Hot ICs

FOR THE ELECTRONICS HOBBYIST

FOR THE ELECTRONICS HOBBYIST

Stan Gibilisco

FIRST EDITION
FIRST PRINTING

© 1993 by **TAB Books**.
TAB Books is a division of McGraw-Hill, Inc.

Printed in the United States of America. All rights reserved. The publisher takes no responsibility for the use of any of the materials or methods described in this book, nor for the products thereof.

Library of Congress Cataloging-in-Publication Data

Gibilisco, Stan.
    Hot ICs for the electronics hobbyist / by Stan Gibilisco.
      p.   cm.
    Includes index.
    ISBN 0-8306-3841-5 (H)   ISBN 0-8306-3839-3 (P)
    1. Integrated circuits—Amateur's manuals. I. Title.
TK9966.G53   1992
621.381'5—dc20   92-7490
                                            CIP

TAB Books offers software for sale. For information and a catalog, please contact TAB Software Department, Blue Ridge Summit, PA 17294-0850.

Acquisitions Editor: Roland S. Phelps
Book Editor: Andrew Yoder
Director of Production: Katherine G. Brown
Book Design: Jaclyn J. Boone and Andrew Yoder
Cover Design: Graphics Plus, Hanover, Pa.

HT3

Contents

Acknowledgments x
Introduction xi

1 Clocks, counters, and timers (CCT) 1
Allegro Microsystems
    2436/37 countdown power timers 1
    5233/38, 5458/63/74, 5604 automotive clocks 3
    5615/16 LCD programmable automotive clocks 6
GEC Plessey Semiconductors
    PDSP1640 FFT address generation 8
LSI Computer Systems
    LS7055/56 6-decade predetermining up/down counter 12
    LS7061 32-bit binary up counter 17
    LS7063 dual 16-bit binary up counter 20
    LS7080/81 quadrature clock converters 23
    LS7083/84 quadrature clock converters 25
    LS7210 programmable digital delay timer 27
    LS7338 touch control light switch 32

2 Communications circuits (COM) 39
Allegro Microsystems
    3845/46 AM noise blankers 39
    3847 dual-conversion AM receiver 45
    3827 FM stereo decoder 46

3330/60/63 optoelectronic switches   49
Analog Devices
    AD507 general purpose operational amplifier   54
    AD517 precision operational amplifier   57
    AD644 dual high-speed BiFET operational amplifier   61
    AD829 video operational amplifier   66
    AD840 wideband operational amplifier   70
GEC Plessey Semiconductors
    MV5087 DTMF generator   76
    MV5089 DTMF generator   78
    ZN478E microphone amplifier for telephone circuits   80
    MV4320 keypad pulse dialer   81
LSI Computer Systems
    LS7501/10 tone activated line isolation device   84
Raytheon
    RC747 general purpose operational amplifier   87
    RC4097 series precision operational amplifiers   91
Silicon Systems
    Application guide: DTMF receivers   95
    75T2089 DTMF transceiver   101
    75T2090 DTMF transceiver and call progress detector   105

3 *Control circuits (CON)* 109

Allegro Microsystems
    8932 voice coil motor driver   109
    8958 voice coil motor driver   111
    Power ICs for motor-drive applications   114
    Relay-driver applications   120
Analog Devices
    AD598 LVDT signal conditioner   121
    AD596/7 thermocouple conditioners and set-point controllers   131
LSI Computer Systems
    LS7237 light switch and motor-speed controller   135
    LS7260/61/62 brushless dc motor commutator/controller   139
    LS7263 brushless dc motor speed controller   145
    LS7270 programmable controller/sequencer   149
Silicon Systems
    32B451 SCSI controller   152
    32C260 PC AT/XT combo controller   156
    32C452 storage controller   160
    32C453 dual-port buffer controller   163

4 Data-conversion and processing circuits (DCP) — 169

Analog Devices
    AD630 balanced modulator/demodulator 169
    AD632 internally trimmed precision multiplier 174

GEC Plessey Semiconductors
    MS2014 digital filter and detector 178
    MV1441 HDB3 encoder/decoder/clock regenerator 181
    MV3506 A-law filter/codec 182
    MV3507 mu-law filter/codec 182
    MV3507A mu-law filter/codec with A/B signalling 182
    MV3508 A-law filter/codec with optional squelch 182
    MV3509 mu-law filter/codec with optional squelch 182
    SL9009 adaptive balance circuit 185
    ZN5683E/J PCM line interface circuit 189

Raytheon
    DAC-08 8-bit high-speed multiplying D/A converter 192
    DAC-10 10-bit high-speed multiplying D/A converter 196
    RC4151/2 voltage-to-frequency converters 204
    RC4153 voltage-to-frequency converter 208

Silicon Systems
    32F8011 programmable electronic filter 210
    32F8020 low-power programmable electronic filter 214

5 Logic circuits (LOG) — 215

Allegro Microsystems
    3611-14 dual 2-input peripheral/power drivers 215
    5800/01 BiMOS II latched drivers 218
    5810 BiMOS II 10-bit serial-input latched source driver 221
    5822 BiMOS II 8-bit serial-input latched driver 222

GEC Plessey Semiconductors
    PDSP16256/A programmable FIR filter 223
    PDSP16330/A/B Pythagoras processor 230
    PDSP16340 polar-to-Cartesian converter 233
    PDSP16350 I/Q splitter/NCO 236
    PDSP16488 single-chip 2D convolver with integral line delays 240

LSI Computer Systems
    LS7220 digital lock circuit 253
    LS7222 keyboard programmable digital lock circuit 257
    LS7223 keyboard programmable digital lock circuit 259
    AN 201 application note for LS7222/3 262

LS7225/6 digital lock circuits with tamper output   263
   LS7228/9 address decoder/two pushbutton digital lock   269
  Raytheon
   RC4805 precision high speed latching comparator   273
   LM139/139A/339/339A single-supply quad comparators   277
   LP165/365 micropower programmable quad comparator   280

6 Microcomputer peripherals (MIC)   285
  Analog Devices
   1B31 wide bandwidth strain gage signal conditioner   285
   1B32 bridge transducer signal conditioner   290
  Raytheon
   DAC-4881 microprocessor-compatible 12-bit D/A converter   296
   DAC-4888 D/A converter with microprocessor
    interface latches   301
  Silicon Systems
   73K222/K222L single-chip modem   306
   73K322L single-chip modem   314
   73K224L single-chip modem   322
   73K324L single-chip modem   330
   Setting DTMF levels for K-series modems   337
   34P570 2-channel floppy disk read/write device   338
   34R575 2-or-4-channel floppy disk read/write device   342
   34B580 port expander floppy disk drive   344
   34D441 data synchronizer and write precompensator device   348

7 Power supplies, test equipment, and instruments (PTI)   353
  Allegro Microsystems
   2429 fluid detector   353
   2453/4/5 automotive lamp monitors   356
   3501 linear-output Hall effect sensor   361
   3503 ratiometric, linear Hall effect sensor   364
   3311/12 precision light sensors   367
  Analog Devices
   AD834 500-MHz four-quadrant multiplier   371
   AD532 internally trimmed multiplier   376
   AD521 precision instrumentation amplifier   380
   AD625 programmable gain instrumentation amplifier   384
  LSI Computer Systems
   LS7100 BCD to 7-segment latch/decoder/driver   388

LS7110 binary addressable latched 8-channel demultiplexer/driver 391
LS7310-13 ac power controllers 394
LS7314/15 ac power controllers 397

Raytheon
RV4143/4 ground fault interrupters 401
RV4145 low-power ground-fault interrupter 405
XR-2207 voltage-controlled oscillator 409

Suggested Additional Reading 415
Index 417

Acknowledgments

THANKS ARE EXTENDED TO THE FOLLOWING MANUFACTURERS, WHO graciously gave their permission for use of their data in this book.

Allegro Microsystems, Inc.
70 Pembroke Road
Concord, NH 03301
(603) 228-5533

Analog Devices, Inc.
2 Technology Way
Norwood, MA 02062
(617) 329-4700

GEC Plessey Semiconductors
9 Parker
Irvine, CA 92718
(714) 472-0303

LSI Computer Systems, Inc.
1235 Walt Whitman Road
Melville, NY 11747
(516) 271-0400

Raytheon Company
350 Ellis Street
Mountain View, CA 94039
(415) 966-7769

Silicon Systems, Inc.
14351 Myford Road
Tustin, CA 92680
(714) 731-7110

Introduction

THIS VOLUME CONTAINS DATA FOR SELECTED INTEGRATED circuits (ICs). The material has been chosen with future trends in mind. Experimenters have always played an important role in new technologies. This book gives a glimpse of not-so-common, cutting-edge devices for serious experimenters, not generally found in other books about ICs.

The information herein is arranged according to application category. The categories and their abbreviations are:

CCT: Clocks, counters, and timers
COM: Communications circuits
CON: Control circuits
DCP: Data-conversion and processing circuits
LOG: Logic circuits
MIC: Microcomputer peripherals
PTI: Power supplies, test equipment, and instruments

Within each category, ICs are subcategorized alphabetically by manufacturer.

A given IC can almost always be put to use in ways that fall into two or three of these categories. In such cases, efforts have been made to locate the data within the category where the IC is most often used.

Emphasis is on specifications, useful characteristics, and applications circuits. This is not a design manual. When involved in serious engineering work, always refer to the manufacturer's data books and sheets.

More detailed data, covering more ICs on the forefront of engineering technology, can also be found in *International Encyclopedia of Integrated Circuits—2nd Edition* (Blue Ridge Summit, PA: TAB Books, 1992).

None of the applications circuits herein are guaranteed. It is expected that some trial and error will be required, as is always the case with experimentation. Applications circuits often must be modified in order to meet specific needs.

These parts are generally available through local distributors, as well as directly from the manufacturer (see the Acknowledgments for addresses and phone numbers of manufacturers whose products are represented here). Consumer retail and direct-mail outlets stock equivalents for many of the ICs for which data appears in this book. Some "digging" is required to find the best bargains.

CHAPTER 1

CLOCKS, COUNTERS, AND TIMERS

**Allegro
2436 and 2437
Countdown Power Timers**

FEATURES
- 28-V/400-mA output switch
- Low-cost ceramic timing capacitor
- Dual-mode timing operation
- −40 °C to +85 °C operation
- 10- to 16-V operation
- Internal stabilizing regulator
- Low-cost 8-pin mini-DIP

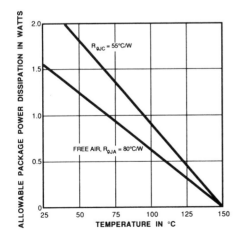

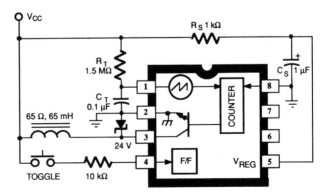

**TEST CIRCUIT AND TYPICAL
REAR-WINDOW DEFOGGER APPLICATION**

APPLICATIONS
- Automotive rear-window defogger timer
- Automotive courtesy light timer
- Appliance power timer
- Power control system

ABSOLUTE MAXIMUM RATINGS at $T_A = +25\,°C$

Supply current, I_{REG}	15 mA
Output voltage, V_{OUT}	28 V
Output current, I_{OUT}	400 mA
Input voltage, V_1 or V_4 (2 min.)	24 V
(continuous)	16 V
Package power dissipation, P_D	See Graph
Operating temperature range, T_A	$-40\,°C$ to $+85\,°C$
Storage temperature range, T_S	$-65\,°C$ to $+150\,°C$

ELECTRICAL SPECIFICATIONS at $T_A = -40\,°C$ to $+85\,°C$, $V_{CC} = 12$ V
(unless otherwise specified)

Characteristic	Test Conditions	Limits Min.	Limits Max.	Units
Regulator Voltage	I_{REG} = 12 mA, Output Off	7.0	9.0	V
Output Saturation Voltage	I_{OUT} = 400 mA, T_A = +25°C	—	2.5	V
	I_{OUT} = 250 mA, T_A = +25°C	—	1.35	V
Output Leakage Current	V_{OUT} = 28 V, V_{CC} = 12 V	—	100	µA
	V_{OUT} = 22 V, V_{CC} = Open Circuit	—	100	µA
Input Threshold Voltage	10 kΩ Series Resistor	1.0	5.0	V
Oscillator Tolerance	T_A = +25°C	—	±3.0	%
	T_A = -40°C to +85°C	—	±6.0	%
Divider Count (V_{CC} = 10 V to 16 V)	Initial Timeout, ULQ2436M	4064	4064	—
	Subsequent Timeouts, ULQ2436M	2032	2032	—
	Initial Timeout, ULQ2437M	992	992	—
	Subsequent Timeouts, ULQ2437M	496	496	—

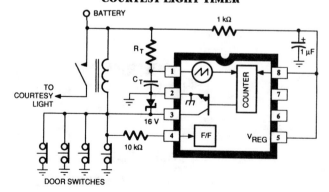

COURTESY LIGHT TIMER

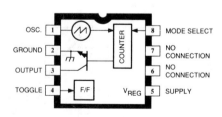

Allegro 5233/38, 5458/63/74, 5604
2-Function 4-Digit VF Automotive Clocks

FEATURES
- Pushbutton "time tone" setting
- Variable pulse-width display dimming
- 12- or 24-hour timekeeping option
- Rollover protection on minutes set
- High noise immunity
- Diode protection on all inputs
- RFI/EMI minimized

ABSOLUTE MAXIMUM RATINGS

Logic supply current, I_{SS}	-7.0 mA
Total supply current, I_{DD}	20 mA
Input voltage range (Ref. V_{SS}), V_{IN}	-0.5 V to $V_{DD}+0.5$V
Input current, I_{IN}	± 10 mA
Package power dissipation, P_D	300 mW
Operating temperature range, T_A	$-40\,°C$ to $+85\,°C$
Storage temperature range, T_S	$-65\,°C$ to $+150\,°C$

Caution: These CMOS devices have static protection but are susceptible to damage if exposed to extremely high static electrical charges.

ELECTRICAL CHARACTERISTICS at $T_A = -40\,°C$ to $+85\,°C$, $V_{SS} = 0$ V, in Typical Application (unless otherwise noted)

Characteristic	Symbol	Test Conditions	Min.	Typ.	Max.	Units
Min. Operating Voltage	V_{DD}	$T_A = +25\,°C$	4.5	—	—	V
		$T_A = -40\,°C$ to $+85\,°C$	5.0	—	—	V
Zener Voltage	V_{Z1}	$I_{SS} = -6.0$ mA, $T_A = +25\,°C$	6.2	6.9	7.5	V
	V_{Z2}	$I_{GND} = -17$ mA, $T_A = +25\,°C$	15	16.5	18	V
Output Breakdown Voltage	$V_{(BR)DS}$	$I_{OUT} = -5$ µA, $T_A = +25\,°C$	15	—	—	V
Segment Output Current	I_{OUT}	$V_{DD} = 6.0$ V, $V_{OUT} = 5.5$ V	-150	—	—	µA
Filament Output Current	I_{FIL}	$V_{DD} = 6.0$ V, $V_{FIL} = 5.0$ V	-2.0	—	—	mA
Oscillator Frequency	f_{OSC}	$T_A = +25\,°C$	—	4.194304	—	MHz
Input Resistance	R_{IN}	S1, S2, or S3 only	300	—	—	kΩ
Switch Debounce Time	t_{DB}		0	31	62.5	ms
Segment Switching Time	t_r	$\Delta V_{OUT} = 5.0$ V	1.0	—	—	µs
Zener Temperature Coefficient	ΔV_{Z1}	$T_A = +25\,°C$	—	+6.0	—	mV/°C
Supply Current	I_{DD}	$V_{DD} = 5.0$ V, $T_A = +25\,°C$	—	—	1.0	mA

NOTE: Negative current is defined as coming out of (sourcing) the specified device terminal.

IGNITION Input	LAMP Input 12 V (ON)	Ground (OFF)
12 V (ON)	50% brightness†, dimmer enabled, S1, S2, S3 enabled	100% brightness, dimmer disabled, S1, S2, S3 enabled
Ground (OFF)	50% brightness†, display command enabled, dimmer enabled	display OFF, display command enabled

† 25% for SCL5604E

Clock Version	S1 (Set Hours)	S2 (Set Minutes)	S3	Colon
SCL5233E	one increment per depression	one increment per depression, start clock	optional ... time tone and display command*	steady
SCL5238E	one increment per depression	one increment per depression	time tone, display command*, start clock	steady
SCL5458E	one increment per depression	one increment per depression, start clock	optional ... time tone and display command*	one second flashing
SCL5463E	one increment per second while depressed	one increment per second while depressed, start clock	optional ... time tone and display command*	one second flashing
SCL5474E SCL5604E	two increments per second while depressed	two increments per second while depressed, start clock	optional ... time tone and display command*	steady

* See IGNITION

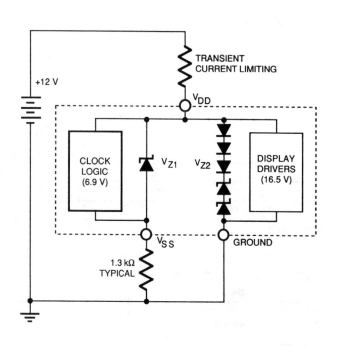

NOTE: b1 = SEGMENT b, DIGIT 1

TYPICAL APPLICATION

TYPICAL INPUT CIRCUITRY

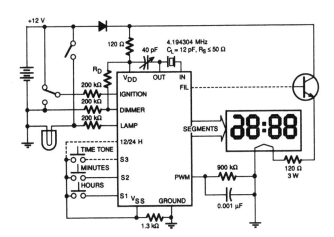

PIN DESIGNATIONS

Pin	SCL5604E	All Other Devices	Pin	SCL5604E	All Other Devices
1	S2	S2	21	PWM	PWM
2	S1	LAMP	22	GROUND	DIMMER
3	S3	adeg1	23	V_{SS}	OSC. IN
4	IGNITION	12/24 H	24	FILAMENT	OSC. OUT
5	NU	c1	25	V_{DD}	b4
6	b1	g2	26	c4	a4
7	f2	e2	27	d4	f4
8	a2	d2	28	e4	b3
9	b2	c2	29	g4	ad3
10	TEST	g3	30	c3	f3
11	COLON	e3	31	e3	COLON
12	f3	c3	32	g3	TEST
13	ad3	g4	33	c2	b2
14	b3	e4	34	d2	a2
15	f4	d4	35	e2	f2
16	a4	c4	36	g2	b1
17	b4	V_{DD}	37	c1	NU
18	OSC. OUT	FILAMENT	38	NU	IGNITION
19	OSC. IN	V_{SS}	39	NU	S3
20	DIMMER	GROUND	40	LAMP	S1

NU = Internal connection, do not use.

Allegro 5615/5616
2-Function 4-Digit LCD Automotive Clocks (Programmable)

FEATURES
- Digital tuning of crystal frequency
- PROM for storing frequency correction information
- 12- or 24-hour timekeeping option (SCL5616EP/HN/HW only)
- Flashing colon
- Two switches control all-setting functions
- High noise immunity
- Internal power-up reset circuitry
- Internal voltage regulation

ABSOLUTE MAXIMUM RATINGS
Supply current, I_{DD} .... 2.0 mA
Input voltage range, V_{IN} (except V_{PP}) ... -0.3 V to V_{DD}
(Programming power voltage, V_{PP}) ... 18.5 V
Input current (except V_{PP}), I_{IN} ... ± 10 mA
Power dissipation, P_D .... 300 mW
Operating temperature range, T_A ... $-40\,°C$ to $+85\,°C$
Storage temperature range, T_S ... $-65\,°C$ to $+150\,°C$

Caution: These CMOS devices have static protection, but are susceptible to damage if exposed to extremely high static electrical charges.

ELECTRICAL CHARACTERISTICS at $T_A = -40\,°C$ to $+85\,°C$, in Typical Application (unless otherwise noted)

Characteristic	Symbol	Test Conditions	Min.	Typ.	Max.	Units
Operating Voltage Range	V_{DD}	$T_A = +25\,°C$	4.5	—	—	V
Zener Voltage	V_{DD}	$I_{DD} = 1.0$ mA	5.5	—	6.8	V
Segment Output Current	I_{OUT}	$V_{DD} = 5.0$ V, $V_{OUT} = 4.8$ V	-20	—	—	µA
		$V_{DD} = 5.0$ V, $V_{OUT} = 0.2$ V	120	—	—	µA
Backplane Output Current	I_{OUT}	$V_{DD} = 5.0$ V, $V_{OUT} = 4.8$ V	-80	—	—	µA
		$V_{DD} = 5.0$ V, $V_{OUT} = 0.2$ V	240	—	—	µA
LCD Drive Signal	V_{DISP}	$V_{DD} \geq 5.0$ V	4.0	—	—	V
Input Current	I_{IN}	S1, S2, DATA, or SELECT	-55	—	-700	µA
Oscillator Frequency	f_{OSC}		—	4.194 304	—	MHz
Oscillator Starting Time	t_{OSC}	V_{DD} = Zener voltage	—	—	200	ms
Oscillator Stability	Δf_{OSC}	$\Delta V_{DD} = \pm 100$ mV	—	—	± 1.0	ppM
Backplane Frequency	f_{BP}		—	64	—	Hz
Switch Debounce Time	t_{DB}		0	—	62.5	ms
Osc. Feedback Resistance	R_{OSC}		—	16	—	MΩ
Osc. Input Capacitance	C_{OSCI}		—	15	—	pF
Osc. Output Capacitance	C_{OSCO}		—	30	—	pF
Supply Current	I_{DD}	$V_{DD} = 5.0$ V	—	—	1.0	mA

NOTE: Negative current is defined as coming out of (sourcing) the specified device terminal.

RECOMMENDED FLASH PROGRAMMING CHARACTERISTICS
at $T_A = +25\,°C$, Logic Levels are V_{DD} and Ground (except PROGRAM High)

Characteristic	Symbol	Min.	Max.	Units
PROGRAM High (18 V) to DATA Low	t_{PHDL}	1.0	—	μs
SELECT Valid to DATA Low	t_{SVDL}	25	—	μs
DATA Low to DATA High	t_{DLH}	1.0	1.5	μs
DATA High to DATA Low	t_{DHL}	1.0	—	μs
DATA Store Pulse Duration	t_{wD}	10	—	ms
DATA High to PROGRAM Low	t_{DHPL}	1.0	—	μs
PROGRAM Low to SELECT Change	t_{PLSX}	1.0	—	μs
SELECT Low (Verify) to DATA Valid	t_{SLDV}	—	1.0	μs
DATA Hold from End of Verify	t_{SHDX}	—	10	ns

DISPLAY FORMAT

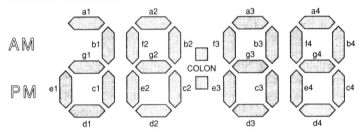

SCL5615HN/HW or SCL5616HN/HW

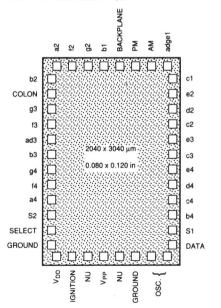

SCL5615EP or SCL5616EP

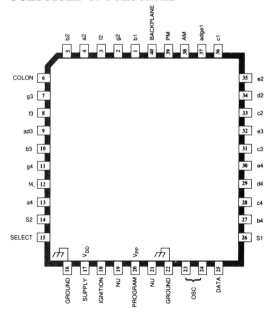

TYPICAL APPLICATION

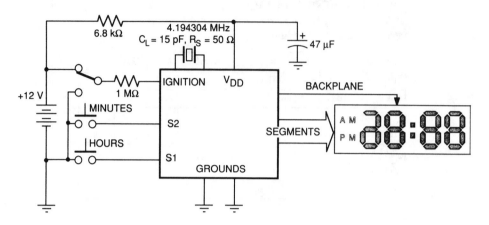

GEC Plessey
FFT Address Generation Using the PDSP1640

Fast Fourier Transforms (FFTs) are used in a wide variety of applications as a means of calculating the Discrete Fourier Transform of a signal, in order to estimate the signal's spectral energy. There are many different algorithms for computing FFTs, each designed to exhibit particular characteristics, but each of which also have certain disadvantages. The in-place radix 2 DIT algorithm puts the outputs from each stage into the same memory location from which the inputs are read. This minimizes the system memory requirements, but has the disadvantage of requiring a different register addressing sequence at each stage and also requires four memory accesses per butterfly cycle, necessitating the use of a very fast RAM.

The main advantage of the "constant geometry" algorithm is that the data memory can be configured so that any particular RAM device only needs to be accessed once during each butterfly cycle, thus allowing the use of slower devices. The algorithm also has the feature that the memory read address sequence and the memory write address sequence remain the same from one stage to the next. However, the algorithm is not in-place and therefore requires twice as much RAM storage as an in-place algorithm and the coefficients are not in a simple order.

These two algorithms are, perhaps, the most common of FFT algorithms and, as such, are used in this application note as examples of the FFT address generation capabilities of the PDSP1640. The variety of FFT algorithms available shows that the data and coefficient address sequencing is a major consideration in the implementation of any one, requiring in nearly all cases some form of programmable address generation. The principles and techniques explained here for the use of the PDSP1640 can be used as the basis for the implementation of address generation for many other algorithms.

The Plessey Semiconductors PDSP1640 is an 8-bit programmable address generator capable of operation at speeds up to 20 MHz. It can be cascaded with other devices to produce wider address fields, for example operating at up to 10 MHz for a 24-bit address and is ideally suited to many situations requiring high-speed address generation, both for FFT computation and for other digital signal-processing applications.

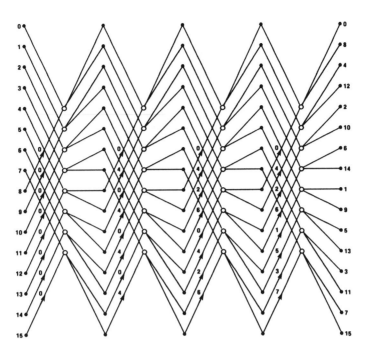

16-point constant geometry radix 2 DIT FFT algorithm

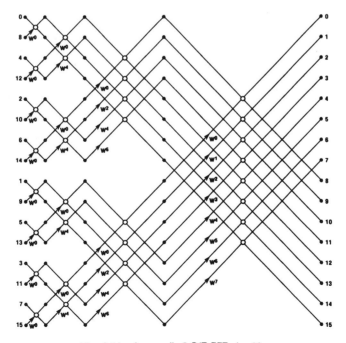

16-point in-place radix 2 DIT FFT algorithm

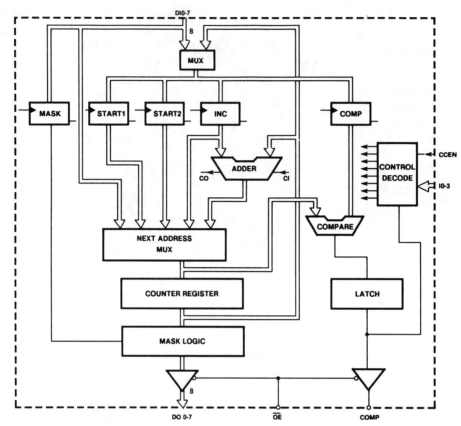

PDSP1640 Address Generator block diagram

	Address sequence	Count sequence				Address sequence	Count sequence		
	0	0	(0000)	000		0	0	(0000)	000
	2	0	(0010)	001		4	0	(0100)	010
	4	0	(0100)	010		8	0	(1000)	100
	6	0	(0110)	011		12	0	(1100)	110
	8	0	(1000)	100		1	1	(0001)	000
	10	0	(1010)	101		5	1	(0101)	010
	12	0	(1100)	110		9	1	(1001)	100
Increment = 11$_H$	14	0	(1110)	111	Increment = 22$_H$	13	1	(1101)	110
	1	1	(0001)	000		2	0	(0010)	000
	3	1	(0011)	001		6	0	(0110)	010
	5	1	(0101)	010		10	0	(1010)	100
	7	1	(0111)	011		14	0	(1110)	110
	9	1	(1001)	100		3	1	(0011)	000
	11	1	(1011)	101		7	1	(0111)	010
	13	1	(1101)	110		11	1	(1011)	100
	15	1	(1111)	111		15	1	(1111)	110
	(a)					(b)			

PDSP1640 address count sequence using bit redundancy

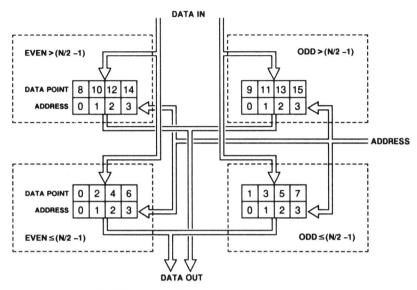

Fig.4 Data storage for constant geometry algorithm

Cycle No.	Mnemonic	Op. code	Data [1,2]	Operation
1	CLRCR	7_H	XX	CLEAR COUNT/MASK REGISTERS
2	LIRDI	C_H	XX	LOAD INCREMENT REGISTER
3	LCPDI	E_H	01_H	LOAD COMPARE REG WITH END ADDRESS
4	LS1DI	8_H	07_H	LOAD START1 REG WITH BRANCH ADDRESS
5	CCJS1	1_H	00_H	COUNT BY INC OR GO TO SR1
6	CCJS1	1_H	XX	COUNT BY INC OR GO TO SR1
(36)	CCJS1	1_H	XX	COUNT BY INC OR GO TO SR1

NOTES
1. XX = don't care
2. Data is input on cycle following relevant instruction

Data addressing instruction sequence, constant geometry algorithm

ADDR CNTR REG	1ST STAGE MASK = FF_H		2ND STAGE MASK = FE_H		3RD STAGE MASK = FC_H		4TH STAGE MASK = $F8_H$	
	OUTPUT ADDR	BIT RVRSD ADDR	OUTPUT ADDR	BIT RVRSD ADDR	OUTPUT ADDR	BIT RVRSD ADDR	OUTPUT ADDR	BIT RVRSD ADDR
00	00	00	00	00	00	00	00	00
01	00	00	01	80	01	80	01	80
02	00	00	00	00	02	40	02	40
03	00	00	01	80	03	C0	03	C0
04	00	00	00	00	00	00	04	20
05	00	00	01	80	01	80	05	A0
06	00	00	00	00	02	40	06	60
07	00	00	01	80	03	C0	07	E0

Coefficient output address sequence for constant geometry algorithm

Cycle No.	Mnemonic	Op. code	Data [1,2]	Operation
1	CLRCR	7_H	XX	CLEAR COUNT/MASK REGISTERS
2	LIRDI	C_H	XX	LOAD INCREMENT REGISTER
3	LCPDI	E_H	01_H	LOAD COMPARE REG WITH END ADDRESS
4	LS1DI	8_H	07_H	LOAD START1 REG WITH BRANCH ADDRESS
5	LMRDI	3_H	00_H	LOAD MASK REGISTER
6	CCJS1	1_H	FF_H	COUNT BY INC OR GO TO SR1
7	CCJS1	1_H	XX	COUNT BY INC OR GO TO SR1
13	CCJS1	1_H	XX	COUNT BY INC OR GO TO SR1
14	LMRDI	3_H	XX	LOAD MASK REGISTER
15	CCJS1	1_H	FE_H	COUNT BY INC OR GO TO SR1
16	CCJS1	1_H	XX	COUNT BY INC OR GO TO SR1
22	CCJS1	1_H	XX	COUNT BY INC OR GO TO SR1
23	LMRDI	3_H	XX	LOAD MASK REGISTER
24	CCJS1	1_H	FC_H	COUNT BY INC OR GO TO SR1
25	CCJS1	1_H	XX	COUNT BY INC OR GO TO SR1
31	CCJS1	1_H	XX	COUNT BY INC OR GO TO SR1
32	LMRDI	3_H	XX	LOAD MASK REGISTER
33	CCJS1	1_H	$F8_H$	COUNT BY INC OR GO TO SR1
34	CCJS1	1_H	XX	COUNT BY INC OR GO TO SR1
40	CCJS1	1_H	XX	COUNT BY INC OR GO TO SR1

NOTES
1. XX = don't care
2. Data is input on cycle following relevant instruction

Coefficient addressing instruction sequence for constant geometry algorithm

---

The information included herein is believed to be accurate and reliable. However, LSI Computer Systems, Inc. assumes no responsibilities for inaccuracies, nor for any infringements of patent rights of others which may result from its use.

LSI
LS7055/LS7056
6-Decade Predetermining Up/Down Counter
(with 3 presettable storage registers)

FEATURES
- Single power-supply operation +4.75 to +15 V
- Preset, presignal, and mainsignal store
- dc to 250-kHz count frequency
- Fully synchronous operation
  Three comparators with output flags
  Automatic or manual preset/reset control
- Thumbwheel interface for storage selects
- Prescale on count input selectable
- Count inhibit
- Up/down control
- Scan rate up to 150 kHz
- Scan oscillator has override capability
- Blanking override for decimal-point operation
- Multiplexed 7-segment and BCD data output
- Output latches
- Reset
- Hysteresis circuit on count input
- CMOS-type noise immunity on all other inputs
- Pull-down resistors on BCD inputs

□ CCT

MAXIMUM RATINGS

PARAMETER	SYMBOL	VALUE	UNITS
Storage temperature	T_{STG}	−65 to +150	°C
Operating temperature	T_A	−25 to +70	°C
Voltage (any pin to V_{SS})	V_{MAX}	−30 to +0.5	V

PIN ASSIGNMENT

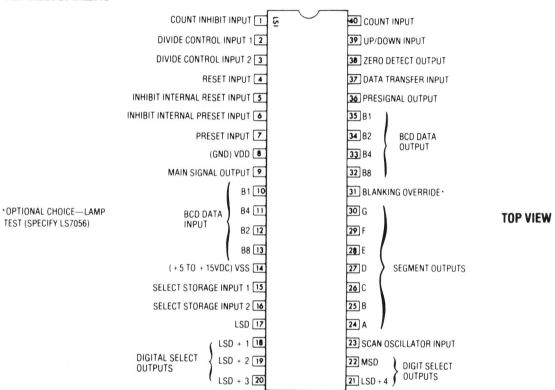

*OPTIONAL CHOICE—LAMP TEST (SPECIFY LS7056)

TOP VIEW

MODES OF OPERATION

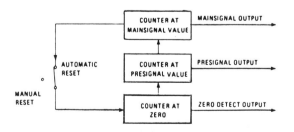

AUTOMATIC OR MANUAL OPERATION IN UP MODE

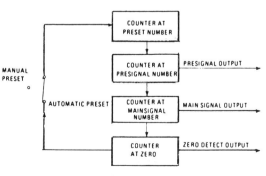

AUTOMATIC OR MANUAL OPERATION IN DOWN MODE

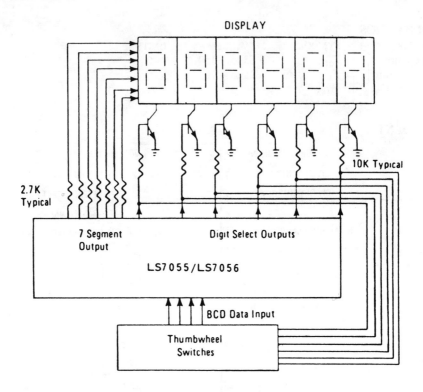

Driving a small LED Display (Typically 1/8") at 12 volt power supply. The 2.7K resistors provide approximately 3 milliamperes segment drive.

Typical resistor/capacitor values for the scan oscillator

Resistor	Capacitor	Typical Frequency
10KΩ	750pF	150KHz
15KΩ	750pF	100KHz
100KΩ	1000pF	10KHz
1.0 MEGΩ	1000pF	1KHz

Driver Requirements for Overriding Scan Oscillator Input

Power Supply(volts)	Sink Current	Source Current
5	1.0mA	0
10	4.5mA	0
15	10.0mA	0

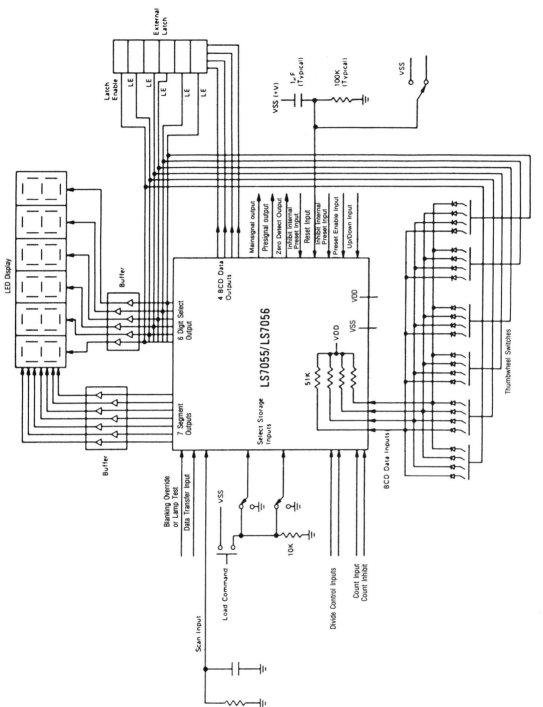

SYSTEM INTERCONNECTION DIAGRAM

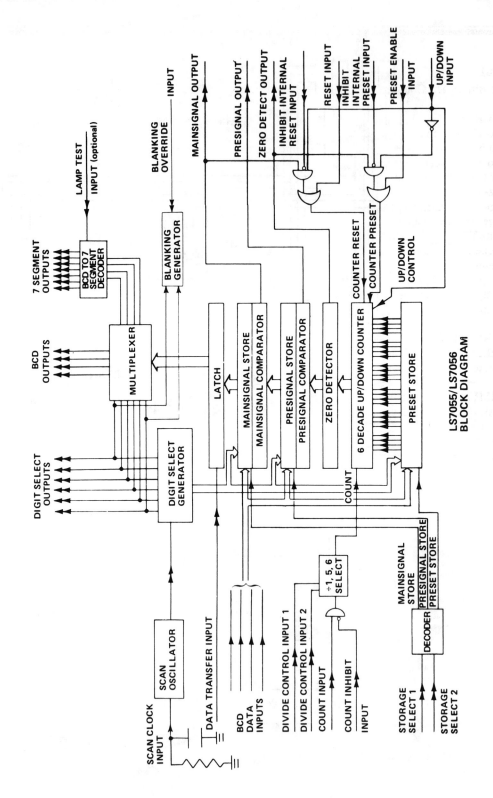

LSI LS7061
32-Bit Binary Up Counter with 40-Bit Latch, Multiplexer, and Three-State Drivers

FEATURES
- dc to 15-MHz count frequency
- 8-bit byte multiplexer
- dc to 1-MHz scan frequency
- Ability to latch external 8 bits of high-speed external prescaler thereby extending count frequency to 3.84 GHz
- Single power-supply operation, +4.75 Vdc to +5.25 Vdc
- Three-state data outputs, bus, and TTL compatible
- Inputs TTL, NMOS, and CMOS compatible
- Unique cascade feature allows multiplexing of successive bytes of data in sequence in multiple counter systems
- Low power dissipation
- All inputs protected
- 24-pin DIP

DESCRIPTION
The LS7061 is a monolithic, ion-implanted MOS silicon gate, 32-bit up counter. The circuit includes 40 latches, multiplexer, eight three-state binary data output drivers and output cascading logic.

MAXIMUM RATINGS

PARAMETER	SYMBOL	VALUE	UNITS
Storage temperature	T_{STG}	−55 to +150	°C
Operating temperature	T_A	0 to +70	°C
Voltage (any pin to V_{SS})	V_{MAX}	+10 to −0.3	V

APPLICATION EXAMPLE: High Speed Differential Energy Analyzer

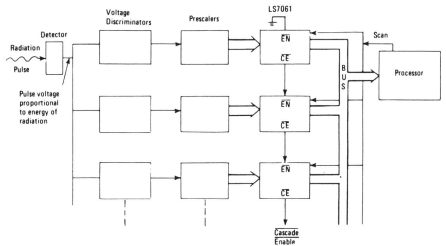

Note: The processor subtracts counts from successive counters to determine the differential energy spectrum.

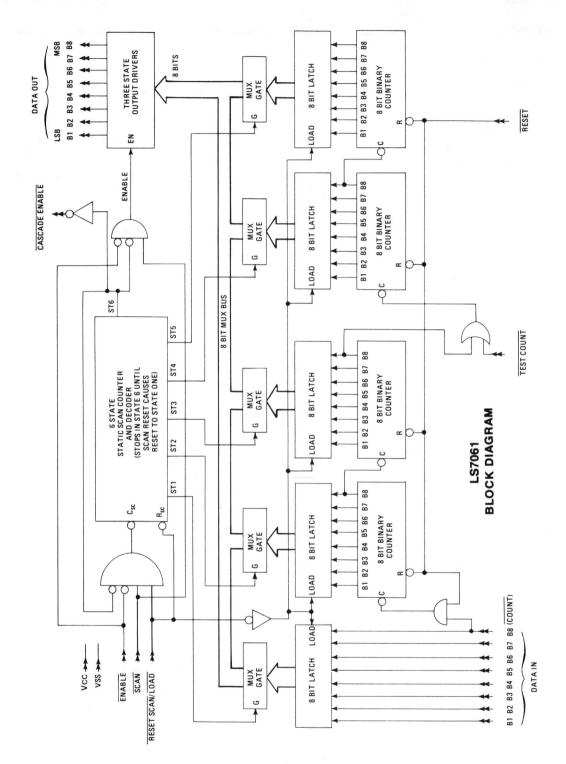

☐ **CCT** LSI LS7061 19

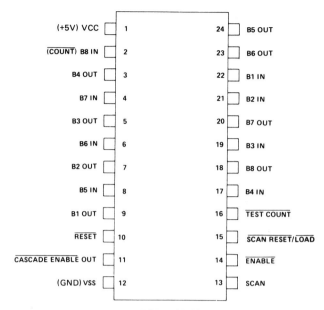

TOP VIEW
PIN ASSIGNMENT

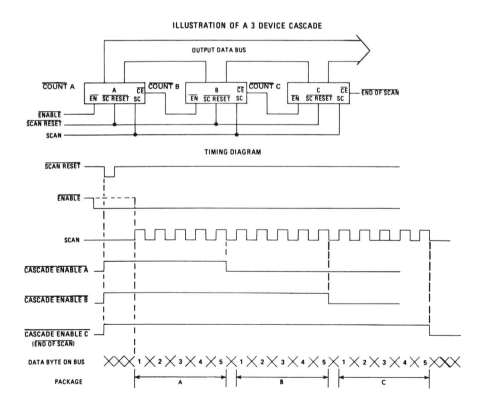

LSI
LS7063
Dual 16-Bit Binary Up Counter with 40-Bit Latch, Multiplexer, and Three-State Drivers

FEATURES
- dc to 15-MHz count frequency
- 8-bit byte multiplexer
- dc to 1-MHz scan frequency
- Ability to latch external 8 bits of high-speed external prescaler thereby extending count frequency to 3.84 GHz
- Single power supply operation, +4.75 to +5.25 Vdc
- Three-state data outputs, bus, and TTL compatible
- Inputs TTL, NMOS, and CMOS compatible
- Unique cascade feature allows multiplexing of successive bytes of data in sequence in multiple counter systems
- Low power dissipation
- All inputs protected
- 24-pin DIP

DESCRIPTION
The LS7063 is a monolithic, ion-implanted MOS silicon gate, dual 16-bit up counter. The circuit includes 40 latches, multiplexer, eight three-state binary data output drivers and output cascading logic.

MAXIMUM RATINGS

PARAMETER	SYMBOL	VALUE	UNITS
Storage temperature	T_{STG}	−55 to +150	°C
Operating temperature	T_A	0 to +70	°C
Voltage (any pin to V_{SS})	V_{MAX}	+10 to −0.3	V

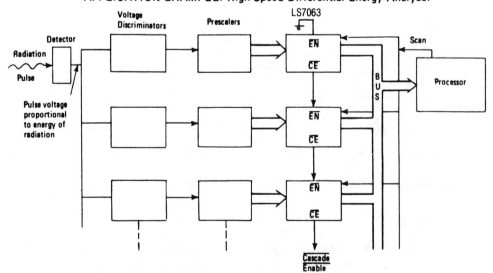

APPLICATION EXAMPLE: High Speed Differential Energy Analyzer

Note: The processor subtracts counts from successive counters to determine the differential energy spectrum.

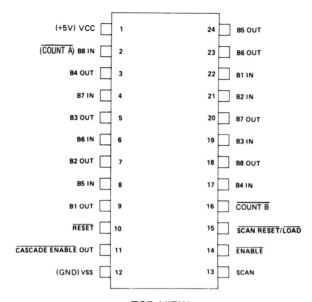

TOP VIEW
LS7063 PIN ASSIGNMENT

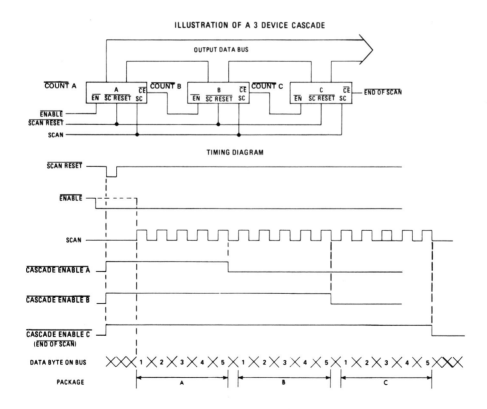

22 LSI LS7063

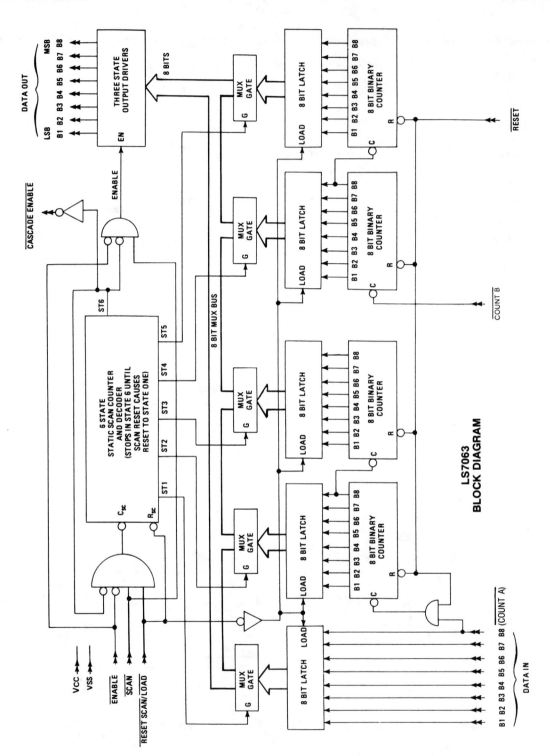

LS7063
BLOCK DIAGRAM

LSI
LS7080/7081
Quadrature Clock Converters

FEATURES
- X4 of X1 mode selection
- Input clock speed up to 1.5 MHz
- Programmable output clock width in X4 mode
- On-chip filtering of input for optical or magnetic encoder application
- TTL and CMOS compatible

DESCRIPTION

The LS7080 and LS7081 are monolithic CMOS quadrature clock converter circuits. Quadrature clocks derived from optical or magnetic encoder type devices, when applied to the A and B inputs of the LS7080/LS7081 are converted to strings of Up Clocks and Down Clocks (LS7080) or to a Clock and an Up/$\overline{\text{Down}}$ direction control (LS7081). These outputs can then be used to drive standard Up/Down counters for direction or position sensing of the encoder.

ABSOLUTE MAXIMUM RATINGS

PARAMETER	SYMBOL	VALUE	UNITS
Voltage at any input	V_{IN}	$V_{SS}-.5$ to $V_{DD}+.5$	V
Operating temperature	T_A	0 to +70	°C
Storage temperature	T_{STG}	−65 to +150	°C

PIN ASSIGNMENT TOP VIEW

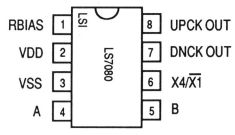

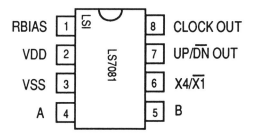

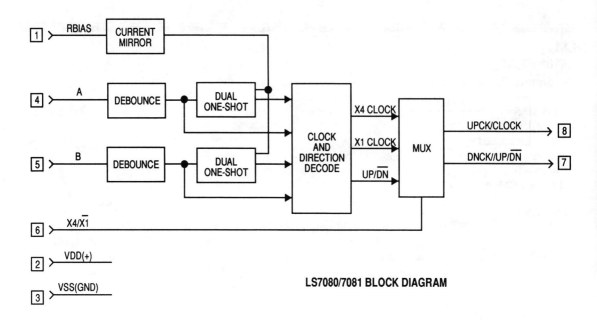

LS7080/7081 BLOCK DIAGRAM

TYPICAL APPLICATION FOR LS7080 IN X4 MODE

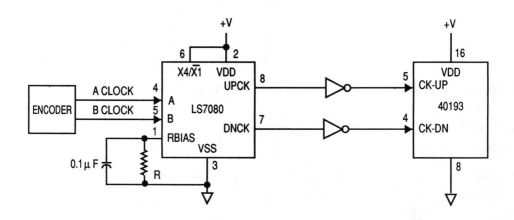

□ CCT　　　　　　　　　　　　　　　　　　　　　　　LSI LS7083/7084　25

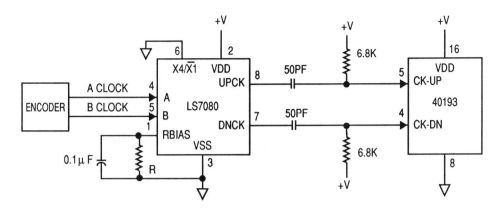

TYPICAL APPLICATION FOR LS7080 IN X1 MODE
NOTE: The Differentiated clock outputs guarantee non-overlapping UP clocks and DN clocks when the direction reverses (X1 mode only).

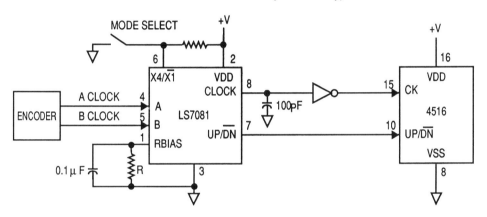

TYPICAL APPLICATION FOR LS7081 WITH X4/$\overline{X1}$ MODE SELECTION
NOTE: The 100pF capacitor guarantees proper counting when coincident UP and DN clocks occur (X1 mode only) when the direction reverses.

LSI
LS7083/7084
Quadrature Clock Converters

FEATURES
- X1 and X4 mode selection
- Up to 16-MHz output clock frequency
- Programmable output clock pulse width
- On-chip filtering of inputs for optical or magnetic encoder applications
- TTL and CMOS compatible

LSI LS7083/7084

DESCRIPTION

The LS7083 and LS7084 are monolithic CMOS quadrature clock converter circuits. Quadrature clocks derived from optical or magnetic encoders, when applied to the A and B inputs of the LS7083/LS7084, are converted to strings of Up Clocks and Down Clocks (LS7083) or to a Clock and an Up/Down direction control (LS7084). These outputs can then be used to drive standard Up/Down counters for direction and position sensing of the encoder.

ABSOLUTE MAXIMUM RATINGS

PARAMETER	SYMBOL	VALUE	UNITS
Voltage at any input	V_{IN}	$V_{SS} - .5$ to $V_{DD} + .5$	V
Operating temperature	T_A	0 to +70	°C
Storage temperature	T_{STG}	−65 to +150	°C

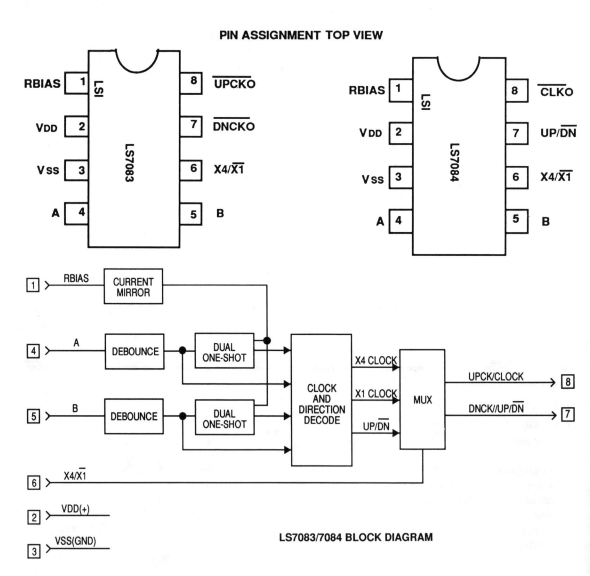

LS7083/7084 BLOCK DIAGRAM

TYPICAL APPLICATION FOR LS7083 IN X4 MODE

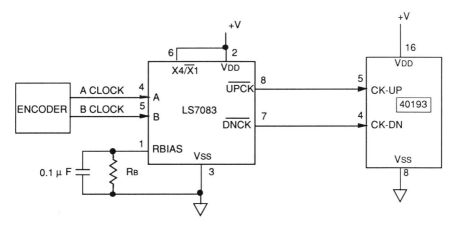

TYPICAL APPLICATION FOR LS7084 WITH X4/X̄1 MODE SELECTION

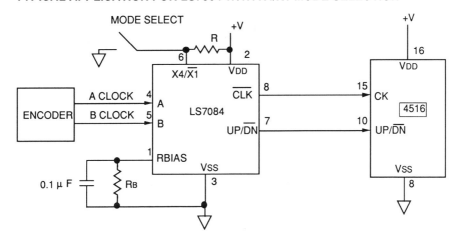

LSI
LS7210
Programmable Digital Delay Timer

FEATURES
- Programmable delay from milliseconds to hours
- Can be cascaded for sequential events or extended delay
- Single power supply operation +4.75 to +15 V
- On-chip oscillator
- Alternate clock input
- On-chip power on reset
- Internal pull-ups on inputs
- Frequency range to 160 kHz
- CMOS-type noise immunity on all inputs
- All inputs are CMOS, PMOS, and TTL compatible

28 LSI LS7210

DESCRIPTION

The LS7210 is a monolithic, ion-implanted MOS programmable digital timer that can generate a delay in the range of 6 ms to infinity. The delay is programmed by five binary weighted input bits in combination with the oscillator provided. The chip can be operated into four different modes: delayed operate, delayed release, dual delay and one-shot. These modes are selected by the control inputs A and B.

ABSOLUTE MAXIMUM RATINGS (All voltages referenced to V_{DD})

	SYMBOL	VALUE	UNITS
dc supply voltage	V_{SS}	+18	V
Voltage (any pin)	V_{IN}	0 to V_{SS}+.3	V
Operating temperature	T_A	−25 to +70	°C
Storage temperature	T_{STG}	−65 to +150	°C

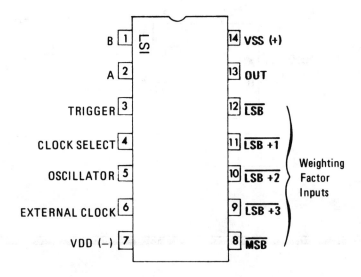

TOP VIEW
STANDARD 14 PIN DIP

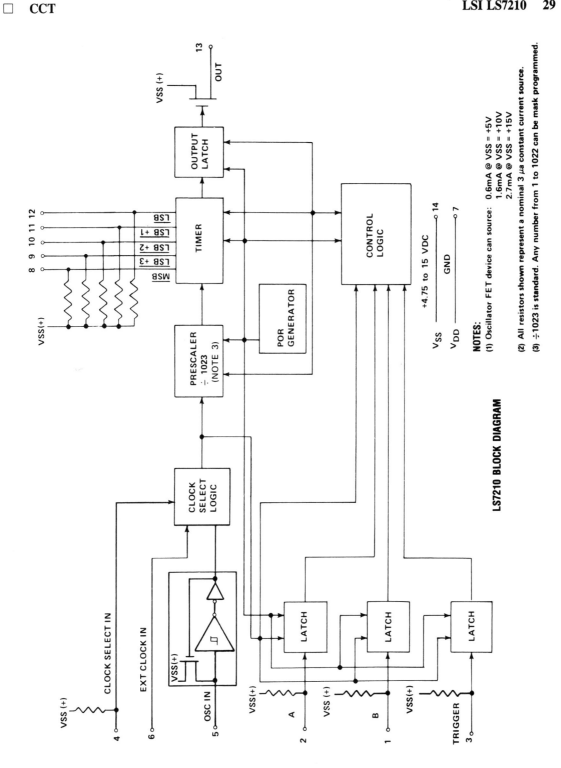

LS7210 BLOCK DIAGRAM

NOTES:
(1) Oscillator FET device can source: 0.6mA @ VSS = +5V
  1.6mA @ VSS = +10V
  2.7mA @ VSS = +15V
(2) All resistors shown represent a nominal 3 μa constant current source.
(3) ÷1023 is standard. Any number from 1 to 1022 can be mask programmed.

30 LSI LS7210

APPLICATION EXAMPLES

SEQUENTIAL TURN ON

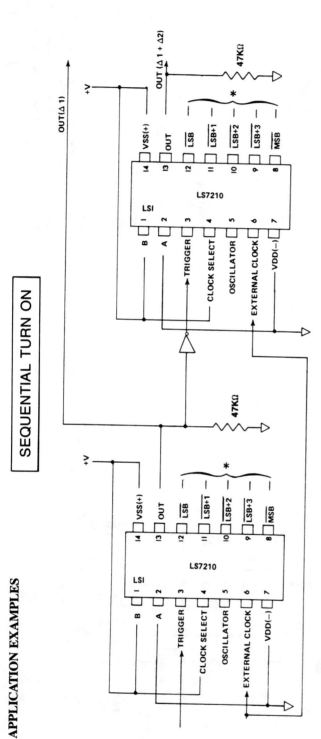

*Connect for desired weighting factor.

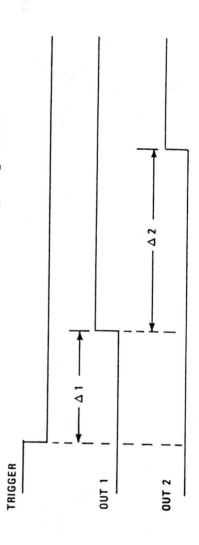

NOTE: Output of LS7210 is open drain FET. A Resistor to ground is required to cause output to go negative.

UNSYMMETRICAL FLASHER

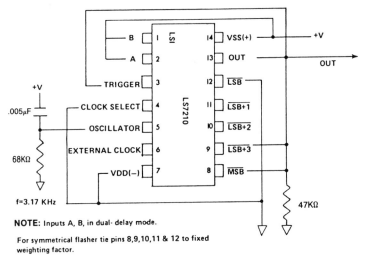

NOTE: Inputs A, B, in dual- delay mode.

For symmetrical flasher tie pins 8,9,10,11 & 12 to fixed weighting factor.

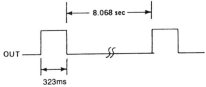

LS7210 IN DELAYED OPERATE MODE TO ACHIEVE ONE TO 31 MINUTE DELAY

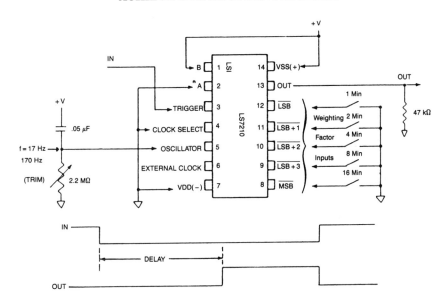

AUTO RESET WATCHDOG CIRCUIT

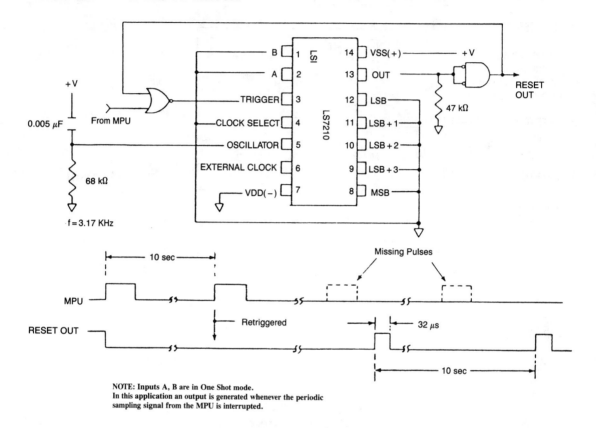

NOTE: Inputs A, B are in One Shot mode.
In this application an output is generated whenever the periodic sampling signal from the MPU is interrupted.

LSI
LS7338
Touch-Control Light Switch With Timed-On and Delayed-Off Modes

FEATURES
- Phase-locked-loop synchronization allows use in wall switch applications
- Timed-On mode provides 99% of full-rated ac power. Time is controlled by external RC with on-chip oscillator and counter
- End of Timed-On mode initializes Delayed-Off mode
- Delayed-Off mode begins at 68% of full-rated ac power and DIMS-to-Off in 209 seconds for 60 Hz (See note 1)
- Positive indication of mode change is provided by the 31% reduction in delivered power
- Modes can be sequenced by activating a control input
- Input for extensions or remote sensors
- Operates at 50-Hz/60-Hz line frequency
- 12 to 18-Vdc supply voltage
- 8-pin plastic DIP

☐ CCT

ABSOLUTE MAXIMUM RATINGS

PARAMETER	SYMBOL	VALUE	UNITS
DC supply voltage	V_{SS}-V_{DD}	+20	Volt
Any input voltage	V_{IN}	V_{SS} -20 to V_{SS} +.5	Volt
Operating temperature	T_A	0 to +80	°C
Storage temperature	T_{stg}	-65 to +150	°C

DC ELECTRICAL CHARACTERISTICS:
(T_A = 25°C, all voltages referenced to VDD)

PARAMETER	SYMBOL	MIN.	TYP	MAX	UNIT	CONDITIONS/REMARKS
Supply voltage	V_{SS}	+12	—	+18	Volts	
Supply current	I_{SS}	—	1.6	2.2	mA	@ V_{SS} = +15V, output off
Input voltage:						
SYNC Lo	V_{IRL}	V_{DD}	—	V_{SS}-9.5	Volts	
SYNC Hi	V_{IRH}	V_{SS}-5.5	—	V_{SS}	Volts	
$\overline{SENSE}$ Lo	V_{IOL}	V_{DD}	—	V_{SS}-8	Volts	
$\overline{SENSE}$ Hi	V_{IOH}	V_{SS}-2	—	V_{SS}	Volts	
SLAVE Lo	V_{IVL}	V_{DD}	—	V_{SS}-8	Volts	
SLAVE Hi	V_{IVH}	V_{SS}-2	—	V_{SS}	Volts	
Input Current:						
SYNC, $\overline{SENSE}$ & SLAVE Hi	I_{IH}	—	—	110	µA	With series 1.5nΩ resistor to 115VAC
SYNC, $\overline{SENSE}$ & SLAVE Lo Voltage	I_{IL}	—	—	100	nA	
$\overline{OUT}$ Hi Voltage	V_{OH}	—	V_{SS}	—	volts	
$\overline{OUT}$ Lo Voltage	V_{OL}	—	V_{SS}-8	—	Volts	@ V_{SS} = +15V
$\overline{OUT}$ Sink Current	I_{OS}	25	—	—	mA	@ V_{SS} = +15V, V_{OL} = V_{SS}-4V

TIMING CHARACTERISTICS
All timings are based on f_S = 60 Hz, unless otherwise specified.

PARAMETER	SYMBOL	MIN.	TYP.	MAX.	UNIT
SYNC Frequency	f_S	40	—	70	Hz
$\overline{SENSE}$/SLAVE Duration	T_{S1}	50	—	Infinite	ms
$\overline{OUT}$ Pulse Width	T_W	—	33	—	µS
STATE 1 Timeout	T_{D1}	—	255RC	—	Sec
STATE 2 Timeout	T_{D2}	—	209	—	Sec

34 LSI LS7338

CCT ☐

PIN ASSIGNMENT TOP VIEW

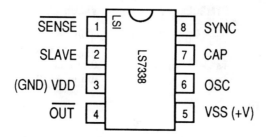

BLOCK DIAGRAM

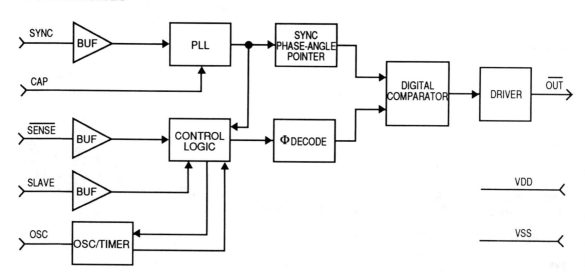

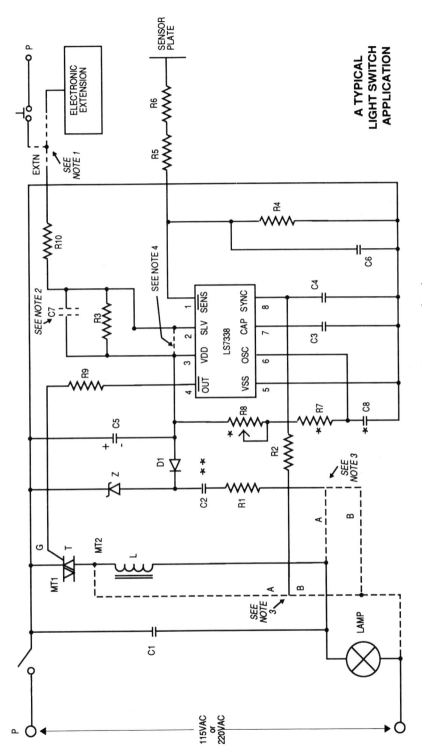

A TYPICAL LIGHT SWITCH APPLICATION

Notes:
1) All circuits connected by broken lines are optional.
2) C7 is used only when electronic extension is connected.
3) Use connection A when neutral is not available.
   Use connection B when neutral is available.
4) When SLAVE is not used, tie Pin 2 to Pin 3 and eliminate R3 and R10.

ELECTRONIC SWITCH EXTENSION

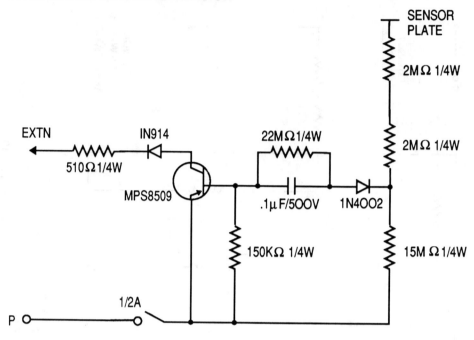

115 VAC

C1 = 0.15µF/150VAC
**C2 = 0.33µF/150VAC
(Connection A)
C3 = 0.047µF/25V
C4 = 470pF/25V
C5 = 47µF/25V
C6 = 680pF/25V
C7 = 0.1µF/25V
R1 = 270Ω/1W
R2 = 1.5M/¼W
R3   1.5MΩ/¼W
R4 = 1MΩ to 5MΩ/¼W
(Select for sensitivity)
R5,R6 = 2.7MΩ/¼W
R9 = 100Ω/¼W
D1 = 1N4148
Z = 15V/1W (Zener)
T = Q4004 L4 Triac
(Typical)
L = 100µH
(RFI Filter)
R10 = 220KΩ/¼W

220 VAC

C1 = 0.15µF/300VAC
**C2 = 0.22µF/300VAC
(Connection A)
C3 = 0.047µF/25V
C4 = 470pF/25V
C5 = 47µF/25V
C6 = 680pF/25V
C7 = 0.1µF/25V
R1 = 1KΩ/2W
R2 = 1.5MΩ/¼W
R3 = 1.5MΩ/¼W
R4 = 1MΩ to 5MΩ/¼W
(Select for sensitivity)
R5,R6 = 4.7MΩ/¼W
R9 = 100Ω/¼W
D1 = 1N4148
Z = 15V/1W (Zener)
T = Q5004L4 Triac
(Typical)
L = 200µH
(RFI Filter)
R10 = 220KΩ/¼W

AGC TOUCH CONTROL WITH LINE PLUG REVERSIBILITY FOR TABLE LAMPS

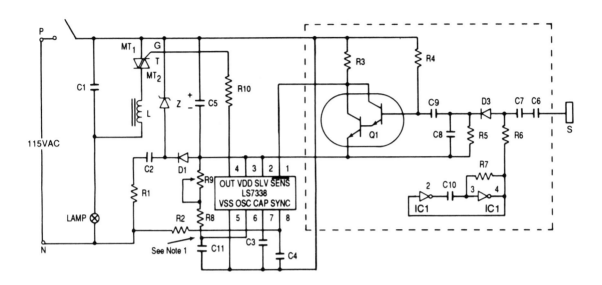

C1 = 0.1µF/150VAC
C2 = 0.47µF/150VAC
C3 = 0.047µF/25V
C4 = 470pF/25V
C5 = 68µF/25V
C6,7 = .001µF/1KVAC
C8 = 0.1µF/25V
C9 = 0.5µF/25V
C10 = 100pF/25V
R1 = 1KΩ/1W
R2 = 1.5MΩ/¼W
R3 = 100KΩ/¼W

R4 = 4.7MΩ/¼W
R5 = 680KΩ/¼W
R6 = 10KΩ/¼W
R7 = 51KΩ/¼W
R10 = 100Ω/¼W
D1 = 1N4148
D3 = 1N4007
IC1 = CD4069
Z = 15V/1W (ZENER)
Q1 = MPSA13
L = 100µH (RFI FILTER)
S = SENSOR PLATE
T = Q4004L4 Triac (TYPICAL)

CHAPTER 2

COMMUNICATIONS CIRCUITS

Allegro 3845 and 3846
AM Noise Blankers

FEATURES
- RF blanking to 30 MHz
- Single-channel or stereo audio blanking
- Adjustable RF and audio blanking time
- Adjustable audio blanking delay
- Sample-and-hold MOS audio gates
- Internal voltage regulation
- Minimum external components

APPLICATIONS
- AM and AM-stereo automotive radios
- CB transmitter/receivers
- Shortwave receivers
- Mobile communications equipment

ABSOLUTE MAXIMUM RATINGS
Supply voltage, V_{CC}	12 V
Package power dissipation, P_D	800 mW
Operating temperature range, T_A	$-40\,°C$ to $+125\,°C$
Storage temperature range, T_S	$-55\,°C$ to $+125\,°C$

Always order by complete part number:

Part Number	Function	Package
ULN3845A	Stereo	18-Pin DIP
ULN3845LW	Noise Blanker	18-Lead SOIC
UNL3846A	Mono	18-Pin DIP
ULN3846LW	Noise Blanker	18-Lead SOIC

Allegro 3845/3846

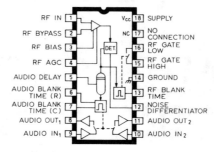

The ULN3845A (dual in-line package) and ULN3845LW (small-outline IC package) are electrically identical and share a common pin number assignment.

TEST CIRCUIT

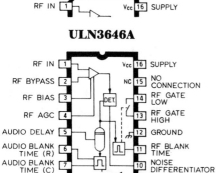
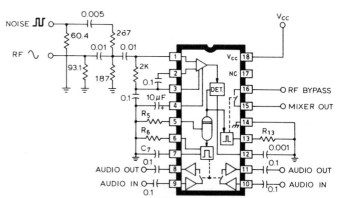

The ULN3846A (dual in-line package) and the ULN3846LW (small-outline IC package) are electrically identical and share a common pin number assignment.

Note that the noise-pulse input is attenuated 20 dB by the test circuit.

FUNCTIONAL BLOCK DIAGRAM
(ULN3845A/LW)

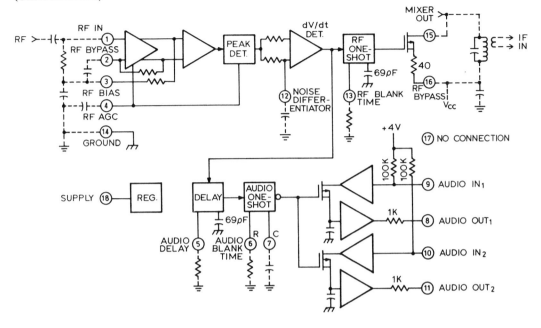

QUIESCENT DC VOLTAGES
(FOR CIRCUIT DESIGN INFORMATION ONLY)

Pin Number		Pin Function	Typical
ULN3845A/LW	ULN3846A/LW		DC Voltage
1	1	RF In	3.1
2	2	RF Bypass	3.1
3	3	RF Bias	3.1
4	4	RF AGC	0.9
5	5	Audio Delay	4.8
6	6	Audio Blank Time (R)	4.8
7	7	Audio Blank Time (C)	4.8
8	8	Audio Out$_x$	4.75
9	9	Audio In$_x$	4.0
10	—	Audio In$_2$	4.75
11	—	Audio Out$_2$	4.0
12	10	Noise Differentiator	4.9
13	11	RF Blank Time	4.8
14	12	Ground	Reference
15	13	RF Gate High	—
16	14	RF Gate Low	—
17	15	No Connection	0
18	16	Supply	V_{CC}

ELECTRICAL CHARACTERISTICS at $T_A = +25°C$, $V_{CC} = 9$ V, $f_{rf} = 1$ MHz, Noise (f_{noise}) = 500 Hz Square Wave, $f_{af} = 1$ kHz

Characteristic	Test Pins*	Test Conditions	Limits Min.	Limits Typ.	Limits Max.	Units
Supply Voltage Range	18	Operating	7.5	9.0	12	V
Quiescent Supply Current	18	$V_{RF} = 0$	—	12	20	mA
RF INPUT AMPLIFIER:						
Trigger Threshold	1	Noise Pulse Amplitude for $V_{RF} = 0$	—	100	—	µV
Modulation Threshold	1	Noise Pulse Modulation for $V_{RF} = 1$ mV	—	85	—	%
Detector Rise Time	12	$C_{12} = 0$	—	500	—	ns
RF SWITCH:						
ON Resistance	15-16		—	50	70	Ω
OFF Resistance	15-16		—	100	—	kΩ
Time Delay	1-15	From Beginning of RF Pulse to Beginning of RF Blanking	—	1.5	3.0	µs
AUDIO SWITCHES:						
Attenuation	9-8, 10-11		60	80	—	dB
Noise	8, 11		—	1.5	6.0	mVpp
Crosstalk	8, 11	ULN3845A/LW Only	—	60	—	dB
Gain	9-8, 10-11		-1.0	-0.5	0	dB
Total Harmonic Distortion	8, 11	$V_{af} = 300$ mV, $V_{noise} = 0$	—	<0.1	0.5	%
Input Impedance	9, 10		—	100	—	kΩ
Output Impedance	8, 11		—	1.0	—	kΩ
BLANKING TIMERS:						
RF Blanking	15	$R_{13} = 350$ kΩ	45	55	65	µs
Audio Delay	8	$R_5 = 350$ kΩ	40	50	62	µs
Audio Blanking	8	$R_6 = 110$ kΩ, $C_7 = 0.0012$ µF	220	290	360	µs

*Pin numbers are for ULN3845A/LW.

TYPICAL PULSE RESPONSE

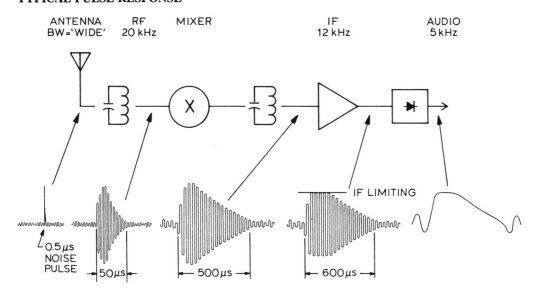

TYPICAL RF FREQUENCY RESPONSE

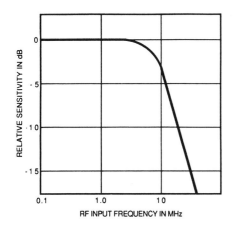

COIL INFORMATION FOR HIGH-PERFORMANCE ETR AM STEREO RECEIVER WITH NOISE BLANKING

	Symbol	Q	N1:N2	N1:N3	Toko Part Number
Antenna	T_1		1:1.6		7HN-60064CY
RF	T_2, T_3	120		10:1	RWOS-6A7894AO, $L = 178\ \mu H$
Local Osc.	T_4	120		5:1	7TRS-A5609AO
Mixer	T_5		2:1	8.9:1	7LC-502112N4, $C_T = 180\ pF$
Detector	L_2	100			A7BRS-T1041Z, $C_T = 1000\ pF$

® Registered trademark of MOTOROLA, INC.

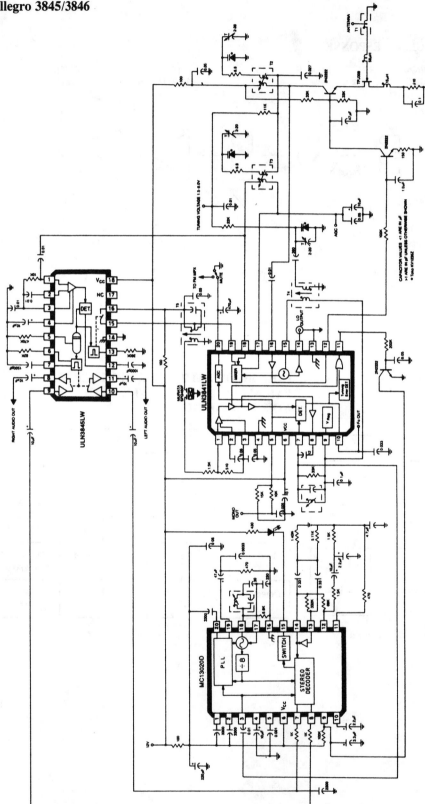

ETR AM STEREO RECEIVER WITH NOISE BLANKING

Allegro 3847
Dual-Conversion AM Receiver

FEATURES
- Low noise figure
- Balanced mixers
- Buffered oscillators
- Very effective stop detector
- Dual wide-band AGC
- Delay AGC
- Narrow-band FM output
- 6.5- to 12-V operating range

APPLICATIONS
- Automobile radios
- High-quality home entertainment receivers

ABSOLUTE MAXIMUM RATINGS
Supply voltage, V_{CC} 12 V
Package power dissipation, P_D 1.2 W
Operating temperature range, T_A $-40\,°C$ to $+85\,°C$
Storage temperature range, T_S $-65\,°C$ to $+150\,°C$

ELECTRICAL CHARACTERISTICS at $T_A = +25\,°C$, $V_{CC} = 10$ V

Characteristic	Symbol	Test Conditions	Min.	Typ.	Max.	Units
Supply Current	I_{CC}	I_2, $V_{in} = 0$	—	45	65	mA
Sensitivity	V_{in}	$V_{out} = 50$ mV	—	6.0	—	µV
Recovered Audio	V_{out}	$V_{in} = 1$ mV	200	250	—	mV
Total Harmonic Dist.	THD	$V_{in} = 1$ mV, Mod = 80%	—	0.4	1.5	%
Oscillator Output	V_o	V_{15}	180	300	—	mV
Stop Output Voltage	V_{STP}	V_7, $V_{in} = 0$	—	4.8	—	V
		V_7, $V_{in} = 1$ mV	—	0.05	—	V
Stop Sensitivity	V_{stp}	$V_{11} = 2.5$ V, Mod = 0%	—	100	—	µV
Stop Bandwidth	BW_{STP}	$V_{in} = 1$ mV, $V_{11} = 1.5$ V, Mod = 0%	—	10.2	—	kHz
Wide-Band AGC	V_{AGC}	$V_{in} = 0$	—	7.5	—	V
		$V_{in} = 18$ mV	—	6.5	—	V
		$V_{in} = 60$ mV	—	1.0	—	V
Overload	V_{in}	$V_{out} = 10\%$ THD, Mod = 80%	—	200	—	mV
-3dB Limiting	V_{in}	Mod = 3 kHz peak deviation	—	12	—	µV
FM Recovered Audio	V_{out}	V_6, Mod = 3 kHz peak deviation	—	380	—	mV
Signal to Noise Ratio	S+N/N	$V_{in} = 250$ µV	—	50	—	dB
		$V_{in} = 10$ mV	—	60	—	dB
AGC Figure of Merit	FOM	Ref. at $V_{in} = 5$ mV, V_{in} or $V_{out} = -10$ dB	—	20	—	µV
Regulator Voltage	V_{REG}	V_5	—	5.1	—	V
		V_5, $V_{24} = 0$ (Muted)	—	0	0.2	V
Reference Voltage	V_{REF}	V_{13}	—	3.7	—	V

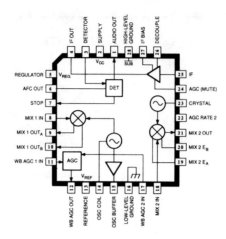

Allegro 3827
FM Stereo Decoder

FEATURES
- Reduced automotive stereo multi-path effects
- Dual-bandwidth phase-locked loop
- No adjustments required
- Improved adjacent-channel rejection
- Good ARI/RDS rejection
- 19-kHz pilot canceling
- Noise-actuated blending and high cut
- Ceramic-resonator controlled oscillator
- Automatic stereo/mono switching

ABSOLUTE MAXIMUM RATINGS
Supply voltage, V_{CC}	12 V
Package power dissipation, P_D	1.0 W
Operating temperature range, T_A	$-40\,°C$ to $+85\,°C$
Storage temperature range, T_S	$-65\,°C$ to $+150\,°C$

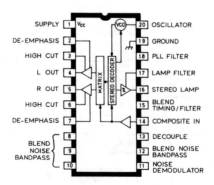

FUNCTIONAL BLOCK DIAGRAM

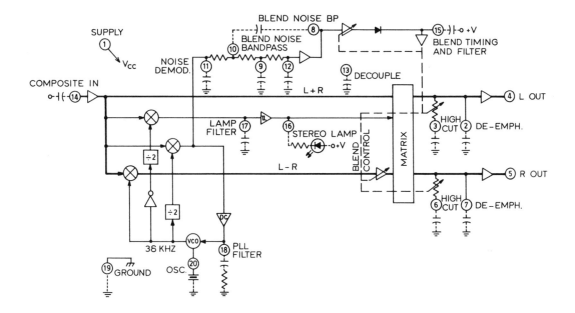

TEST CIRCUIT AND TYPICAL APPLICATION

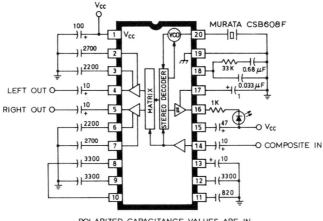

POLARIZED CAPACITANCE VALUES ARE IN µF, NON-POLARIZED CAPACITANCE VALUES ARE IN pF, UNLESS OTHERWISE INDICATED.

The typical application and circuit constants herein are included only as an example and provide no guarantee for designing equipment to be mass-produced. The information herein is believed to be accurate and reliable. However, no responsibility is assumed by Allegro MicroSystems for its use, nor for any infringements of patents or other rights of third parties which may result from its use.

ELECTRICAL CHARACTERISTICS at $T_A = 25°C$, $V_{CC} = 10$ V, Composite Input = 400 mVrms (L=R, pilot OFF), Pilot Level = 40 mVrms, $f_m = 1$ kHz, unless otherwise specified

Characteristic	Test Conditions	Limits			Units
		Min.	Typ.	Max.	
Supply Voltage Range	Functional	8.5	10	12	V
Max. Composite Input	THD = 1.0 %	600	800	—	mVrms
Input Impedance		—	25	—	kΩ
Output Impedance		—	1.0	—	kΩ
Stereo Channel Separation	f_m = 100 Hz	—	50	—	dB
	f_m = 1.0 kHz	30	50	—	dB
	f_m = 10 kHz	—	40	—	dB
Monaural Gain	19 kHz Pilot Level = 0	0.4	0.6	1.6	dB
Monaural Channel Balance	19 kHz Pilot Level = 0	—	0	±1.0	dB
Total Harmonic Distortion (100 Hz to 1 kHz)	19 kHz Pilot = 0	—	0.05	0.5	%
	L or R only	—	0.1	0.5	%
Ultrasonic Frequency Rejection	19 kHz	36	51	—	dB
	38 kHz	35	45	—	dB
SCA Rejection	67 kHz (Note 2)	—	65	—	dB
Spurious Response	114 kHz, 10% modulation	—	65	—	dB
	190 kHz, 10% modulation	—	65	—	dB
PLL Bandwidth	Loop Locked	—	20	—	Hz
Stereo Switch Level	19 kHz Pilot Only, Lamp ON	10	15	22	mVrms
	19 kHz Pilot Only, Lamp OFF	6.0	11	16	mVrms
Stereo Lamp Hysteresis	Lamp OFF to Lamp ON	—	3.0	—	dB
Capture Range	Pilot = 6.0 mV	—	300	—	Hz
Lock Range	Pilot = 20 mV	—	300	—	Hz
Blend Threshold	$S+N/N$	—	36	—	dB
Stereo Lamp Output Current	Short Circuit, Lamp ON	5.0	20	40	mA
	Lamp OFF, V_{CC} = 12 V	—	0	3.0	µA
Quiescent Supply Current	Lamp OFF	—	22	35	mA

NOTES: 1) Typical values are given for circuit design information only.
2) Measured with a stereo composite signal of 80% stereo, 10% pilot, and 10% SCA.

PIN FUNCTIONS

Pin	DC Voltage	Function	Notes
1	10	Supply	—
2	4.0	Left De-Emphasis	R_s = 27.8 kΩ
3	4.3	Left High Cut	R_s = 20 kΩ
4	3.4	Left Output	R_s = 1 kΩ
5	3.4	Right Output	R_s = 1 kΩ
6	4.3	Right High Cut	R_s = 20 kΩ
7	4.0	Right De-Emphasis	R_s = 27.8 kΩ
8	3.0	Blend Capacitor	—
9	3.7	Blend Capacitor	—
10	4.4	Blend Capacitor	—
11	5.1	Blend Capacitor	—
12	5.4	Blend Capacitor	—
13	5.4	Blend Decouple	—
14	3.6	Composite Input	R_{IN} = 25 kΩ [1]
15	8.3	Blend Timing & Filter	[2]
16	9.0	Stereo Indicator	Locked = 0.1 V
17	5.0	Lock Detector Filter	Locked = 4.7 V
18	5.6	Loop Filter	Locked = 4.7 [3]
19	0.0	Ground	—
20	9.0	608 kHz Resonator	—

NOTES: 1) The decoder matrix does not account for FM detector frequency roll-off. An input RC network can be used to correct for this if separation is not sufficient.

2) Blend threshold can be increased to about 42 dB (but separation will be reduced at lower levels) by adding 470 kΩ to V_{CC}. Smaller values will cause blending when it is not desired.

3) The loop filter capacitor should be low-leakage current because the phase detector output current is very low.

Allegro
3330, 3360, 3363
Optoelectronic Switches

The ULN3330T/TA, ULN3360T, and ULN3363T optoelectronic switches provide light detection and low-level signal processing in single three-lead packages. The monolithic integrated circuits, requiring no external components, meet the need for cost-effective, light-sensing devices in consumer and industrial applications. Their high sensitivity makes them ideal for low-level light detection in optically noise-free environments.

Each optoelectronic IC includes a 0.030″×0.030″ (0.76×0.76 mm) photodiode, a high-gain current amplifier, a comparator with 12% hysteresis, output driver stage, and voltage regulator.

The ULN3330T/TA and ULN3360T switches turn on as illumination of the photodiode falls below 55 μW/cm² at 880 nm. An internal latch provides hysteresis: the output turns off when illumination exceeds the turn-on threshold by approximately 12%.

The ULN3363T switch has an inverted output characteristic. It turns off as illumination falls below 55 μW/cm² at 880 nm; it remains off until increasing illumination at the photodiode typically reaches 62 μW/cm².

The ULN3330T/TA and ULN3363T have buffered open-collector outputs for current-sink applications. Typical loads include incandescent lamps, LEDs, sensitive relays, or dc motors.

The output circuitry for the ULN3360T includes an internal 5.4-kΩ pull-up resistor that enables its direct use with microprocessors and TTL logic.

Allegro 3330/3360/3363

FEATURES
- Photodiode with:
  - On-chip amplifier
  - On-chip comparator with hysteresis
  - On-chip power driver
  - On-chip voltage regulator
- Sensitive switch points
- Operation to 30 kHz

ABSOLUTE MAXIMUM RATINGS

Supply voltage, V_{CC}	15 V
Output voltage, V_{OUT}	15 V
Output current, I_{OUT}	25 mA
Operating temperature range, T_A	$-40\,°C$ to $+70\,°C$
Storage temperature range, T_S	$-55\,°C$ to $+110\,°C$

ULN3330/60/63T

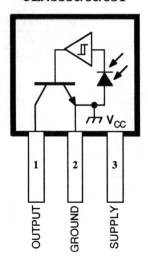

ULN3330TA

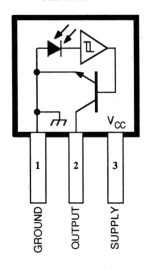

FUNCTIONAL BLOCK DIAGRAM

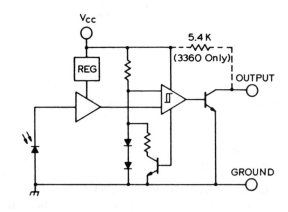

Device Type	Output	Package	Pinout (1-2-3)
ULN3330T	Open Collector	T	OUT-GND-V_{CC}
ULN3330TA	Open Collector	TA	GND-OUT-V_{CC}
ULN3360T	5.4 kΩ Pull-Up	T	OUT-GND-V_{CC}
ULN3363T	Inv. Open Collector	T	OUT-GND-V_{CC}

ELECTRICAL CHARACTERISTICS at $T_A = +25°C$, $V_{CC} = 6.0$ V, $\lambda = 800$ nm

Characteristic	Symbol	Test Conditions	Limits Min.	Limits Typ.	Limits Max.	Units
Supply Voltage Range	V_{CC}		4.0	6.0	15	V
Supply Current	I_{CC}		—	4.0	8.0	mA
Light Threshold Level	E_{ON}	Output ON	45	55	65	$\mu W/cm^2$
	E_{OFF}	Output OFF	—	62	—	$\mu W/cm^2$
Hysteresis	ΔE	$(E_{OFF} - E_{ON})/E_{OFF}$	10	13	16	%
Output ON Voltage	V_{OUT}	$I_{OUT} = 15$ mA	—	300	500	mV
		$I_{OUT} = 25$ mA	—	500	800	mV
Output OFF Current	I_{OUT}	$V_{OUT} = 15$ V	—	—	1.0	μA
Output Fall Time	t_f	90% to 10%	—	200	500	ns
Output Rise Time	t_r	10% to 90%	—	200	500	ns

TYPICAL TRANSFER CHARACTERISTICS

ULN3330T/TA AND ULN3360T

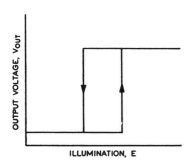

ULN3363T

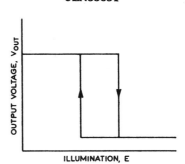

SENSOR LOCATIONS

SUFFIX 'T'

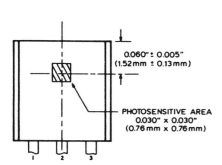

SUFFIX 'TA'

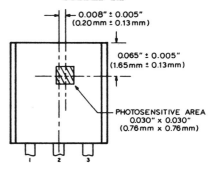

RELATIVE SPECTRAL RESPONSE AT $T_A = +25°C$ AS A FUNCTION OF WAVELENGTH OF LIGHT

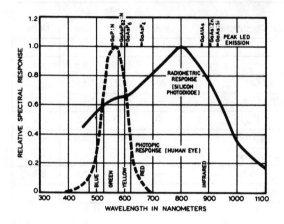

OUTPUT SATURATION VOLTAGE AS A FUNCTION OF OPERATING TEMPERATURE

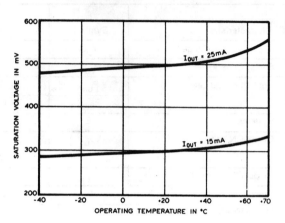

PROPAGATION DELAY AS A FUNCTION OF LIGHT LEVEL

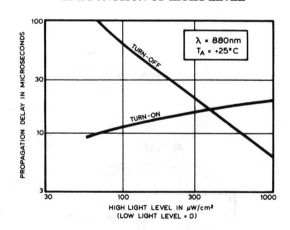

LIGHT-THRESHOLD CHANGE AS A FUNCTION OF OPERATING TEMPERATURE

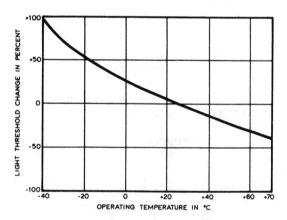

TYPICAL APPLICATIONS*

BAR CODE READER

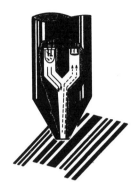

OPTICAL ISOLATOR

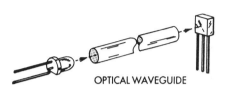

SHEET DETECTOR

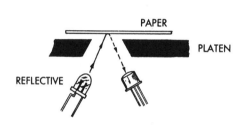

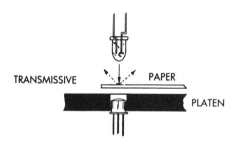

OPTICAL ENCODER

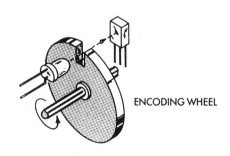

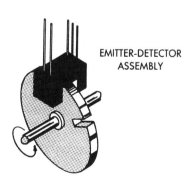

*Optics and ambient light shields omitted for clarity.

Analog Devices
AD507
IC Wideband Fast-Slewing General-Purpose Operational Amplifier

FEATURES
- Gain bandwidth: 100 MHz
- Slew rate: 20 V/μs min
- I_B: 15 nA max (AD507K)
- V_{OS}: 3 mV max (AD507K)
- V_{OS} drift: 15 μV/°C max (AD507K)
- High capacitive drive

PRODUCT HIGHLIGHTS
1. Excellent dc and ac performance combined with low cost.
2. The AD507 will drive several hundred pF of output capacitance without oscillation.
3. All guaranteed dc parameters, including offset voltage drift, are 100% tested.
4. To ensure compliance with gain bandwidth and slew rate specifications, all devices are tested for ac-performance characteristics.
5. To take full advantage of the inherent high reliability of ICs, every AD507S receives a 24-hour stabilization bake at +150°C.

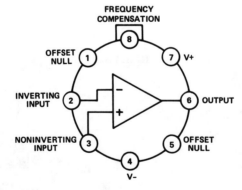

AD507 PIN CONFIGURATION

Analog Devices AD507

SPECIFICATIONS (typical at +25°C and ±15 Vdc, unless otherwise noted)

PARAMETER	AD507J	AD507K	AD507S(AD507S/883)**
OPEN LOOP GAIN			
$R_L = 2k\Omega$, $C_L = 50pF$	80,000 min (150,000 typ)	100,000 min (150,000 typ)	100,000 min (150,000 typ)
@ T_{min} to T_{max}	70,000 min	85,000 min	70,000 min
OUTPUT CHARACTERISTICS			
Voltage @ $R_L = 2k\Omega$, $C_L = 50pF$, T_{min} to T_{max}	±10V min (±12V typ)	*	±10V min (±12V typ)
Current @ $V_o = \pm10V$	±10mA min (±20mA typ)	*	±15mA min (±22mA typ)
Short Circuit Current	25mA	*	25mA
FREQUENCY RESPONSE			
Unity Gain, Small Signal			
@ A = 1 (open loop)	35MHz	*	*
@ A = 100 (closed loop)	1MHz	*	*
Full Power Response	320kHz min (600kHz typ)	400kHz min (600kHz typ)	400kHz min (600kHz typ)
Slew Rate	±20V/µs min (±35V/µs typ)	±25V/µs min (±35V/µs typ)	20V/µs min (±35V/µs typ)
Settling Time (to 0.1%)	900ns	*	*
INPUT OFFSET VOLTAGE			
Initial	5.0mV max (3.0mV typ)	3.0mV max (1.5mV typ)	4mV max (0.5mV typ)
Avg vs Temp, T_{min} to T_{max}	15µV/°C	15µV/°C max (8µV/°C typ)	20µV/°C max (8µV/°C typ)
vs Supply, T_{min} to T_{max}	200µV/V max	100µV/V max	100µV/V max
INPUT BIAS CURRENT			
Initial	25nA max	15nA max	15nA max
T_{min} to T_{max}	40nA max	25nA max	35nA max
INPUT OFFSET CURRENT			
Initial	25nA max	15nA max	15nA max
T_{min} to T_{max}	40nA max	25nA max	35nA max
Avg vs Temp, T_{min} to T_{max}	0.5nA/°C	0.2nA/°C	0.2nA/°C
INPUT IMPEDANCE			
Differential	40MΩ min (300MΩ typ)	*	65MΩ min (500MΩ typ)
Common Mode	1000MΩ	*	*
INPUT VOLTAGE NOISE			
f = 10Hz	100nV/√Hz	*	*
f = 100Hz	30nV/√Hz	*	*
f = 100kHz	12nV/√Hz	*	*
INPUT VOLTAGE RANGE			
Differential, Max Safe	±12.0V	*	*
Common Mode Voltage Range, T_{min} to T_{max}	±11.0V	*	*
Common Mode Rejection @ ±5V, T_{min} to T_{max}	74dB min (100dB typ)	80dB min (100dB typ)	80dB min (100dB typ)
POWER SUPPLY			
Rated Performance	±15V	*	*
Operating	±(5 to 20)V	*	*
Current, Quiescent	4.0mA max (3.0mA typ)	*	*
TEMPERATURE RANGE			
Rated Performance	0 to +70°C	*	-55°C to +125°C
Operating	-25°C to +85°C	*	-65°C to +150°C
Storage	-65°C to +150°C	*	*
PACKAGE OPTION			
H-08A	AD507JH	AD507KH	AD507SH

NOTES
*Specifications same as AD507J.
**AD507S/883 minimum order 10 pieces.
Specifications subject to change without notice.

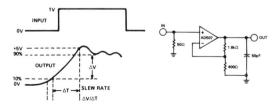

Slew Rate Definition and Test Circuit

Analog Devices AD507

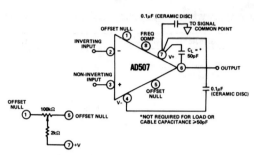

General Purpose Configuration to Closed Loop Gain > 10

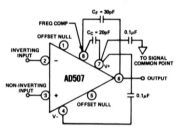

Configuration for Unity Gain Applications

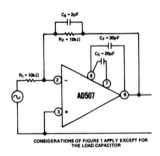

Fast Settling Time Configuration

TYPICAL PERFORMANCE CURVES

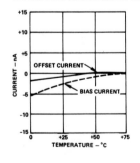

Input Bias Current and Offset Current vs Temperature

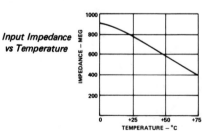

Input Impedance vs Temperature

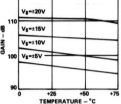

Open Loop Voltage Gain vs Temperature

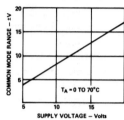

Common Mode Voltage Range vs Supply Voltage

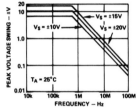

Output Voltage Swing vs Frequency

Power Supply Current vs Temperature

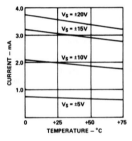

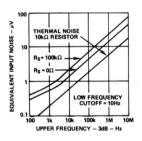

Broadband Input Noise Characteristics

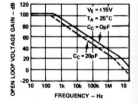

Open Loop Gain vs Frequency

Analog Devices
AD517
Low-Cost Laser-Trimmed Precision IC Op Amp

FEATURES
- Low input bias current: 1 nA max (AD517L)
- Low input offset current: 0.25 nA max (AD517L)
- Low V_{OS}: 50 μV max (AD517L), 150 μV max (AD517J)
- Low V_{OS} drift: 1.3 μV/°C (AD517L)
- Internal compensation
- MIL-standard parts available
- 8-pin TO-99 hermetic metal can
- Available in chip form

AD517 PIN CONFIGURATION

TOP VIEW

PRODUCT HIGHLIGHTS
1. Offset voltage is 100% tested and guaranteed on all models.
2. The AD517 exhibits extremely low input bias currents without sacrificing CMRR (over 100 dB) or offset voltage stability.
3. The AD517 inputs are protected (to $\pm V_S$), preventing offset-voltage and bias-current degradation as a result of reverse breakdown of the input transistors.
4. Internal compensation is provided, eliminating the need for additional components (often required by high-accuracy IC op amps).
5. The AD517 can directly replace 725, 108, and AD510 amplifiers. In addition, it can replace 741-type amplifiers if the offset-nulling potentiometer is removed.
6. Thermally-balanced layout ensures high open-loop gain independent of thermal gradients induced by output loading, offset nulling, and power-supply variations.
7. Chips are available.

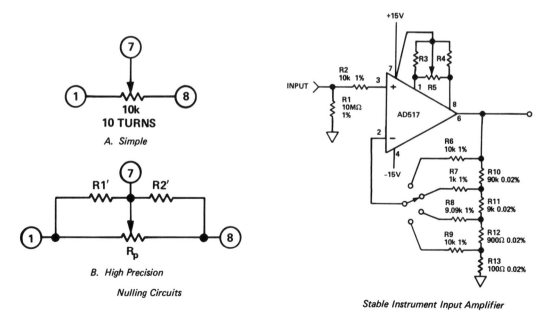

A. Simple

B. High Precision

Nulling Circuits

Stable Instrument Input Amplifier

Analog Devices AD517

SPECIFICATIONS (@ +25°C and $V_S = \pm 15$ Vdc)

Model	AD517J Min	AD517J Typ	AD517J Max	AD517K Min	AD517K Typ	AD517K Max	AD517L Min	AD517L Typ	AD517L Max	AD517S Min	AD517S Typ	AD517S Max	Units
OPEN LOOP GAIN													
$V_O = \pm 10V$, $R_L \geq 2k\Omega$	10^6			10^6			10^6			10^6			V/V
T_{min} to T_{max}, $R_L = 2k\Omega$	500,000			500,000			500,000			250,000			V/V
OUTPUT CHARACTERISTICS													
Voltage @ $R_L = 2k\Omega$, T_{min} to T_{max}	± 10			± 10			± 10			± 10			V
Load Capacitance		1000			1000			1000			1000		pF
Output Current	10			10			10			10			mA
Short Circuit Current		25			25			25			25		mA
FREQUENCY RESPONSE													
Unity Gain Small Signal		250			250			250			250		kHz
Full Power Response		1.5			1.5			1.5			1.5		kHz
Slew Rate, Unity Gain		0.10			0.10			0.10			0.10		V/μs
INPUT OFFSET VOLTAGE													
Initial Offset			150			75			50			75	μV
Input Offset vs. Temp.			3.0			1.8			1.3			1.8	μV/°C
Input Offset vs. Supply			25			10			10			10	μV/V
T_{min} to T_{max}			40			15			15			20	μV/V
INPUT BIAS CURRENT													
Initial			5			2			1.0			2.0	nA
T_{min} to T_{max}			8			3.5			1.5			10	nA
vs. Temp, T_{min} to T_{max}			±20			±10			±4			±10	pA/°C
INPUT OFFSET CURRENT													
Initial			1.0			0.75			0.25			2.0	nA
T_{min} to T_{max}			1.5			1.25			0.4			10	nA
INPUT IMPEDANCE													
Differential		15∥1.5			20∥1.5			20∥1.5			20∥1.5		MΩ∥pF
Common Mode		$2.0 \times 10^{''}$			$2.0 \times 10^{''}$			$2.0 \times 10^{''}$			$2.0 \times 10^{''}$		Ω
INPUT VOLTAGE RANGE													
Differential		$\pm V_S$			$\pm V_S$			$\pm V_S$			$\pm V_S$		V
Common Mode Rejection	94			110			110			110			dB
Common Mode Rejection T_{min} to T_{max}	94			110			100			100			dB
INPUT NOISE													
Voltage, 0.1Hz to 10Hz		2			2			2			2		μV p-p
f = 10Hz		35			35			35			35		nV/√Hz
f = 100Hz		25			25			25			25		nV/√Hz
f = 1kHz		20			20			20			20		nV/√Hz
Current, f = 10kHz		0.05			0.05			0.05			0.05		pA/√Hz
f = 100Hz		0.03			0.03			0.03			0.03		pA/√Hz
f = 1kHz		0.03			0.03			0.03			0.03		pA/√Hz
POWER SUPPLY													
Rated Performance		±15			±15			±15			±15		V
Operating	±5		±18	±5		±18	±5		±18	±5		±18	V
Quiescent Current			4			3			3			3	mA
TEMPERATURE RANGE													
Operating, Rated Performance	0		+70	0		+70	0		+70	−55		+125	°C
Storage	−65		+150	−65		+150	−65		+150	−65		+150	°C
PACKAGE OPTION TO-99 Style (H-08B) J and S Grade Chips Also Available		AD517JH			AD517KH			AD517LH			AD517SH			

NOTES

Specifications subject to change without notice.
Specifications shown in boldface are tested on all production units at final electrical test. Results from those tests are used to calculate outgoing quality levels. All min and max specifications are guaranteed, although only those shown in boldface are tested on all production units.

TYPICAL PERFORMANCE CURVES

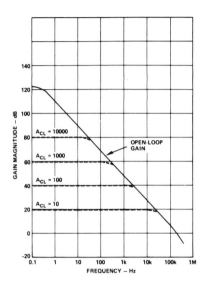

Small-Signal Gain vs. Frequency

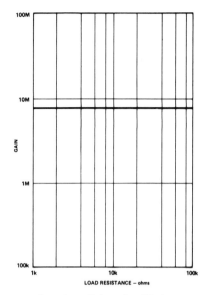

Open-Loop Gain vs. Load Resistance

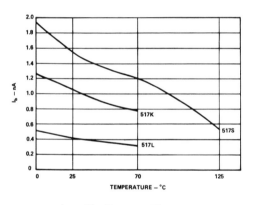

Input Bias Current vs. Temperature

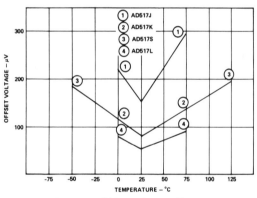

Untrimmed Offset Voltage vs. Temperature

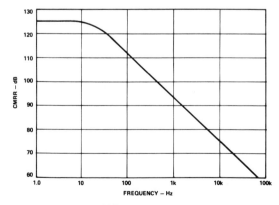

CMRR vs. Frequency

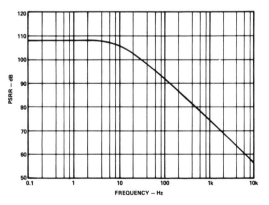

PSRR vs. Frequency

60 Analog Devices AD517

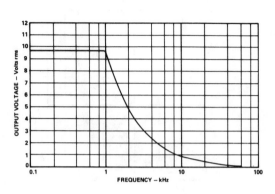

Maximum Undistorted Output vs. Frequency (Distortion ≤ 1%)

Output Voltage vs. Load Resistance

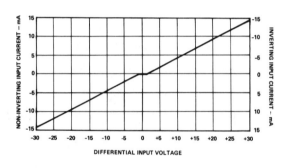

Input Current vs. Differential Input Voltage

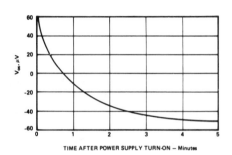

Warm-Up Offset Voltage Drift

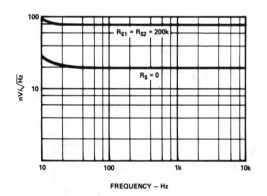

Total Input Noise Voltage vs. Frequency

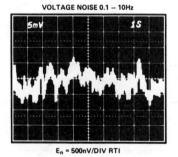

Low Frequency Voltage Noise (0.1 to 10Hz)

Analog Devices
AD644
Dual High-Speed Implanted BiFET Op Amp

FEATURES
- Matched offset voltage
- Matched offset voltage over temperature
- Matched bias currents
- Crosstalk −124 dB at 1 kHz
- Low bias current: 35 pA max warmed up
- Low offset voltage: 500 μV max
- Low input voltage noise: 2 μV p-p
- High slew rate: 13 V/μs
- Low quiescent current: 4.5 mA max
- Fast setting to ±0.01%: 3 μs
- Low total harmonic distortion: 0.0015% at 1 kHz
- Standard dual amplifier pin out
- Available in hermetic metal can package and chip form
- MIL-STD-883B processing available
- Single version available: AD544

PRODUCT HIGHLIGHTS
1. The AD644 has tight side-to-side matching specifications to ensure high performance without matching individual devices.
2. Analog Devices, unlike some manufacturers, specifies each device for the maximum bias current at either input in the warmed-up condition, thus assuring the user that the AD644 will meet its published specifications in actual use.
3. Laser-wafer-trimming reduces offset voltage as low as 0.5 mV max matched side to side to 0.25 mV (AD644L), thus eliminating the need for external nulling.
4. Improved bipolar and JFET processing on the AD644 result in the lowest matched bias current available in a high-speed monolithic FET op amp.
5. Low voltage noise (2 μV p-p) and high open-loop gain enhance the AD644's performance as a precision op amp.
6. The high slew rate (13.0 V/μs) and fast settling time to 0.01% (3.0 μs) make the AD644 ideal for D/A, A/D, sample-and-hold circuits and dual high-speed integrators.
7. Low harmonic distortion (0.0015%) and low crosstalk (−124 dB) make the AD644 an ideal choice for stereo audio applications.
8. The standard dual amplifier pin out allows the AD644 to replace lower performance duals without redesign.
9. The AD644 is available in chip form.

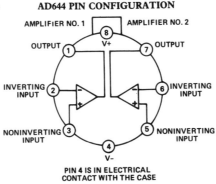

Analog Devices AD644

SPECIFICATIONS (@ +25°C and $V_S = \pm 15$ Vdc)

Model	AD644J Min	AD644J Typ	AD644J Max	AD644K Min	AD644K Typ	AD644K Max	AD644L Min	AD644L Typ	AD644L Max	AD644S Min	AD644S Typ	AD644S Max	Units
OPEN LOOP GAIN													
$V_O = \pm 10V$, $R_L \geq 2k\Omega$	30,000			50,000			50,000			50,000			V/V
T_{min} to T_{max}, $R_L = 2k\Omega$	20,000			40,000			40,000			20,000			V/V
OUTPUT CHARACTERISTICS													
Voltage @ $R_L = 2k\Omega$, T_{min} to T_{max}	±10	±12		±10	±12		±10	±12		±10	±12		V
Voltage @ $R_L = 10k\Omega$, T_{min} to T_{max}	±12	±13		±12	±13		±12	±13		±12	±13		V
Short Circuit Current		25			25			25			25		mA
FREQUENCY RESPONSE													
Unity Gain Small Signal		2.0			2.0			2.0			2.0		MHz
Full Power Response		200			200			200			200		kHz
Slew Rate, Unity Gain	8.0	13.0		8.0	13.0		8.0	13.0		8.0	13.0		V/µs
Total Harmonic Distortion		0.0015			0.0015			0.0015			0.0015		%
INPUT OFFSET VOLTAGE[1]													
Initial Offset			2.0			1.0			0.5			1.0	mV
Input Offset Voltage T_{min} to T_{max}			3.5			2.0			1.0			3.5	mV
Input Offset Voltage vs. Supply, T_{min} to T_{max}			200			100			100			100	µV/V
INPUT BIAS CURRENT[2]													
Either Input		10	75		10	35		10	35		10	35	pA
Offset Current		10			5			5			5		pA
MATCHING CHARACTERISTICS[3]													
Input Offset Voltage			1.0			0.5			0.25			0.5	mV
Input Offset Voltage T_{min} to T_{max}			3.5			2.0			1.0			3.5	mV
Input Bias Current			35			25			25			35	pA
Crosstalk		−124			−124			−124			−124		dB
INPUT IMPEDANCE													
Differential		$10^{12}\|6$			$10^{12}\|6$			$10^{12}\|6$			$10^{12}\|6$		MΩ|pF
Common Mode		$10^{12}\|3$			$10^{12}\|3$			$10^{12}\|3$			$10^{12}\|3$		MΩ|pF
INPUT VOLTAGE RANGE													
Differential[4]		±20			±20			±20			±20			V
Common Mode	±10	±12		±10	±12		±10	±12		±10	±12		V
Common Mode Rejection	76			80			80			80			dB
INPUT NOISE													
Voltage 0.1Hz to 10Hz		2			2			2			2		µV p-p
f = 10Hz		35			35			35			35		nV/√Hz
f = 100Hz		22			22			22			22		nV/√Hz
f = 1kHz		18			18			18			18		nV/√Hz
f = 10kHz		16			16			16			16		nV/√Hz
POWER SUPPLY													
Rated Performance		±15			±15			±15			±15		V
Operating	±5		±18	±5		±18	±5		±18	±5		±18	V
Quiescent Current		3.5	4.5		3.5	4.5		3.5	4.5		3.5	4.5	mA
TEMPERATURE RANGE													
Operating, Rated Performance	0		+70	0		+70	0		+70	−55		+125	°C
Storage	−65		+150	−65		+150	−65		+150	−65		+150	°C
PACKAGE OPTION													
TO-99 Style (H-08B)		AD644JH			AD644KH			AD644LH			AD644SH		

NOTES
[1] Input Offset Voltage specifications are guaranteed after 5 minutes of operation at $T_A = +25°C$.
[2] Bias Current specifications are guaranteed at maximum at either input after 5 minutes of operation at $T_A = +25°C$. For higher temperatures, the current doubles every 10°C.
[3] Matching is defined as the difference between parameters of the two amplifiers.
[4] Defined as voltage between inputs, such that neither exceeds ±10V from ground.

Specifications subject to change without notice.

Specifications shown in boldface are tested on all production units at final electrical test. Results from those tests are used to calculate outgoing quality levels. All min and max specifications are guaranteed, although only those shown in boldface are tested on all production units.

Analog Devices AD644

TYPICAL CHARACTERISTICS

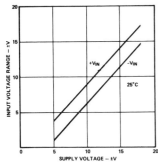

Figure 1. Input Voltage Range vs. Supply Voltage

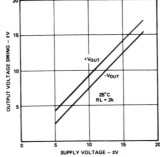

Figure 2. Output Voltage Swing vs. Supply Voltage

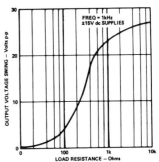

Figure 3. Output Voltage Swing vs. Load Resistance

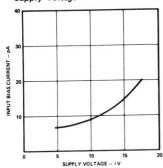

Figure 4. Input Bias Current vs. Supply Voltage

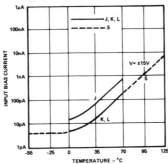

Figure 5. Input Bias Current vs. Temperature

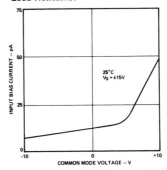

Figure 6. Input Bias Current vs. CMV

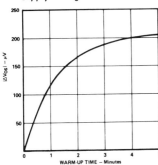

Figure 7. Change in Offset Voltage vs. Warm-Up Time

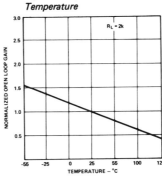

Figure 8. Open Loop Gain vs. Temperature

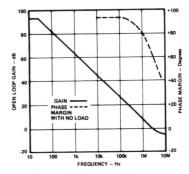

Figure 9. Open Loop Frequency Response

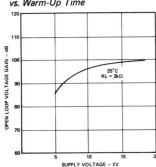

Figure 10. Open Loop Voltage Gain vs. Supply Voltage

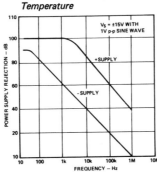

Figure 11. Power Supply Rejection vs. Frequency

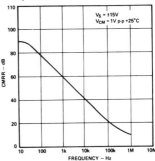

Figure 12. Common Mode Rejection Ratio vs. Frequency

64 Analog Devices AD644

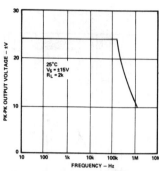

Figure 13. Large Signal Frequency Response

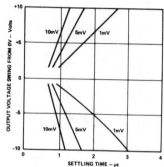

Figure 14. Output Swing and Error vs. Settling Time (Circuit of Figure 23a)

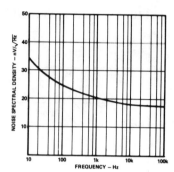

Figure 15. Noise Spectral Density

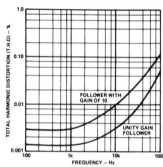

Figure 16. Total Harmonic Distortion vs. Frequency

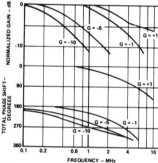

Figure 17. Closed Loop Gain & Phase vs. Frequency

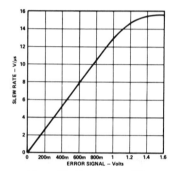

Figure 18. Slew Rate vs. Error Signal

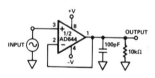

a. Unity Gain Follower

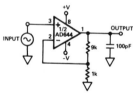

b. Follower with Gain = 10

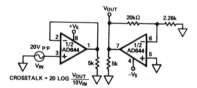

Figure 20. Crosstalk Test Circuit

Figure 19. T.H.D. Test Circuits

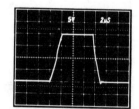

Figure 21a. Unity Gain Follower Pulse Response (Large Signal)

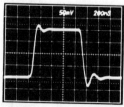

Figure 21b. Unity Gain Follower Pulse Response (Small Signal)

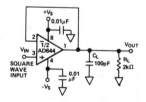

Figure 21c. Unity Gain Follower

Analog Devices AD644

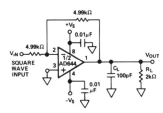

Figure 22a. Unity Gain Inverter

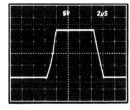

Figure 22b. Unity Gain Inverter Pulse Response (Large Signal)

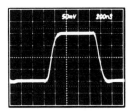

Figure 22c. Unity Gain Inverter Pulse Response (Small Signal)

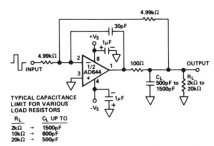

Circuit for Driving a Large Capacitive Load

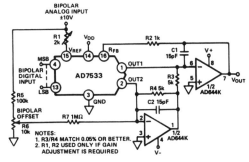

AD644 Used as DAC Output Amplifiers

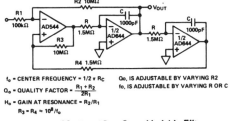

f_o = CENTER FREQUENCY = $1/2 \pi RC$

Q_o = QUALITY FACTOR = $\frac{R_1 + R_2}{2R_1}$

H_o = GAIN AT RESONANCE = R_2/R_1

$R_3 = R_4 \approx 10^8/f_o$

Q_o IS ADJUSTABLE BY VARYING R_2

f_o IS ADJUSTABLE BY VARYING R OR C

Figure 28. Band Pass State Variable Filter

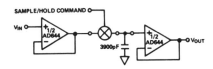

Sample and Hold Circuit

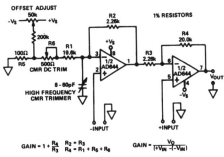

GAIN = $1 + \frac{R_4}{R_3}$ $R_2 = R_3$ $R_4 = R_1 + R_5 + R_6$

GAIN = $\frac{V_O}{(+V_{IN}) - (-V_{IN})}$

INSTRUMENTATION AMPLIFIER WITH GAIN OF TEN

Wide Bandwidth Instrumentation Amplifier

Analog Devices
AD829
High-Speed Low-Noise Video Op Amp

FEATURES
- High speed
  - 120-MHz bandwidth, gain = −1
  - 230-V/µs slew rate
  - 90-ns settling time to 0.1%
- Ideal for video applications
  - 0.02% differential gain
  - 0.04% differential phase
- Low noise
  - 2-nV/$\sqrt{\text{Hz}}$ input voltage noise
  - 1.5-pA/$\sqrt{\text{Hz}}$ input current noise
- Excellent dc precision
  - 1 mV max input offset voltage (over temp)
  - 0.3-µV/°C input offset drift
- Flexible operation
  - Specified for ±5- to ±15-V operation
  - ±3-V output swing into a 150-Ω load
  - External compensation for gains 1 to 20
  - 5-mA supply current

PRODUCT HIGHLIGHTS

1. Input voltage noise of 2 nV/$\sqrt{\text{Hz}}$, current noise of 1.5 pA/$\sqrt{\text{Hz}}$ and 50-MHz bandwidth, for gains of 1 to 20, make the AD829 an ideal preamp.
2. Differential phase error of 0.04° and 0.02% differential-gain error, at the 3.58-MHz NTSC and 4.43-MHz PAL and SECAM color subcarrier frequencies, make it an outstanding video performer for driving reverse-terminated 50- and 75-Ω cables to ±1 V (at their terminated end).
3. The AD829 can drive heavy capacitive loads.
4. Performance is fully specified for operation from ±5- to ±15-V supplies.
5. Available in plastic, cerdip, and small outline packages. Chips and MIL-STD-883B parts are also available.

SPECIFICATIONS (@ T_A = +25 °C and V_S = ±15 Vdc, unless otherwise noted)

Model	Conditions	V_S	AD829J Min	AD829J Typ	AD829J Max	AD829 A/S Min	AD829 A/S Typ	AD829 A/S Max	Units
INPUT OFFSET VOLTAGE		±5 V, ±15 V		0.2	1		0.1	0.5	mV
	T_{min} to T_{max}				1			0.5	mV
Offset Voltage Drift		±5 V, ±15 V		0.3			0.3		µV/°C
INPUT BIAS CURRENT		±5 V, ±15 V		3.3	7		3.3	7	µA
	T_{min} to T_{max}				8.2			9.5	µA
INPUT OFFSET CURRENT		±5 V, ±15 V		50	500		50	500	nA
	T_{min} to T_{max}				500			500	nA
Offset Current Drift		±5 V, ±15 V		0.5			0.5		nA/°C
OPEN-LOOP GAIN	$V_O = ±2.5$ V	±5 V							
	$R_{LOAD} = 500$ Ω		30	65		30	65		V/mV
	T_{min} to T_{max}		20			20			V/mV
	$R_{LOAD} = 150$ Ω			40			40		V/mV
	$V_{OUT} = ±10$ V	±15 V							
	$R_{LOAD} = 1$ kΩ		50	100		50	100		V/mV
	T_{min} to T_{max}		20			20			V/mV
	$R_{LOAD} = 500$ Ω			85			85		V/mV
DYNAMIC PERFORMANCE									
Gain Bandwidth Product		±5 V		600			600		MHz
		±15 V		750			750		MHz
Full Power Bandwidth[1, 2]	$V_O = 2$ V p-p								
	$R_{LOAD} = 500$ Ω	±5 V		25			25		MHz

Model	Conditions	V_S	AD829J Min	AD829J Typ	AD829J Max	AD829 A/S Min	AD829 A/S Typ	AD829 A/S Max	Units
DYNAMIC PERFORMANCE	$V_O = 20$ V p-p								
	$R_{LOAD} = 1$ kΩ	± 15 V		3.6			3.6		MHz
Slew Rate[2]	$R_{LOAD} = 500\ \Omega$	± 5 V		150			150		V/μs
	$R_{LOAD} = 1$ kΩ	± 15 V		230			230		V/μs
Settling Time to 0.1%	$A_V = -1$,								
	-2.5 V to $+2.5$ V	± 5 V		65			65		ns
	10 V Step	± 15 V		90			90		ns
Phase Margin[2]	$C_{LOAD} = 10$ pF	± 15 V							
	$R_{LOAD} = 1$ kΩ			60			60		Degrees
DIFFERENTIAL GAIN ERROR[3]	$R_{LOAD} = 100\ \Omega$	± 15 V							
	$C_{COMP} = 30$ pF			0.02			0.02		%
DIFFERENTIAL PHASE ERROR[3]	$R_{LOAD} = 100\ \Omega$	± 15 V							
	$C_{COMP} = 30$ pF			0.04			0.04		Degrees
COMMON-MODE REJECTION	$V_{CM} = \pm 2.5$ V	± 5 V	100	120		100	120		dB
	$V_{CM} = \pm 12$ V	± 15 V	100	120		100	120		dB
	T_{min} to T_{max}		96			96			dB
POWER SUPPLY REJECTION	$V_S = \pm 4.5$ V to ± 18 V		98	120		98	120		dB
	T_{min} to T_{max}		94			94			dB
INPUT VOLTAGE NOISE	$f = 1$ kHz	± 15 V		2			2		nV/$\sqrt{Hz}$
INPUT CURRENT NOISE	$f = 1$ kHz	± 15 V		1.5			1.5		pA/$\sqrt{Hz}$
INPUT COMMON-MODE VOLTAGE RANGE		± 5 V		+4.3			+4.3		V
				−3.8			−3.8		V
		± 15 V		+14.3			+14.3		V
				−13.8			−13.8		V
OUTPUT VOLTAGE SWING	$R_{LOAD} = 500\ \Omega$	± 5 V	3.0	3.6		3.0	3.6		$\pm$V
	$R_{LOAD} = 150\ \Omega$	± 5 V	2.5	3.0		2.5	3.0		$\pm$V
	$R_{LOAD} = 50\ \Omega$	± 5 V		1.4			1.4		$\pm$V
	$R_{LOAD} = 1$ kΩ	± 15 V	12	13.3		12	13.3		$\pm$V
	$R_{LOAD} = 500\ \Omega$	± 15 V	10	12.2		10	12.2		$\pm$V
Short Circuit Current		± 5 V, ± 15 V		32			32		mA
INPUT CHARACTERISTICS									
Input Resistance (Differential)				13			13		kΩ
Input Capacitance (Differential)[4]				5			5		pF
Input Capacitance (Common Mode)				1.5			1.5		pF
CLOSED-LOOP OUTPUT RESISTANCE	$A_V = +1$, $f = 1$ kHz			2			2		MΩ
POWER SUPPLY									
Operating Range			± 4.5		± 18	± 4.5		± 18	V
Quiescent Current		± 5 V		5	6.5		5	6.5	mA
	T_{min} to T_{max}				8.0			8.2/8.7	mA
		± 15 V		5.3	6.8		5.3	6.8	mA
	T_{min} to T_{max}				8.3			8.5/9.0	mA
TRANSISTOR COUNT	Number of Transistors			46			46		

NOTES
[1] Full Power Bandwidth = Slew Rate/2 π V_{PEAK}.
[2] Tested at Gain = +20, $C_{COMP} = 0$ pF.
[3] 3.58 MHz (NTSC) and 4.43 MHz (PAL & SECAM).
[4] Differential input capacitance consists of 1.5 pF package capacitance plus 3.5 pF from the input differential pair.

Specifications subject to change without notice.

Analog Devices AD829

AD829 CONNECTION DIAGRAM

8-Pin Plastic Mini-DIP (N),
Cerdip (Q) and SOIC (R) Packages

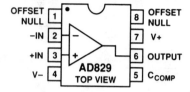

ORDERING GUIDE

Model	Temperature Range	Package Description
AD829JN	0 to +70°C	8-Pin Plastic Mini-DIP
AD829JR	0 to +70°C	8-Pin Plastic SOIC
AD829AQ	−40°C to +85°C	8-Pin Cerdip
AD829SQ	−55°C to +125°C	8-Pin Cerdip
AD829SQ/883B	−55°C to +125°C	8-Pin Cerdip

ABSOLUTE MAXIMUM RATINGS[1]

Supply Voltage .... ±18 V
Internal Power Dissipation[2]
    Plastic (N) .... 1.3 Watts
    Small Outline (R) .... 0.9 Watts
    Cerdip (Q) .... 1.3 Watts
Input Voltage .... ±Vs
Differential Input Voltage[3] .... ±6 Volts
Output Short Circuit Duration .... Indefinite
Storage Temperature Range Q .... −65°C to +150°C
Storage Temperature Range N, R .... −65°C to +125°C
Operating Temperature Range
    AD829J .... 0 to +70°C
    AD829A .... −40°C to +85°C
    AD829S .... −55°C to +125°C
Lead Temperature Range (Soldering 60 sec) .... +300°C

NOTES
[1]Stresses above those listed under "Absolute Maximum Ratings" may cause permanent damage to the device. This is a stress rating only and functional operation of the device at these or any other conditions above those indicated in the operational section of this specification is not implied. Exposure to absolute maximum rating conditions for extended periods may affect device reliability.
[2]Maximum internal power dissipation is specified so that T_J does not exceed +175°C at an ambient temperature of +25°C.
Thermal characteristics:
    8-pin plastic package: θ_{JA} = 100°C/watt (derate at 8.7 mW/°C)
    8-pin cerdip package: θ_{JA} = 110°C/watt (derate at 8.7 mW/°C)
    8-pin small outline package: θ_{JA} = 155°C/watt (derate at 6 mW/°C).
[3]If the differential voltage exceeds 6 volts, external series protection resistors should be added to limit the input current.

TYPICAL PERFORMANCE CHARACTERISTICS

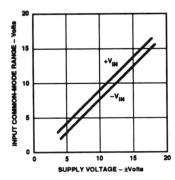

Figure 1. Input Common-Mode Range vs. Supply Voltage

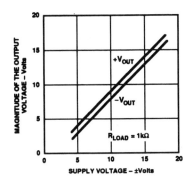

Figure 2. Output Voltage Swing vs. Supply Voltage

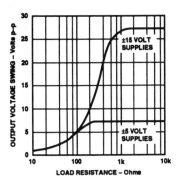

Figure 3. Output Voltage Swing vs. Resistive Load

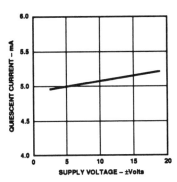

Figure 4. Quiescent Current vs. Supply Voltage

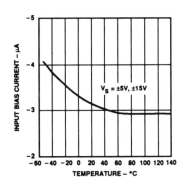

Figure 5. Input Bias Current vs. Temperature

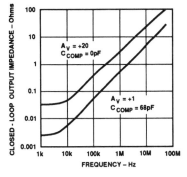

Figure 6. Closed-Loop Output Impedance vs. Frequency

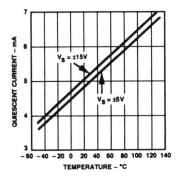

Figure 7. Quiescent Current vs. Temperature

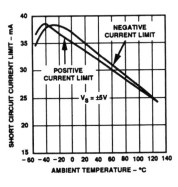

Figure 8. Short Circuit Current Limit vs. Temperature

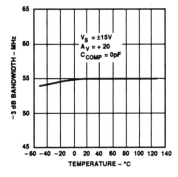

Figure 9. −3 dB Bandwidth vs. Temperature

Analog Devices
AD840
Wideband, Fast-Settling Op Amp

FEATURES
- Wideband ac performance
  Gain bandwidth product: 400 MHz (Gain ≥ 10)
  Fast settling: 100 ns to 0.01% for a 10-V step
  Slew rate: 400 V/μs
  Stable at gains of 10 or greater
  Full power bandwidth: 6.4 MHz for 20 V p-p into a 500-Ω load
- Precision dc performance
  Input offset voltage: 0.3 mV max.
  Input offset drift: 3 μV/°C typ.

Input voltage noise: 4 nV/√Hz
Open-loop gain: 130 V/mV into a 1-kΩ load
Output current: 50 mA min.
Supply current: 12 mA max.

APPLICATIONS
Video and pulse amplifiers
DAC and ADC buffers
Line drivers
Available in 14-pin plastic DIP, hermetic cerdip packages, and chip form
MIL-STD-883B processing available

APPLICATION HIGHLIGHTS
1. The high slew rate and fast settling time of the AD840 make it ideal for DAC and ADC buffers, line drivers, and all types of video instrumentation circuitry.
2. The AD840 is truly a precision amplifier. It offers a 12-bit accuracy to 0.01% or better and wide bandwidth, performance previously available only in hybrids.
3. The AD840's thermally balanced layout and the high speed of the CB process allow the AD840 to settle to 0.01% in 100 ns without the long "tails" that occur with other fast op amps.
4. Laser wafer trimming reduces the input offset voltage of 0.3 mV max on the K grade, thus eliminating the need for external offset nulling in many applications. Offset null pins are provided for additional versatility.
5. Full differential inputs provide outstanding performance in all standard high-frequency op amp applications where circuit gain will be 10 or greater.
6. The AD840 is an enhanced replacement for the HA2540.

SPECIFICATIONS (@ +25°C and ±15 Vdc, unless otherwise noted)

Model	Conditions	AD840J Min	AD840J Typ	AD840J Max	AD840K Min	AD840K Typ	AD840K Max	AD840S Min	AD840S Typ	AD840S Max	Units
INPUT OFFSET VOLTAGE[1]			0.2	1		0.1	0.3		0.2	1	mV
	$T_{min} - T_{max}$			1.5			0.7			2	mV
Offset Drift			5			3			5		μV/°C
INPUT BIAS CURRENT			3.5	8		3.5	5		3.5	8	μA
	$T_{min} - T_{max}$			10			6			12	μA
INPUT OFFSET CURRENT			0.1	0.4		0.1	0.2		0.1	0.4	μA
	$T_{min} - T_{max}$			0.5			0.3			0.6	μA
INPUT CHARACTERISTICS	Differential Mode										
Input Resistance			30			30			30		kΩ
Input Capacitance			2			2			2		pF
INPUT VOLTAGE RANGE											
Common Mode	$V_{CM} = \pm 10$ V	±10	12		±10	12		±10	12		V
Common-Mode Rejection		90	110		106	115		90	110		dB
	$T_{min} - T_{max}$	85			90			85			dB

Model	Conditions	AD840J Min	AD840J Typ	AD840J Max	AD840K Min	AD840K Typ	AD840K Max	AD840S Min	AD840S Typ	AD840S Max	Units
INPUT VOLTAGE NOISE	$f = 1$ kHz		4			4			4		nV/$\sqrt{Hz}$
Wideband Noise	10 Hz to 10 MHz		10			10			10		µV rms
OPEN LOOP GAIN	$V_O = \pm 10$ V										
	$R_{LOAD} = 1$ kΩ	100	130		100	130		100	130		V/mV
	$T_{min} - T_{max}$	50	80		75	100		50	80		V/mV
	$R_{LOAD} = 500$ Ω	75			100			75			V/mV
	$T_{min} - T_{max}$	50			75			50			V/mV
OUTPUT CHARACTERISTICS											
Voltage	$R_{LOAD} \geq 500$ Ω										
	$T_{min} - T_{max}$	±10			±10			±10			V
Current	$V_{OUT} = \pm 10$ V	50			50			50			mA
Output Resistance	Open Loop		15			15			15		Ω
FREQUENCY RESPONSE											
Gain Bandwidth Product	$V_{OUT} = 90$ mV p-p										
	$A_V = -10$		400			400			400		MHz
Full Power Bandwidth[2]	$V_O = 20$ V p-p										
	$R_{LOAD} \geq 500$ Ω	5.5	6.4		5.5	6.4		5.5	6.4		MHz
Rise Time	$A_V = -10$		10			10			10		ns
Overshoot[3]	$A_V = -10$		20			20			20		%
Slew Rate[3]	$A_V = -10$	350	400		350	400		350	400		V/µs
Settling Time[3] −10 V Step	$A_V = -10$										
	to 0.1%		80			80			80		ns
	to 0.01%		100			100			100		ns
OVERDRIVE RECOVERY	−Overdrive		190			190			190		ns
	+Overdrive		350			350			350		ns
DIFFERENTIAL GAIN	$f = 4.4$ MHz		0.025			0.025			0.025		%
DIFFERENTIAL PHASE	$f = 4.4$ MHz		0.04			0.04			0.04		Degree
POWER SUPPLY											
Rated Performance			±15			±15			±15		V
Operating Range		±5		±18	±5		±18	±5		±18	V
Quiescent Current			10.5	12		10.5	12		10.5	12	mA
	$T_{min} - T_{max}$			14			14			16	mA
Power Supply Rejection Ratio	$V_S = \pm 5$ V to ± 18 V	90	100		94	100		90	100		dB
	$T_{min} - T_{max}$	80			86			80			dB
TEMPERATURE RANGE											
Rated Performance[4]		0		+75	0		+75	−55		+125	°C
PACKAGE OPTIONS											
Cerdip (Q-14)			AD840JQ			AD840KQ			AD840SQ		
Plastic (N-14)			AD840JN			AD840KN					
LCC (E-20A)									AD840SE		
J and S Grade Chips Also Available											

NOTES
[1] Input offset voltage specifications are guaranteed after 5 minutes at $T_A = +25°C$.
[2] Full power bandwidth = slew rate/$2\pi V_{PEAK}$.
[3] Refer to Figures 22 and 23.
[4] "S" grade $T_{min}-T_{max}$ specifications are tested with automatic test equipment at $T_A = -55°C$ and $T_A = +125°C$.

AD840 CONNECTION DIAGRAMS

Plastic DIP (N) Package and Cerdip (Q) Package

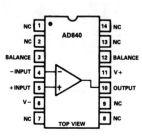

LCC (E) Package

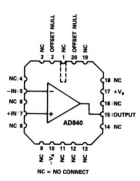

NC = NO CONNECT

ABSOLUTE MAXIMUM RATINGS[1]

Supply Voltage . . . . ±18V
Internal Power Dissipation[2]
   Plastic (N) 1.5W
   Cerdip (Q) 1.3W
Input Voltage . . . . ±V_S
Differential Input Voltage . . . . ±6V
Storage Temperature Range
   Q . . . . −65°C to +150°C
   N . . . . −65°C to +125°C
Junction Temperature (T_J) . . . . +175°C
Lead Temperature Range (Soldering 60sec) . . . . . . +300°C

NOTES

[1] Stresses above those listed under "Absolute Maximum Ratings" may cause permanent damage to the device. This is a stress rating only, and functional operation of the device at these or any other conditions above those indicated in the operational section of this specification is not implied. Exposure to absolute maximum rating conditions for extended periods may affect device reliability.

[2] Maximum internal power dissipation is specified so that T_J does not exceed +175°C at an ambient temperature of +25°C.

Thermal Characteristics:

	θ_{JC}	θ_{JA}	θ_{SA}	
Cerdip Package	35°C/W	110°C/W	38°C/W	Recommended Heat Sink:
Plastic Package	30°C/W	95°C/W		Aavid Engineering ©#602B

Analog Devices AD840

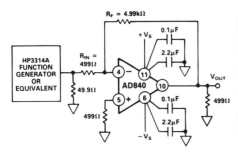
Inverting Amplifier Configuration (DIP Pinout)

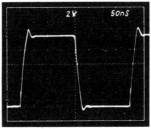
Inverter Large Signal Pulse Response

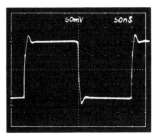
Inverter Small Signal Pulse Response

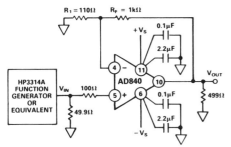
Noninverting Amplifier Configuration (DIP Pinout)

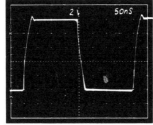
Noninverting Large Signal Pulse Response

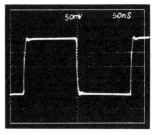
Noninverting Small Signal Pulse Response

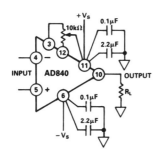
Offset Nulling (DIP Pinout)

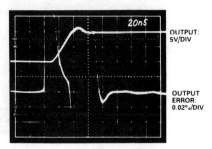

AD840 0.01% Settling Time

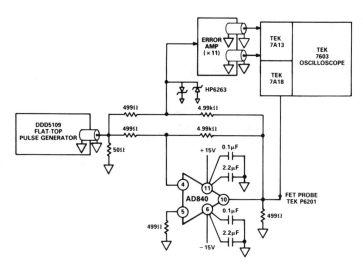

Settling Time Test Circuit

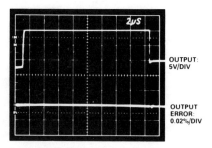

AD840 Settling Demonstrating No Settling Tails

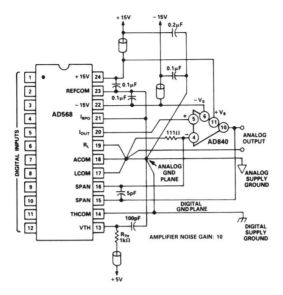

0 to +10.24 V DAC Output Buffer

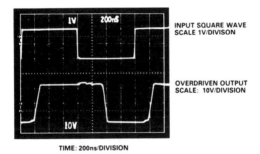

Overdrive Recovery

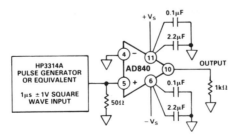

Overdrive Recovery Test Circuit

GEC Plessey
MV5087
DTMF Generator

FEATURES
- Pin-for-pin replacement for MK5087
- Low standby power
- Minimum external parts count
- 3.5- to 10-V operation
- 2-of-8 keyboard or calculator-type single-contact (form A) keyboard input
- On-chip regulation of output tone
- Mute and transmitter drivers on-chip
- High-accuracy tones provided by 3.58-MHz crystal oscillator
- Pin-selectable inhibit of single-tone generation

APPLICATIONS
DTMF signalling for:
- Telephone sets
- Mobile radio
- Remote control
- Point-of-sale and banking terminals
- Process control

ABSOLUTE MAXIMUM RATINGS

	MIN.	MAX.		MIN.	MAX.
$V_{DD} - V_{SS}$	-0.3V	10.5V	Power dissipation		850 mW
Voltage on any pin	V_{SS} - 0.3V	V_{DD} + 0.3V	Derate 16 mW/°C above 75°C		
Current on any pin		10 mA	(All leads soldered to PCB)		
Operating temperature	-40°C	+85°C			
Storage temperature	-65°C	+150°C			

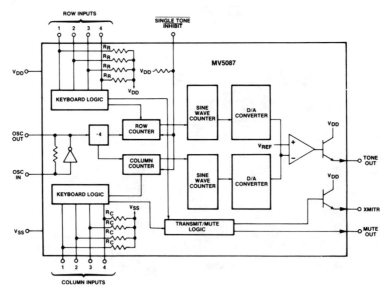

Functional block diagram

Plessey MV5087

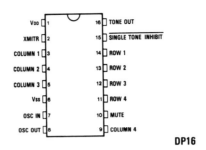

Pin connections - top view

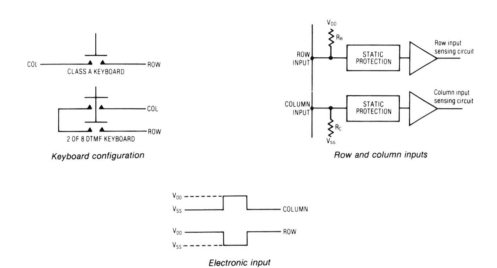

Keyboard configuration

Row and column inputs

Electronic input

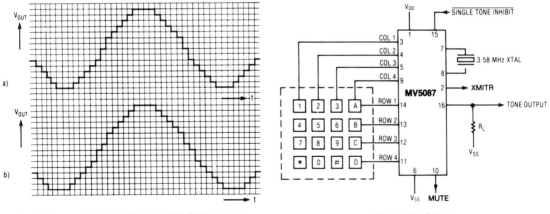

Typical sinewave output (a) Row tones (b) Column tones

Connection diagram

GEC Plessey
MV5089
DTMF Generator

FEATURES
- Pin-for-pin replacement for MK5089
- Low standby power
- Minimum external parts count
- 2.75 V- to 10-V operation
- 2-of-8 keyboard input
- High-accuracy tones provided by 3.58-MHz crystal oscillator
- Pin-selectable inhibit of single-tone generation

APPLICATIONS
DTMF signalling for:
- Telephone sets
- Mobile radio
- Remote control
- Point-of-sale and banking terminals
- Process control

ABSOLUTE MAXIMUM RATINGS

	MIN.	MAX.		MIN.	MAX.
$V_{DD} - V_{SS}$	-0.3V	10.5V	Power dissipation		850 mW
Voltage on any pin	V_{SS} - 0.3V	V_{DD} + 0.3V	Derate 16 mW/°C above 75°C		
Current on any pin		10 mA	(All leads soldered to PCB)		
Operating temperature	-40°C	+85°C			
Storage temperature	-65°C	+150°C			

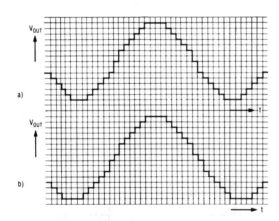

Typical sinewave output (a) Row tones (b) Column tones

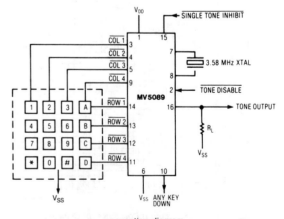

connection diagram

COM

Plessey MV5089

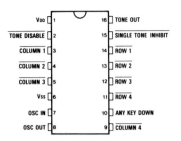

Pin connections - top view

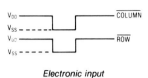

Electronic input

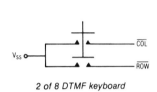

2 of 8 DTMF keyboard

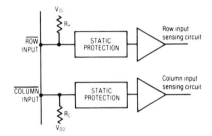

Row and Column inputs

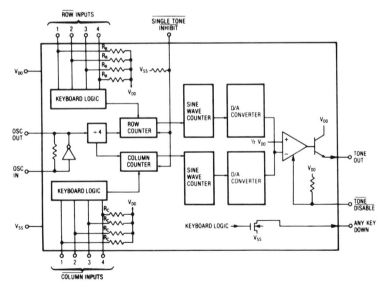

Functional block diagram

GEC Plessey
ZN478E
Microphone Amplifier for Telephone Circuits

FEATURES
- Low working voltage
- Designed to match electrets with FET buffers
- Gain adjustable by external resistor
- Operates from 1 to 100-mA line current
- Low noise
- Low distortion
- Operates on telephone supply lines
- Minimum external components in telephone circuits

DESCRIPTION
The ZN478E was developed specifically for use with low-impedance transducers, such as electret microphones (with FET buffers), to replace the carbon transmitter in telephone handsets. The ZN478E is especially useful where a low operating voltage is required.

The amplifier gain can be adjusted over a wide range by an external resistor to suit a variety of different low impedance transducer sensitivities.

This is a single-polarity device and care should be taken over line connection.

ABSOLUTE MAXIMUM RATINGS
Supply current   100 mA continuous
Operating temp. range -20 to $+80\,°C$
Storage temp. range -55 to $+125\,°C$

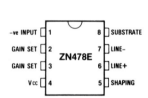

Pin connections - top view

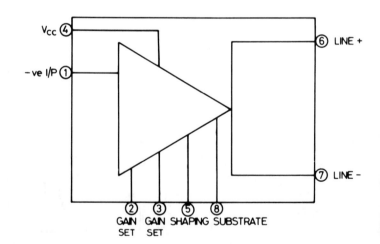

System Diagram

□ COM

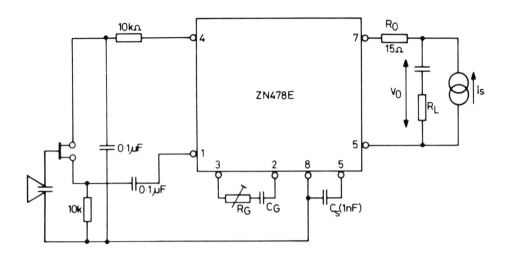

Typical Application Circuit

GEC Plessey
MV4320
Keypad Pulse Dialer

FEATURES
- Pin-for-pin replacement for the DF320
- 2.5- to 5.5-V supply voltage operating range
- 375-μW dynamic power dissipation at 3 V
- Uses inexpensive 3.58-MHz ceramic resonator or crystal
- Stores up to 20 digits
- Selectable outpulsing mark/space ratio
- Selectable dialing speeds of 10, 16, 20, and 932 Hz
- Low cost

APPLICATIONS
- Pushbutton telephones
- Tone-to-pulse converters
- Mobile telephones
- Repertory dialers

ABSOLUTE MAXIMUM RATINGS
The absolute maximum ratings are limiting values above which operating life might be shortened or specified parameters might be degraded.

	MIN.	MAX.
V_{DD}-V_{SS}	-0.3 V	10 V
Voltage on any pin	$V_{SS}-0.3$ V	$V_{DD}+0.3$ V
Current at any pin		10 mA
Operating temperature	-40°C	$+85$°C
Storage temperature	-65°C	$+150$°C
Power dissipation		1000 mW

Derate 16 mW/°C above 75°C. All leads soldered to PC board.

82 Plessey MV4320

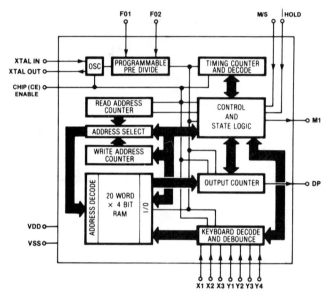

Pin connections (top view)

MV4320 functional block diagram

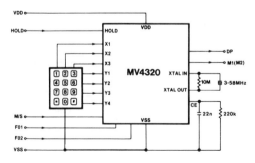

Application diagram

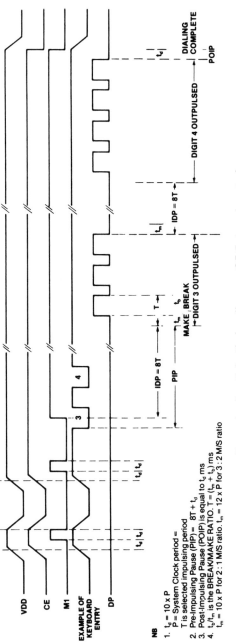

Keypad pulse dialer timing diagram, CE-External control

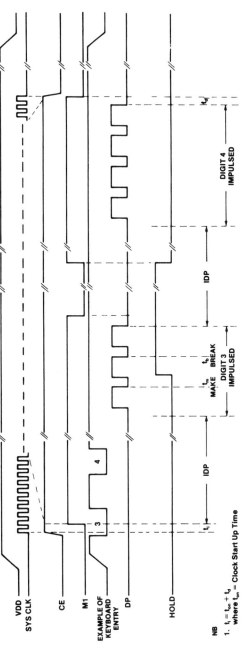

Keypad pulse dialer timing diagram, CE-Internal control

The information included herein is believed to be accurate and reliable. However, LSI Computer Systems, Inc. assumes no responsibilities for inaccuracies, nor for any infringements of patent rights of others which may result from its use.

LSI
LS7501/LS7510
Tone-Activated Line-Isolation Device

FEATURES
- Low-power CMOS design
- On-chip oscillator (32,768 Hz external crystal required)
- Tone input can be low-level sinusoid (as low as -30 dBm) or fully digital.
- Mask programmable available frequencies: 11 to 4095 Hz (in 1-Hz steps)
- Sample interval -4.5 seconds (Mask programmable 0.5 to 8.0 seconds).

MAXIMUM RATINGS (VOLTAGES REFERENCED TO V_{SS}):

RATING	SYMBOL	VALUE	UNIT
dc supply voltage	V_{DD}	$+2.5$ to $+6.0$	Vdc
Operating temperature range	T_A	-25 to $+70$	°C
Storage temperature range	T_{STG}	-65 to $+150$	°C

DESCRIPTION
The LS7501/LS7510 are frequency discriminator circuits that respond to a standard frequency input if the input is maintained within ± 10 Hz during a 4.5-second continuous sample interval. During this interval, the input is being sampled every 0.5 seconds. If it is valid for the sample interval, then the circuit can be used to pulse a relay that disconnects the line to be tested. After 20 seconds of disconnect time, the relay is reset and the line is restored. There are 10 standard frequency versions of this circuit. These are indicated in the table with their associated input discriminator frequencies.

PART NO.	FREQUENCY (HZ)
LS7501	2683
LS7502	2713
LS7503	2743
LS7504	2773
LS7505	2833
LS7506	2863
LS7507	2893
LS7508	2923
LS7509	2953
LS7510	2983

CONNECTION DIAGRAM TOP VIEW
STANDARD 16 PIN PLASTIC DIP

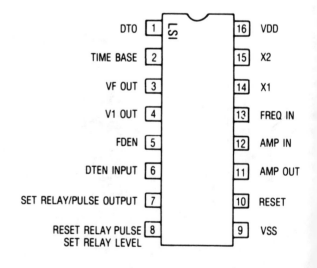

CIRCUIT BLOCK DIAGRAM

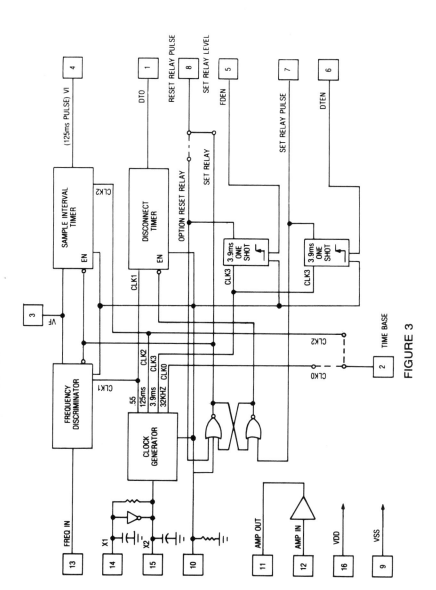

FIGURE 3

NOTE (1) All devices shown on the LS7501 through the LS7510 are configured with the set relay output on Pin 8. The reset option can be substituted by optional mask change.

NOTE (2) All devices shown with the exception of the LS7502 are configured with the clock-0, 32KHz output on Pin 2. The LS7502 is configured with the clock-2 time base output of 8Hz. These outputs may be changed with the same optional mask change referred to in Note 1.

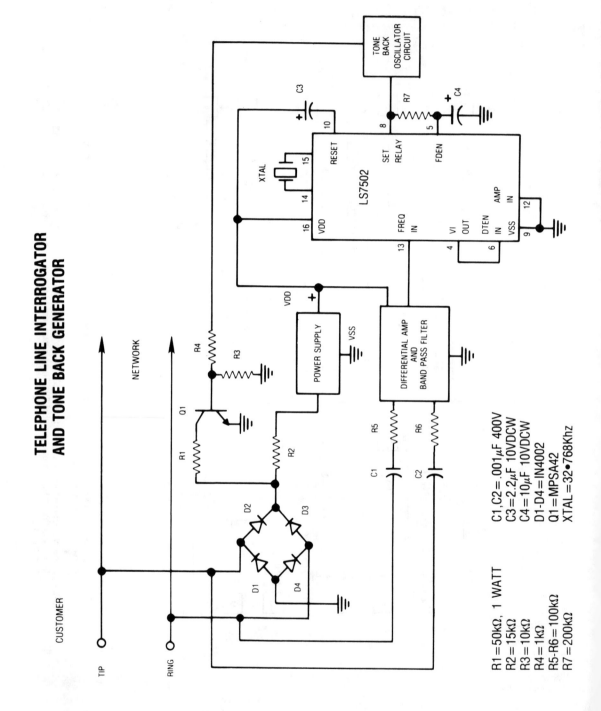

Raytheon RC747
General-Purpose Operational Amplifier

FEATURES
- Short-circuit protection
- No frequency compensation required
- No latch-up
- Large common-mode and differential-voltage ranges
- Low power consumption
- Parameter tracking over temperature range
- Gain and phase match between amplifiers

ORDERING INFORMATION

Part Number	Package	Operating Temperature Range
RC747N	N	0°C to +70°C
RC747T	T	0°C to +70°C
RM747D	D	-55°C to +125°C
RM747D/883B*	D	-55°C to +125°C
RM747T	T	-55°C to +125°C
RM747T/883B*	T	-55°C to +125°C

Notes:
*/883B suffix denotes Mil-Std-883, Level B processing
N = 14-lead plastic DIP
D = 14-lead ceramic DIP
T = 10-lead metal can TO-99
Contact a Raytheon sales office or representative for ordering information on special package/temperature range combinations.

ABSOLUTE MAXIMUM RATINGS
Supply Voltage
 RM747 ....±22V
 RC747 ....±18V
Differential Input Voltage 30V
Input Voltage* ....±15V
Output Short-Circuit Duration** ....Indefinite
Storage Temperature
 Range ....-65°C to +150°C
Operating Temperature Range
 RM747 ....-55°C to +125°C
 RC747 ....0°C to +70°C
Lead Soldering Temperature
 (60 sec) ....+300°C

*For supply voltages less than ±15V, the absolute maximum input voltage is equal to the supply voltage.
**Short-circuit may be to ground or either supply. Rating applies to +125°C case temperature or +75°C ambient temperature for RC747.

CONNECTION INFORMATION

10-Lead TO-100 Metal Can
(Top View)

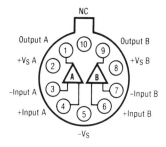

14-Lead Dual In-Line Package
(Top View)

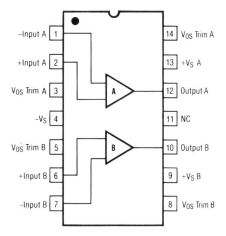

$+V_s$ A is internally connected to $+V_s$ B for the 747S

TYPICAL APPLICATIONS

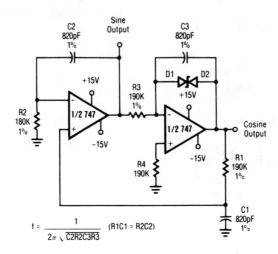

Quadrature Oscillator

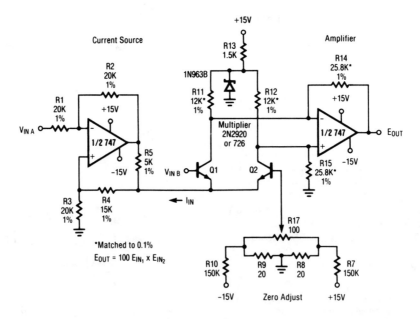

Analog Multiplier

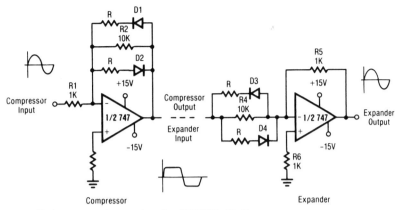

Compressor/Expander Amplifiers

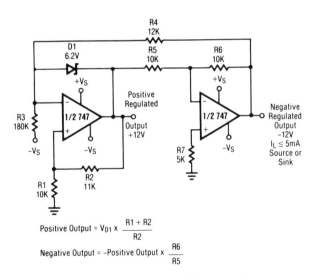

Positive Output = $V_{D1} \times \dfrac{R1 + R2}{R2}$

Negative Output = $-\text{Positive Output} \times \dfrac{R6}{R5}$

Tracking Positive and Negative Voltage References

Raytheon RC747

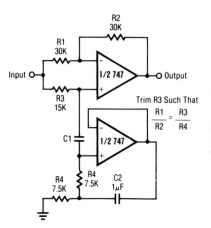
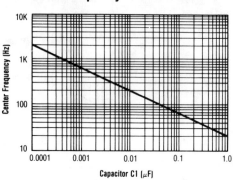

Notch Filter Using the 747 as a Gyrator

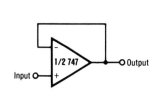

$R_{IN} = 400M\Omega$
$C_{IN} = 1pF$
$R_{OUT} \ll 1\Omega$
$BW = 1MHz$

Unity Gain Voltage Follower

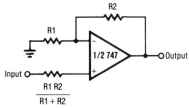

$\dfrac{R1\ R2}{R1 + R2}$

Gain	R1	R2	B.W.	R_{IN}
10	1kΩ	9kΩ	100kHz	400MΩ
100	100Ω	9.9kΩ	10kHz	280MΩ
1000	100Ω	99.9kΩ	1kHz	80MΩ

Non-Inverting Amplifier

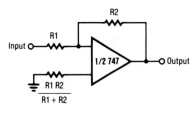

$\dfrac{R1\ R2}{R1 + R2}$

Gain	R1	R2	B.W.	R_{IN}
1	10kΩ	10kΩ	1MHz	10kΩ
10	1kΩ	10kΩ	100kHz	1kΩ
100	1kΩ	100kΩ	10kHz	1kΩ
1000	100Ω	100kΩ	1kHz	100Ω

Inverting Amplifier

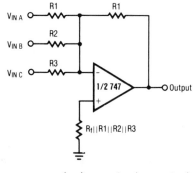

$-V_O = V_{IN\ A} \left(\dfrac{R_f}{R1}\right) - V_{IN\ B} \left(\dfrac{R_f}{R2}\right) - V_{IN\ C} \left(\dfrac{R_f}{R3}\right)$

Weighted Averaging Amplifier

Raytheon RC4097 Series
Low-Power High-Precision Operational amplifiers

FEATURES

- Low input offset voltage: 15 μV max.
- Low V_{OS} drift: 0.3 μV/°C max.
- Low input bias current:
  +25°C, 100 pA max.
  −55°C to +125°C, 600 pA max.
- High gain: 1000 V/mV min.
- High CMRR: 120 dB min.
- High PSRR 114 dB min.
- Low supply current: 600 μA max.
- Low noise: 0.5 μV_{p-p} (0.1 to 10 Hz)
- Replaces OP-97, LT1012

ORDERING INFORMATION

Part Number	Package	Operating Temperature Range
RC4097AN	N	0°C to +70°C
RC4097EN	N	0°C to +70°C
RC4097FN	N	0°C to +70°C
RC4097EM	M	0°C to +70°C
RC4097FM	M	0°C to +70°C
RV4097ET	T	-25°C to +85°C
RV4097FT	T	-25°C to +85°C
RV4097ED	D	-25°C to +85°C
RV4097FD	D	-25°C to +85°C
RM4097AT	T	-55°C to +125°C
RM4097AT/883B	T	-55°C to +125°C
RM4097AD	D	-55°C to +125°C
RM4097AD/883B	D	-55°C to +125°C

Notes:
/883B suffix denotes Mil-Std-883, Level B processing
N = 8-lead plastic DIP
D = 8-lead ceramic DIP
T = 8-lead metal can (TO-99)
M = 8-lead plastic SOIC
Contact a Raytheon sales office or representative for ordering information on special package/temperature range combinations.

ABSOLUTE MAXIMUM RATINGS

Supply Voltage .... ±22V
Input Voltage* .... ±22V
Differential Input Voltage 30V
Internal Power Dissipation** .... 500 mW
Output Short Circuit Duration .... Indefinite
Storage Temperature
   Range .... -65°C to +150°C
Operating Temperature Range
   RM4097A .... -55°C to +125°C
   RV4097E,F (Hermetic) .... -25°C to +85°C
   RC4097A,E,F (Plastic) .... 0°C to +70°C
Lead Soldering Temperature
   (SO-8, 10 sec) .... +260°C
   (DIP, TO-99; 60 sec) .... +300°C

*For supply voltages less than ±22V, the absolute maximum input voltage is equal to the supply voltage.
**Observe package thermal characteristics.

CONNECTION INFORMATION

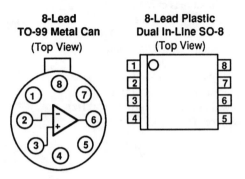

Pin	Function
1	V_{os} Trim
2	-Input
3	+Input
4	$-V_s$
5	NC
6	Output
7	$+V_s$
8	V_{os} Trim

TYPICAL APPLICATIONS

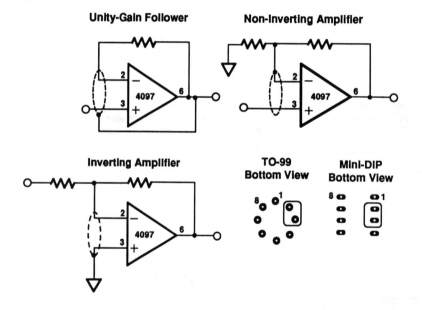

Guard Ring Layout and Connections

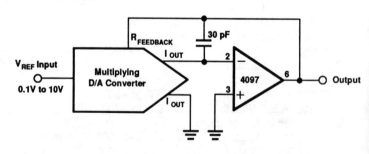

Wide Dynamic Range Multiplying DAC

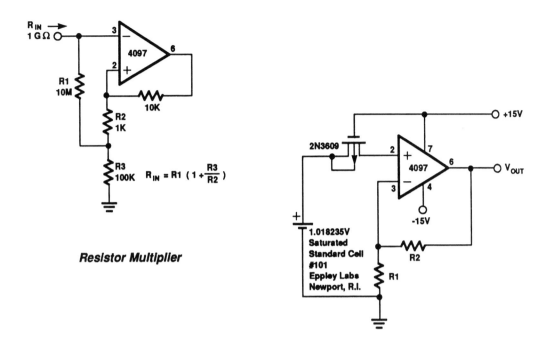

Resistor Multiplier

Long-Life Standard Cell Amplifier

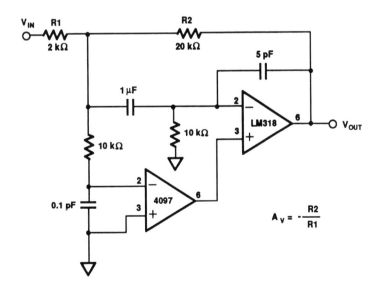

Composite High-Speed, Precision Amplifier

94 Raytheon RC4097

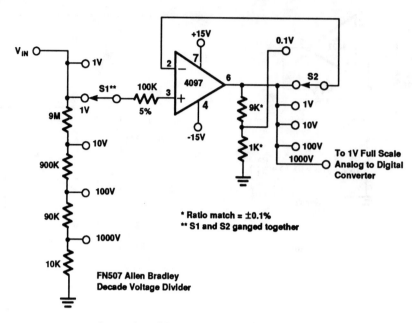

Input Amplifier for 4-1/2 Digit Voltmeter

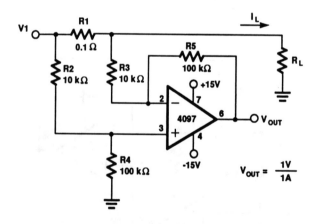

Precision Current Monitor

Silicon Systems
Monolithic Dual-Tone Multi-Frequency (DTMF) Receivers

HOW THE SILICON SYSTEMS DTMF CIRCUITS WORK

The task of a DTMF receiver is to detect the presence of a valid DTMF signal on a telephone line or other transmission medium. The presence of a valid DTMF signal indicates a single dialed digit; to generate a valid digit sequence, each DTMF signal must be separated by a valid pause.

The table gives the established Bell system standards for a valid DTMF signal and a valid pause. The SSI DTMF receivers meet or exceed these standards.

Similar device architecture is used in all SSI DTMF receivers. This architecture is implemented in all Silicon Systems single-chip receivers, as well as in SSI transceivers. In general terms, the detection scheme is as follows: The input signal is prefiltered and then split into two bands, each of which contains only one DTMF tone group. The output of each band-split filter is amplified and limited by a zero-crossing detector. The limited signals, in the form of square waves, are passed through tone frequency bandpass filters. Digital logic is then used to provide detector sampling and determine detection validity, to present the digital output data in the correct format, and to provide device timing and control.

SPEECH IMMUNITY AND NOISE TOLERANCE

The two largest problems confronting a DTMF receiver are:

1. Distinguishing between valid DTMF tone pairs and other speech or stray signals that contain DTMF tone pair frequencies. This is referred to as *speech immunity*.

2. Detecting valid tone pairs in the presence of noise, which is typically found in the telephone (or other transmission medium) environment. This is referred to as *noise tolerance*.

The SSI DTMF receivers use several techniques to distinguish between valid tone pairs and other stray signals. These techniques are explained in later sections. Briefly, the techniques are:

1. Prefiltering of audio signal. Removes supply noise and dial tone from input audio signal and emphasizes the voice frequency domain.

2. Zero-cross detection. Limits the acceptable level of noise during detection of a tone pair. Important for speech rejection.

3. Valid tone pair/pause sampling.

PARAMETER	VALUE
One Low-Group Tone, and	697, 770, 852 or 941 Hz
One High-Group Tone	1209, 1336, 1477 or 1633 Hz
Frequency Tolerance	$f_o \pm (1.5\% + 2 \text{ Hz})$
Amplitude Range	$-24 \text{ dB} \leq A \leq 6 \text{ dBm} @ 600\Omega$ (Dynamic Range 30 dB)
Relative Amplitude (Twist)	$-8 \text{ dB} \leq \dfrac{\text{High Group Tone}}{\text{Low Group Tone}} \leq +4 \text{ dB}$
Duration	40 ms or longer
Inter-tone Pauses	40 ms or longer

Bell System Standards

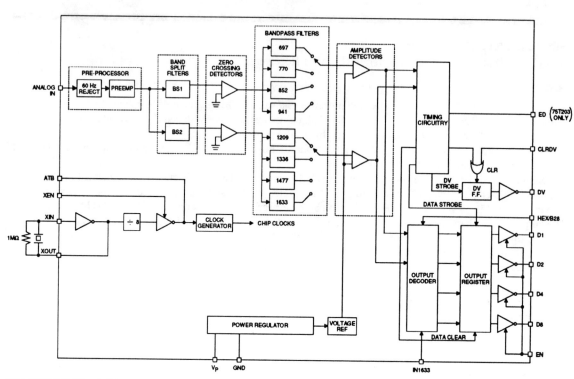

SSI 75T202 Block Diagram

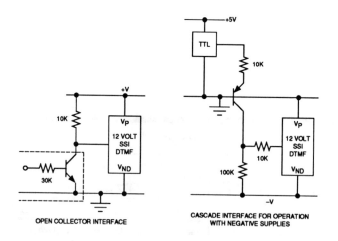

Interface Circuits for Conversion from TTL Output Levels to 12V SSI DTMF Input Levels

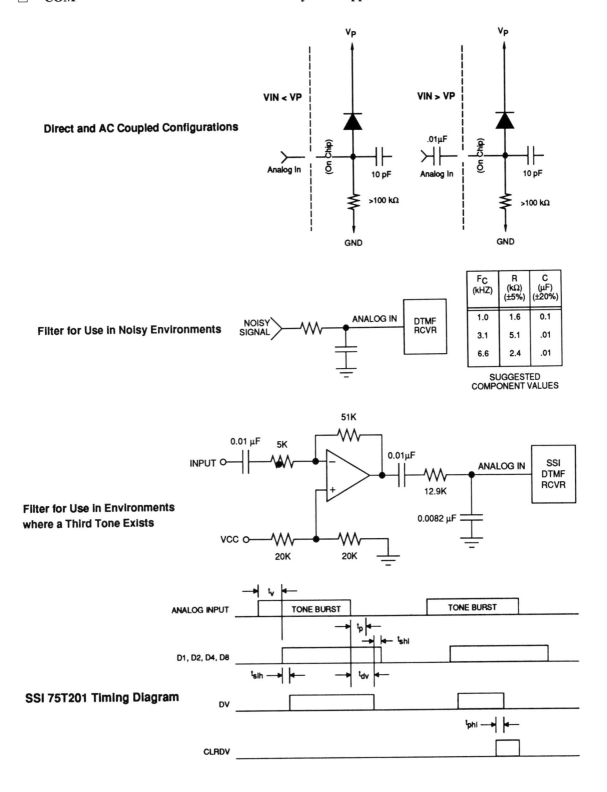

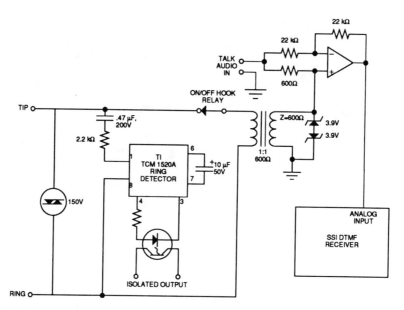

Full Featured Phone Line Interface

PARAMETER		CONDITIONS	MIN	NOM	MAX	UNITS
t_v	Tone Detection Time		20	25	40	ms
t_{slh}	Data Overlap of DV Rising Edge	CLRDV = VND, EN = VP	7			µs
t_p	Pause Detection Time		25	32	40	ms
t_{dv}	Time between end of Tone and Fall of DV		40	45	50	ms
t_{shl}	Data overlap of DV Falling Edge		4	4.56	4.8	ms
t_{phl}	Prop. Delay: Rise of CLRDV to fall of DV	CI = 300 pF Measured at 50% points			1	µs
	Output Enable Time	CI = 300 pF, RI = 10K See Note 1			1	µs
	Output Disable Time	CI = 300 pF, RI = 1K ΔV = 1V, See Note 2			1	µs
	Output 10-90% Transition Time	CI = 300 pF			1	µs

Note 1: Measured from 50% point of Rising Edge of EN to the 50% point of the data output with RI to opposite rail.
Note 2: Measured from 50% point of Falling Edge of EN to time at which output has changed 1V with RI to opposite rail.

SSI 75T201 Timing Specifications (−40°C ≤ TA ≤ +85°C, VP − VND = VP − VNA = 12V ± 10%)

Silicon Systems Application Guide: DTMF Receivers

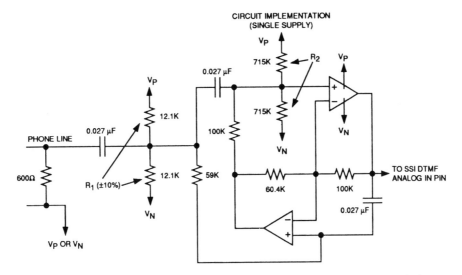

Note: All resistors 1%, all caps 5%, unless noted, op-amps: 1/2 LM1458 or equivalent

Dial Tone Reject Filter

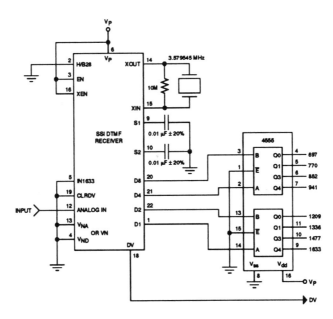

Touch-Tone™ to 2-of-8 Output Converter

100 Silicon Systems Application Guide: DTMF Receivers

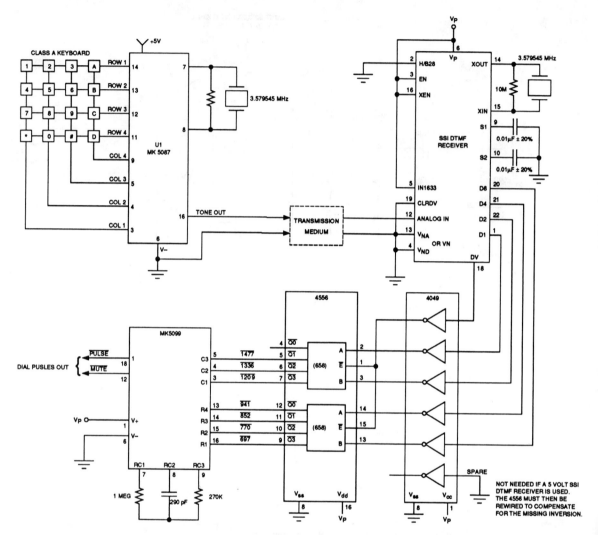

Touch-Tone™ to Rotary Dial Pulse Converter Adding Rotary Dial Pulse Detection Capabilities

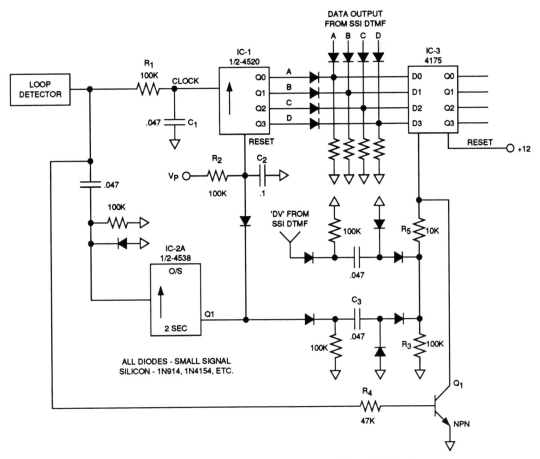

Adding Pulse Detection and Counting to the SSI DTMF Receiver

Silicon Systems SSI 75T2089 DTMF Transceiver

FEATURES

- DTMF generator and receiver on one-chip
- 22-pin, 400-mil plastic DIP
- Low-power 5-V CMOS
- DTMF receiver exhibits excellent speech immunity
- Three-state outputs (4-bit hexadecimal) from the DTMF receiver
- ac coupled, internally biased analog input
- Latched DTMF generator inputs
- Analog input range from -32 to -2 dBm (ref 600 Ω)
- DTMF output typ. -8 dBm (low band) and -5.5 dBm (high band)
- Uses inexpensive 3.579545-MHz crystal for reference
- Easy interface for microprocessor dialing

CAUTION: Use handling procedures necessary for a static sensitive component.

Silicon Systems SSI 75T2089

ORDERING INFORMATION

PART DESCRIPTION	ORDER NO.	PKG. MARK
SSI 75T2089 22-Pin Plastic DIP	SSI 75T2089 - IP	75T2089 - IP

No responsibility is assumed by Silicon Systems for use of this product nor for any infringements of patents and trademarks or other rights of third parties resulting from its use. No license is granted under any patents, patent rights or trademarks of Silicon Systems. Silicon Systems reserves the right to make changes in specifications at any time without notice. Accordingly, the reader is cautioned to verify that the data sheet is current before placing orders.

BLOCK DIAGRAM

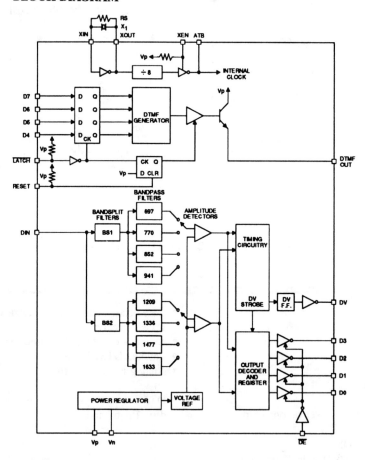

PACKAGE PIN DESIGNATIONS
(TOP VIEW)

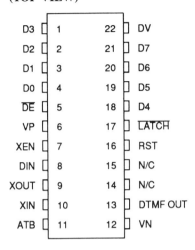

22-Pin DIP

ABSOLUTE MAXIMUM RATINGS
Operating above absolute maximum ratings may damage the device.

PARAMETER	RATING	UNIT
DC Supply Voltage (Vp - Vn)	+7	V
Voltage at any Pin (Vn = 0)	-0.3 to Vp + 0.3	V
DIN Voltage	Vp + 0.5 to Vp - 10	V
Current through any Protection Device	±20	mA
Operating Temperature Range	-40 to +85	°C
Storage Temperature	-65 to 150	°C

RECOMMENDED OPERATING CONDITIONS

PARAMETER	CONDITIONS	MIN	NOM	MAX	UNITS
Supply Voltage		4.5		5.5	V
Power Supply Noise (wide band)				10	mV pp
Ambient Temperature		-40		+85	°C
Crystal Frequency (F Nominal = 3.579545MHz)		-0.01		+0.01	%
Crystal Shunt Resistor		0.8		1.2	MΩ
DTMF OUT Load Resistance		100			Ω

Silicon Systems SSI 75T2089

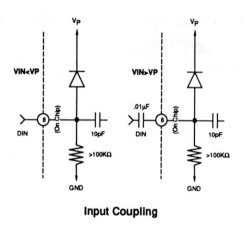

Input Coupling

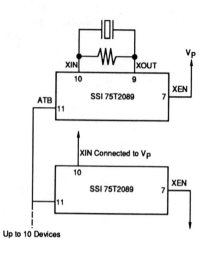

Crystal Connections

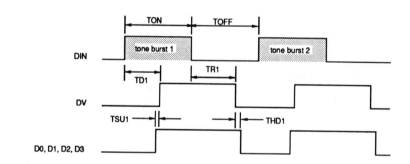

DTMF DECODER

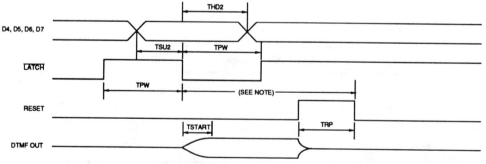

NOTE: THE INDICATED TIME MAY BE AS SMALL AS 0 SEC., MEANING THAT THE LATCH AND RESET LINES MAY BE TIED TOGETHER.

DTMF GENERATOR

Silicon Systems
SSI 75T2090
DTMF Transceiver and Call Progress Detection

FEATURES
- DTMF generator and receiver on one-chip
- 22-pin, 400 mil plastic DIP
- Low-power 5-V CMOS
- DTMF receiver exhibits excellent speech immunity
- Three-state outputs (4-bit hexadecimal) from DTMF receiver
- ac coupled, internally biased analog input
- Latched DTMF generator inputs
- Analog input range from -32 to -2 dBm (ref 600 Ω)
- DTMF output typ. -8 dBm (low band) and -5.5 dBm (high band)
- Uses inexpensive 3.579545-MHz crystal for reference
- Easy interface for microprocessor dialing
- Call progress detection

CAUTION: Use handling procedures necessary for a static sensitive component.

ORDERING INFORMATION

PART DESCRIPTION	ORDER NO.	PKG. MARK
SSI 75T2090 22-Pin DIP	SSI 75T2090 - IP	75T2090 - IP

No responsibility is assumed by Silicon Systems for use of this product nor for any infringements of patents and trademarks or other rights of third parties resulting from its use. No license is granted under any patents, patent rights or trademarks of Silicon Systems. Silicon Systems reserves the right to make changes in specifications at any time without notice. Accordingly, the reader is cautioned to verify that the data sheet is current before placing orders.

PACKAGE PIN DESIGNATIONS
(TOP VIEW)

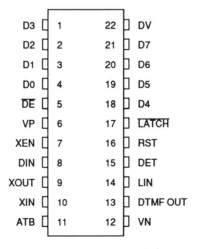

22-Pin DIP

BLOCK DIAGRAM

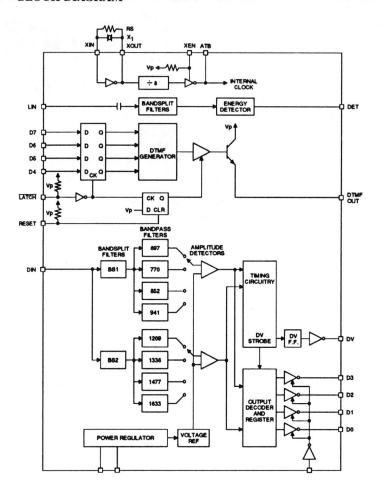

ABSOLUTE MAXIMUM RATINGS
Operating above absolute maximum ratings may damage the device.

PARAMETER	RATING	UNIT
DC Supply Voltage (Vp - Vn)	+7	V
Voltage at any Pin (Vn = 0)	-0.3 to Vp + 0.3	V
DIN Voltage	Vp + 0.5 to Vp - 10	V
Current through any Protection Device	±20	mA
Operating Temperature Range	-40 to + 85	°C
Storage Temperature	-65 to 150	°C

RECOMMENDED OPERATING CONDITIONS

PARAMETER	CONDITIONS	MIN	NOM	MAX	UNITS
Supply Voltage		4.5		5.5	V
Power Supply Noise (wide band)				10	mV pp
Ambient Temperature		-40		+85	°C
Crystal Frequency (F Nominal = 3.579545MHz)		-0.01		+0.01	%
Crystal Shunt Resistor		0.8		1.2	MΩ
DTMF OUT Load Resistance		100			Ω

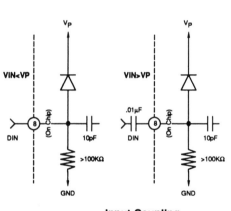

Input Coupling

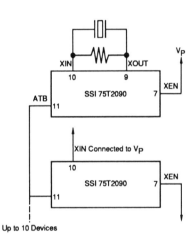

Crystal Connections

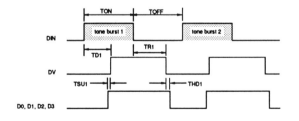

DTMF Decoder

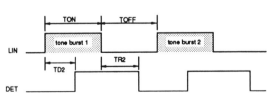

Call Progress Detector

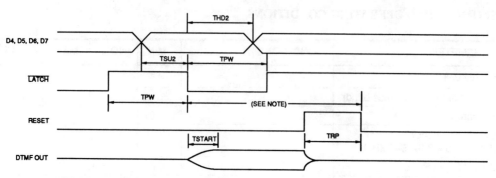

NOTE: THE INDICATED TIME MAY BE AS SMALL AS 0 SEC., MEANING THAT THE LATCH AND RESET LINES MAY BE TIED TOGETHER.

DTMF Generator

CHAPTER 3
CONTROL CIRCUITS

Allegro
8932
Voice-Coil Motor Driver
Preliminary Information

FEATURES
- Internal back-EMF velocity loop option
- Lossless current sensing
- Zero deadband
- High transconductance bandwidth
- User-adjustable transconductance gain
- Digital transconductance gain switch (4:1 ratio)
- 5-V monitor with selectable UV trip point
- Retract circuitry functional to 0 V
- Chip-enable/sleep-mode function
- 1 V at 500 mA output saturation voltage
- Internal thermal shutdown circuitry

ABSOLUTE MAXIMUM RATINGS

Supply voltages, V_{CC} and V_{DD}	6.0 V
Output current, I_{OUT} (peak)	±600 mA
(continuous)	±400 mA
Analog input voltage range, V_{IN}	−0.3 V to V_{CC}
Logic input voltage range, V_{IN}	−0.3 V to +6.0 V
Package power dissipation, P_D	See Graph
Operating temperature range, T_A	0 °C to +70 °C
Junction temperature, T_J	+150 °C†
Storage temperature range, T_S	−55 °C to +150 °C

†Fault conditions that produce excessive junction temperature will activate device thermal shutdown circuitry. These conditions can be tolerated, but should be avoided.

Output current rating may be restricted to a value determined by system concerns and factors. These include: system duty cycle and timing, ambient temperature, and use of any heatsinking and/or forced cooling. For reliable operation the specified maximum junction temperature should not be exceeded.

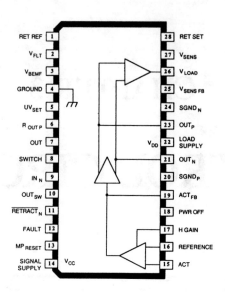

FUNCTIONAL BLOCK DIAGRAM

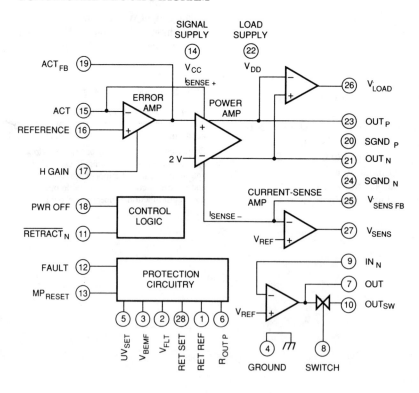

TEST CIRCUIT AND TYPICAL APPLICATION

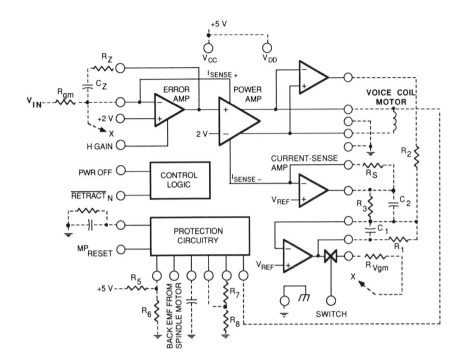

Allegro 8958
Voice-Coil Motor Driver
Preliminary Information

FEATURES
- Controlled-velocity head parking
- 4- to 15-V operation
- Zero deadband
- High transconductance bandwidth
- User-adjustable transconductance gain
- ±800-mA load current
- Dual under-voltage monitors with flag and user-selectable trip points
- Internal thermal shutdown circuitry
- Replaces UC3175

ABSOLUTE MAXIMUM RATINGS at $T_A = 25°C$

Supply voltages, V_{BB} and V_{CC}	16 V
Output current, I_{OUT}	±1.0 A
Park drive output current, I_{PARK}	
Continuous	250 mA
Peak	1.0 A
Amplifier input voltage range, V_{IN}	−2.0 V to V_{CC}
Sense input voltage range, $V_{SENSE\ IN}$	−0.3 V to V_{CC}
Comparator and digital inputs, V_{IN}	−0.3 V to 10 V
I_{IN}	±10 mA
Power OK output, V_{CEX}	20 V
I_C	30 mA
Output clamp diode current, I_F (pulsed)	1.0 A
Package power dissipation, P_D	See Graph
Operating temperature range, T_A	−20°C to +85°C
Junction temperature, T_J	150°C*
Storage temperature range, T_S	−55°C to +150°C

*Fault conditions that produce excessive junction temperature will activate device thermal shutdown circuitry. These conditions can be tolerated but should be avoided.

112 Allegro 8958

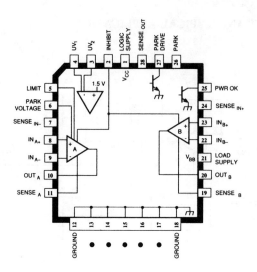

FUNCTIONAL BLOCK DIAGRAM

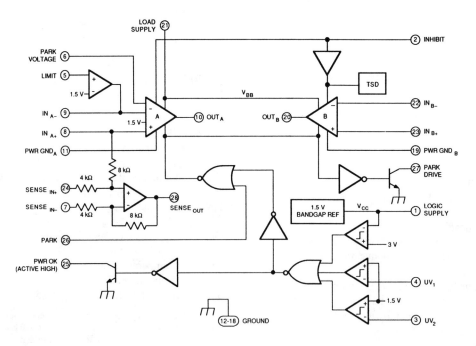

CURRENT SENSING

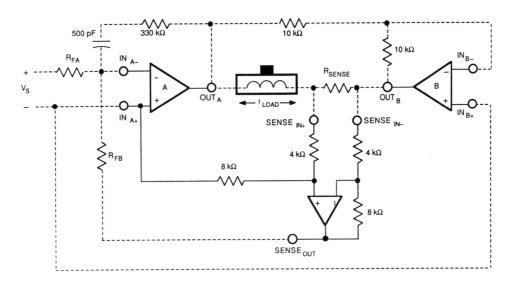

$$I_{LOAD} = g_m V_S = \frac{R_{FB} V_S}{R_{FA} A_{VS} R_{SENSE}}$$

where $A_{VS} = 2$

PARKING FUNCTION

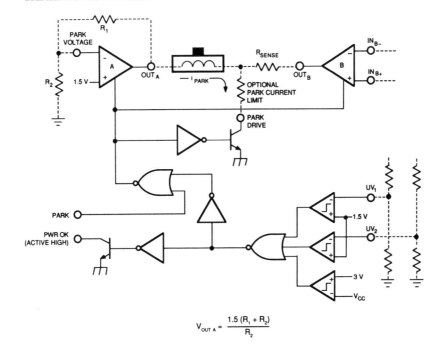

$$V_{OUT\,A} = \frac{1.5\,(R_1 + R_2)}{R_2}$$

Allegro
Applications Information
Power Integrated Circuits
For Motor-Drive Applications

UNIPOLAR STEPPER-MOTOR TRANSLATOR/DRIVER

The UCN5804B integrated circuit drives permanent magnet stepper motors rated to 1.25 A and 35 V with a minimum of external components.

Internal step logic activates one or two of the four output sink drivers to step the load from one position to the next. The logic is activated when STEP INPUT (pin 11) is allowed to go HIGH. Single-phase (A-B-C-D), two-phase (DA-AB-BC-CD), or half-step (A-AB-B-BC-C-CD-D-DA) operation, and step-inhibit are selected by connections at pins 9 and 10. The sequence of states is determined by the DIRECTION CONTROL (pin 14).

RECOMMENDED MAXIMUM OPERATING CONDITIONS

Output voltage, V_{OUT} — 35 V
Output current, I_{OUT} — 1.25 A
Logic supply voltage, V_{CC} — 4.5 V to 5.5 V
Input voltage, V_{IN} — 5.5 V

Drive Format	Pin 9	Pin 10
Two-Phase	L	L
One-Phase	H	L
Half-Step	L	H
Step-Inhibit	H	H

UCN5804B

L/R STEPPER-MOTOR DRIVE

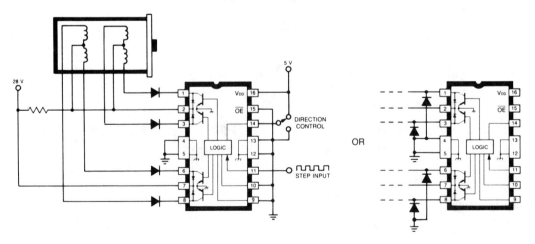

FULL-BRIDGE MOTOR DRIVERS

The UDN2953B and UDN2954W are designed for bidirectional, chopped-mode current control of dc motors with peak start-up currents as high as 3.5 A. The output-current limit is determined by the user's selection of a sensing resistor. The pulse duration is set by an external RC timing network. The chopped mode of operation is characterized by low power-dissipation levels and maximum efficiency.

Internal circuit protection includes thermal shutdown with hysteresis, output transient-suppression diodes, and crossover current protection.

The UDN2953B is supplied in a 16-pin DIP with heatsink contact tabs. The UDN2954W, with increased allowable package power dissipation, is supplied in a 12-lead single in-line power tab package. In both case styles, the heatsink is at ground potential and needs no insulation.

RECOMMENDED MAXIMUM OPERATING CONDITIONS

Motor supply voltage, V_{BB}	7.5 V to 50 V
Continuous output current, I_{OUT}	±2.0 A
Peak output current, I_{OP}	±3.5 A
Logic supply voltage, V_{CC}	4.5 V to 5.5 V
Input voltage, V_{IN}	24 V

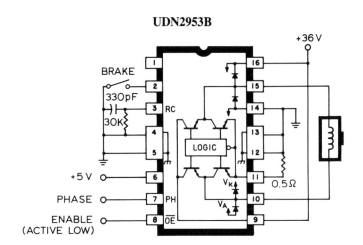

UDN2953B

QUAD DARLINGTON SWITCHES

The UDN2878W and UDN2879W drive motor windings at up to 200 watts per channel. The integrated circuits include transient-suppression diodes and input logic that is compatible with most TTL, LS TTL, and 5 V CMOS. The 12-pin single in-line power-tab package allows maximum power-handling capability.

RECOMMENDED MAXIMUM OPERATING CONDITIONS

Load voltage, V_{CC} (UDN2878W)	35 V
(UDN2879W)	50 V
Continuous output current, I_C	4 A
Peak output current, I_{CP}	5 A
Logic supply voltage range, V_S	4.5 V to 7.0 V
Input voltage, V_{IN}	V_S

STEPPER-MOTOR DRIVE

2-PHASE, UNIPOLAR INPUT WAVEFORMS

UDN2878W

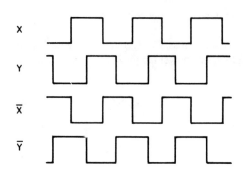

DUAL FULL-BRIDGE MOTOR DRIVER

The UDN2993B motor driver contains two independent full-bridges capable of operating with load currents of up to 600 mA. An internally generated deadtime prevents potentially destructive crossover currents when changing load phase. Internal transient-suppression diodes are included for use with inductive loads. Emitter outputs allow for current sensing in pulse-width modulated applications.

RECOMMENDED MAXIMUM OPERATING CONDITIONS

Load voltage range, V_{BB} 10 to 40 V
Output current, I_{OUT} ±500 mA
Logic voltage range, V_{DD} 4.5 to 5.5 V

UDN2993B

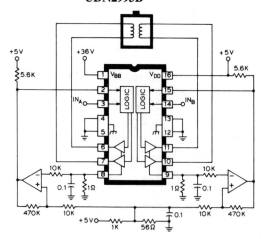
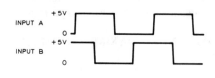

LINEAR MOTOR DRIVERS

Power operational amplifiers are useful in driving voice-coil motors, linear servo motors, and ac and dc motors in a linear mode where motor speed or position is a direct function of a linear input signal. The operational amplifiers listed here are standard "building block" circuits providing almost unlimited application. The high-gain high-impedance op-amp configuration allows many specialized input, output, and feedback arrangements.

All devices feature high-output voltage swings, high-input common-mode range, high PSRR and CMRR. The unity-gain stable versions need no external compensation. Internal thermal-shutdown circuitry protects these devices against output overloads. The dual amplifiers include programmable-output current-sensing capability.

PART NUMBER	TYPE	MAX. ΔV_S	CONT. I_{OUT}	PEAK I_{OP}	FEATURES	PACKAGE
ULN3751Z	Single	28 V	±2.5 A	3.5 A	Unity-Gain Stable Internal Compensation	5-Lead SIP
ULN3755W	Dual	40 V	±2.5 A	±3.5 A	Bootstrapped Output, Unity-Gain Stable,	12-Lead SIP
ULN3755B			±2.5 A	±3.5 A	Prog. Current Sense	16-Pin DIP

POSITION SERVO

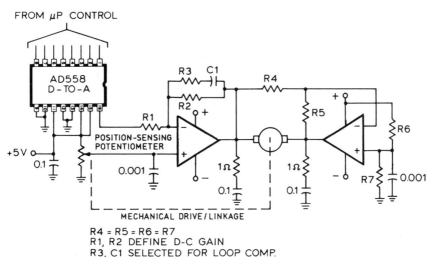

R4 = R5 = R6 = R7
R1, R2 DEFINE D-C GAIN
R3, C1 SELECTED FOR LOOP COMP.

TWO-PHASE, 60 Hz OSCILLATOR/MOTOR DRIVER

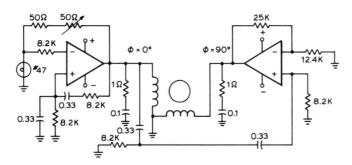

THREE-PHASE, 400 Hz OSCILLATOR/MOTOR DRIVER

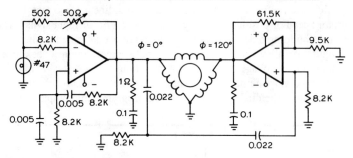

BiMOS UNIPOLAR MOTOR DRIVERS

Driving unipolar motors is one of many successful applications for the UCN5800A, UCN5801A, UCN5830B, and UCN5831B BiMOS II latched sink drivers.

All devices contain CMOS data latches, CMOS-control circuitry, high-voltage, high-current bipolar Darlington outputs, and output transient protection diodes for use with inductive loads.

The UCN5800A is a direct replacement for the original UCN4401A. The UCN5801A replaces the UCN4801A. With a 5-V supply, BiMOS II devices typically operate at data input rates above 5 MHz; at 12 V, significantly higher speeds are obtainable. BiMOS III drivers, with output voltage ratings to 150 V, are available as UCN5900A and UCN5901A.

Device	Package	Drivers	Features
UCN5800A	14-pin DIP	4	Clear, Strobe, Output Enable
UCN5801A	22-pin DIP	8	Clear, Strobe, Output Enable
UCN5830B	16-pin DIP	4	Strobe and Output Enable
UCN5831B	16-pin DIP	4	Strobe, Output Enable, Saturated Outputs

RECOMMENDED MAXIMUM OPERATING CONDITIONS

Output Voltage, V_{OUT}
    UCN5800A & UCN5801A . . . . 35 V
    UCN5830B & UCN5831B . . . . 35 V
Continuous Output Current, I_{OUT}
    UCN5800A & UCN5801A . . . . 350 mA
    UCN5830B & UCN5831B . . . . 1.0 A
Logic Supply Voltage, V_{DD}
    UCN5800A & UCN5801A . . . . 4.5 V to 12 V
    UCN5830B & UCN5831B . . . . 4.5 V to 5.5 V

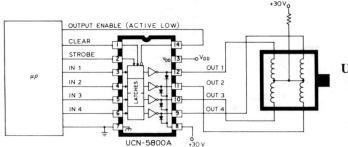

UNIPOLAR STEPPER-MOTOR DRIVE

HIGH-CURRENT BIPOLAR HALF-BRIDGE MOTOR DRIVERS

The UDN2935Z and UDN2950Z ICs are monolithic half-bridge motor drivers in power tab TO-220 style packages. The circuits combine sink and source drivers with diode protection, gain and level shifting systems, and a voltage regulator for single-supply operation. They are designed for servo-motor drive applications using pulse-width modulation.

The PWM drive mode is characterized by minimal power dissipation requirements and allows the output to switch currents of 2 amperes. Output dc current accuracies of better than 10% at 100 kHz can be obtained. The UDN2935Z and UDN2950Z can be used in pairs (full-bridge) to drive dc stepper motors or brushless dc motors.

RECOMMENDED MAXIMUM OPERATING CONDITIONS

Supply voltage, V_S	8.0 V to 35 V
Continuous output current, I_{OUT}	±2.0 A
Peak output current, I_{OP}	±3.5 A
Input voltage, V_{IN}	5.5 V

SINGLE-WINDING DC OR STEPPER-MOTOR DRIVE

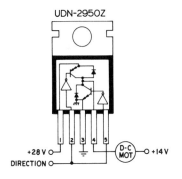

FULL-BRIDGE DC SERVO-MOTOR DRIVE

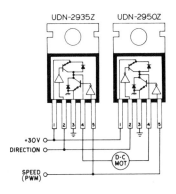

3-PHASE BRUSHLESS DC MOTOR CONTROL

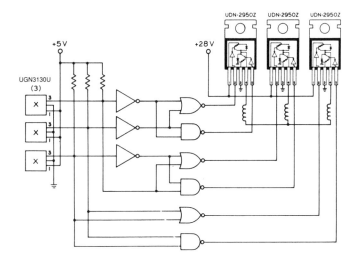

RELAY-DRIVER APPLICATIONS

Series UDN 2580A, 8-channel source drivers, and type UDN2957A, 5-channel source driver, provide current/voltage translation from TTL, positive CMOS, or negative CMOS logic to -48-V telecommunication relays requiring less than 350 mA. All devices have internal inductive-load transient-suppression diodes.

Type UDN 2580A-1 is best driven from negative-reference CMOS or NMOS logic (-5- or -12-V swing) in order to provide a -48-V swing at the output. The active-low input type UDN2588A-1 can be driven from positive logic TTL ($+5$-V swing) or CMOS ($+12$-V swing) levels. The active-high input UDN2957A is similar to the UDN2588A-1, but it also has a chip-enable function that requires a minimum number of drive lines to control outputs from several packages in a simple multiplex scheme.

RECOMMENDED MAXIMUM OPERATING CONDITIONS

Supply voltage, V_{EE} -50 V
Continuous output current, I_{OUT}
  (per output) -350 mA

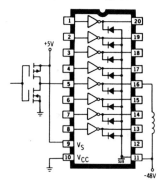

TELECOMMUNICATIONS RELAY DRIVER
(Positive logic)

TELECOMMUNICATIONS RELAY DRIVER
(Negative logic)

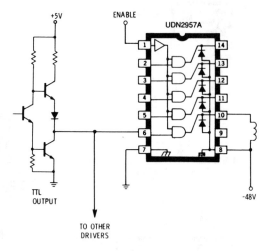

MULTIPLEXED RELAY DRIVER

Analog Devices
AD598
LVDT Signal Conditioner

FEATURES
- Single-chip solution, contains internal oscillator and voltage reference
- No adjustments required
- Insensitive to transducer null voltage
- Insensitive to primary to secondary phase shifts
- dc output proportional to position
- 20-Hz to 20-kHz frequency range
- Single- or dual-supply operation
- Unipolar or bipolar output
- Will operate a remote LVDT at up to 300 feet
- Position output can drive up to 1000 feet of cable
- Will also interface to an RVDT
- Outstanding performance
  Linearity: 0.05% of FS max.
  Output voltage: ±11 V min.
  Gain drift: 50 ppm/°C of FS max.
  Offset drift: 50 ppm/°C of FS max.

PRODUCT HIGHLIGHTS

1. The AD598 offers a monolithic solution to LVDT and RVDT signal conditioning problems; few extra passive components are required to complete the conversion from mechanical position to dc voltage and no adjustments are required.

2. The AD598 can be used with many different types of LVDTs because the circuit accommodates a wide range of input and output voltages and frequencies; the AD598 can drive an LVDT primary with up to 24 V rms and accept secondary input levels as low as 100 mV rms.

3. The 20-Hz to 20-kHz LVDT excitation frequency is determined by a single external capacitor. The AD598 input signal need not be synchronous with the LVDT primary drive. This means that an external primary excitation, such as the 400-Hz power mains in aircraft, can be used.

4. The AD598 uses a ratiometric decoding scheme such that primary to secondary phase shifts and transducer null voltage have absolutely no effect on overall circuit performance.

5. Multiple LVDTs can be driven by a single AD598, either in series or parallel as long as power dissipation limits are not exceeded. The excitation output is thermally protected.

6. The AD598 may be used in telemetry applications or in hostile environments where the interface electronics are remote from the LVDT. The AD598 can drive an LVDT at the end of 300 feet of cable, because the circuit is not affected by phase shifts or absolute signal magnitudes. The position output can drive as much as 1000 feet of cable.

7. The AD598 may be used as a loop integrator in the design of simple electromechanical servo loops.

SPECIFICATIONS

(typical @ +25°C and ±15 Vdc, C1 = 0.015 μF, R2 = 80 kΩ, R_L = 2 kΩ, unless otherwise noted. See Figure 7.)

Model	AD598J Min	AD598J Typ	AD598J Max	AD598A Min	AD598A Typ	AD598A Max	Unit
TRANSFER FUNCTION[1]		$V_{OUT} = \dfrac{V_A - V_B}{V_A + V_B} \times 500\,\mu A \times R2$					V
OVERALL ERROR[2]							
T_{min} to T_{max}		0.6	2.35		0.6	1.65	% of FS
SIGNAL OUTPUT CHARACTERISTICS							
Output Voltage Range (T_{min} to T_{max})	±11			±11			V
Output Current (T_{min} to T_{max})	8			6			mA
Short Circuit Current		20			20		mA
Nonlinearity[3] (T_{min} to T_{max})		75	±500		75	±500	ppm of FS
Gain Error[4]		0.4	±1		0.4	±1	% of FS
Gain Drift		20	±100		20	±50	ppm/°C of FS
Offset[5]		0.3	±1		0.3	±1	% of FS
Offset Drift		7	±200		7	±50	ppm/°C of FS
Excitation Voltage Rejection[6]		100			100		ppm/dB
Power Supply Rejection (±12 V to ±18 V)							
PSRR Gain (T_{min} to T_{max})	300	100		400	100		ppm/V
PSRR Offset (T_{min} to T_{max})	100	15		200	15		ppm/V
Common Mode Rejection (±3 V)							
CMRR Gain (T_{min} to T_{max})	100	25		200	25		ppm/V
CMRR Offset (T_{min} to T_{max})	100	6		200	6		ppm/V
Output Ripple[7]		4			4		mV rms
EXCITATION OUTPUT CHARACTERISTICS (@ 2.5 kHz)							
Excitation Voltage Range	2.1		24	2.1		24	V rms
Excitation Voltage							
(R1 = Open)[8]	1.2		2.1	1.2		2.1	V rms
(R1 = 12.7 kΩ)[8]	2.6		4.1	2.6		4.1	V rms
(R1 = 487 Ω)[8]	14		20	14		20	V rms
Excitation Voltage TC[9]		600			600		ppm/°C
Output Current	30			30			mA rms
T_{min} to T_{max}	12			12			mA rms
Short Circuit Current		60			60		mA
DC Offset Voltage (Differential, R1 = 12.7 kΩ)							
T_{min} to T_{max}		30	±100		30	±100	mV
Frequency	20		20k	20		20k	Hz
Frequency TC, (R1 = 12.7 kΩ)		200			200		ppm/°C
Total Harmonic Distortion		−50			−50		dB
SIGNAL INPUT CHARACTERISTICS							
Signal Voltage	0.1		3.5	0.1		3.5	V rms
Input Impedance		200			200		kΩ
Input Bias Current (AIN and BIN)		1	5		1	5	μA
Signal Reference Bias Current		2	10		2	10	μA
Excitation Frequency	0		20	0		20	kHz
POWER SUPPLY REQUIREMENTS							
Operating Range	13		36	13		36	V
Dual Supply Operation (±10 V Output)	±13			±13			V
Single Supply Operation							
0 to +10 V Output	17.5			17.5			V
0 to −10 V Output	17.5			17.5			V
Current (No Load at Signal and Excitation Outputs)		12	15		12	15	mA
T_{min} to T_{max}			16			18	mA

Model	AD598J Min	AD598J Typ	AD598J Max	AD598A Min	AD598A Typ	AD598A Max	Unit
TEMPERATURE RANGE							
JR (SOIC)	0		70				°C
AD (DIP)				−40		+85	°C
PACKAGE OPTION							
SOIC (R-20)		AD598JR					
Side Brazed DIP (D-20)					AD598AD		

NOTES
[1] V_A and V_B represent the Mean Average Deviation (MAD) of the detected sine waves. Note that for this Transfer Function to linearly represent positive displacement, the sum of V_A and V_B of the LVDT must remain constant with stroke length. See "Theory of Operation." Also see Figures 7 and 12 for R2.
[2] From T_{min} to T_{max} the overall error due to the AD598 alone is determined by combining gain error, gain drift and offset drift. For example, the worst case overall error for the AD598AD from T_{min} to T_{max} is calculated as follows: overall error = gain error at +25°C (±1% full scale) + gain drift from −40°C to +25°C (50 ppm/°C of FS × +65°C) + offset drift from −40°C to +25°C (50 ppm/°C of FS× 65°C) = ±1.65% of full scale. Note that 1000 ppm of full scale equals 0.1% of full scale. Full scale is defined as the voltage difference between the maximum positive and maximum negative output.
[3] Nonlinearity of the AD598 only, in units of ppm of full scale. Nonlinearity is defined as the maximum measured deviation of the AD598 output voltage from a straight line. The straight line is determined by connecting the maximum produced full-scale negative voltage with the maximum produced full-scale positive voltage.
[4] See Transfer Function.
[5] This offset refers to the $(V_A-V_B)/(V_A+V_B)$ input spanning a full-scale range of ±1. [For $(V_A-V_B)/(V_A+V_B)$ to equal +1, V_B must equal zero volts; and correspondingly for $(V_A-V_B)/(V_A+V_B)$ to equal −1, V_A must equal zero volts. Note that offset errors do not allow accurate use of zero magnitude inputs; practical inputs are limited to 100 mV rms.] The ±1 span is a convenient reference point to define offset referred to input. For example, with this input span a value of R2 = 20 kΩ would give V_{OUT} span a value of ±10 volts. Caution, most LVDTs will typically exercise less of the $(V_A-V_B)/(V_A+V_B)$ input span and thus require a larger value of R2 to produce the ±10 V output span. In this case the offset is correspondingly magnified when referred to the output voltage. For example, a Schaevitz E100 LVDT requires 80.2 kΩ for R2 to produce a ±10.69 V output and $(V_A-V_B)/(V_A+V_B)$ equals 0.27. This ratio may be determined from the graph shown in Figure 18, $(V_A-V_B)/(V_A+V_B)$ = (1.71 V rms−0.99 V rms)/(1.71 V rms+0.99 V rms). The maximum offset value referred to the ±10.69 V output may be determined by multiplying the maximum value shown in the data sheet (±1% of FS by 1/0.27 which equals ±3.7% maximum. Similarly, to determine the maximum values of offset drift, offset CMRR and offset PSRR when referred to the ±10.69 V output, these data sheet values should also be multiplied by (1/0.27). For this example, for the AD598AD the maximum values of offset drift, PSRR offset and CMRR offset would be: 185 ppm/°C of FS; 741 ppm/V and 741 ppm/V respectively when referred to the ±10.69 V output.
[6] For example, if the excitation to the primary changes by 1 dB, the gain of the system will change by typically 100 ppm.
[7] Output ripple is a function of the AD598 bandwidth determined by C2, C3 and C4. See Figures 16 and 17.
[8] R1 is shown in Figures 7 and 12.
[9] Excitation voltage drift is not an important specification because of the ratiometric operation of the AD598.

Specifications subject to change without notice.
Specifications shown in **boldface** are tested on all production units at final electrical test. Results from those tested are used to calculate outgoing quality levels. All min and max specifications are guaranteed, although only those shown in **boldface** are tested on all production units.

THERMAL CHARACTERISTICS

	θ_{JC}	θ_{JA}
SOIC Package	22°C/W	80°C/W
Side Brazed Package	25°C/W	85°C/W

ABSOLUTE MAXIMUM RATINGS

Total Supply Voltage +V_S to −V_S . . . . 36 V
Storage Temperature Range
  R Package . . . . −65°C to +150°C
  D Package . . . . −65°C to +150°C
Operating Temperature Range
  AD598JR . . . . 0 to +70°C
  AD598AD . . . . −40°C to +85°C
Lead Temperature Range (Soldering 60 Seconds) . . . . +300°C
Power Dissipation Up to +65°C . . . . 1.2 W
Derates Above +65°C . . . . 12 mW/°C

CONNECTION DIAGRAM

Plastic SOIC (R) Package
and
Side Brazed Ceramic DIP (D) Package

```
       -V_S  [1]      [20] +V_S
       EXC 1 [2]      [19] OFFSET 1
       EXC 2 [3]      [18] OFFSET 2
     LEVEL 1 [4]      [17] SIGNAL REFERENCE
     LEVEL 2 [5] AD598 [16] SIGNAL OUTPUT
      FREQ 1 [6] TOP VIEW [15] FEEDBACK
             (Not to Scale)
      FREQ 2 [7]      [14] OUTPUT FILTER
   B1 FILTER [8]      [13] A1 FILTER
   B2 FILTER [9]      [12] A2 FILTER
         V_B [10]     [11] V_A
```

AD598 FUNCTIONAL BLOCK DIAGRAM

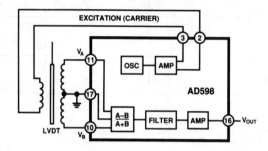

TYPICAL CHARACTERISTICS (at +25°C and $V_S = \pm 15$ V unless otherwise noted)

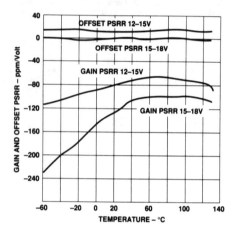

Gain and Offset PSRR vs. Temperature

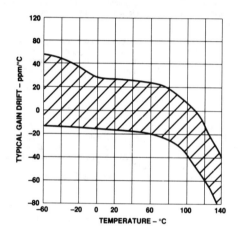

Typical Gain Drift vs. Temperature

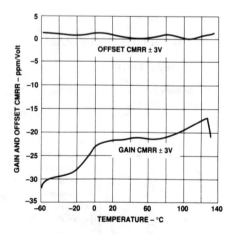

Gain and Offset CMRR vs. Temperature

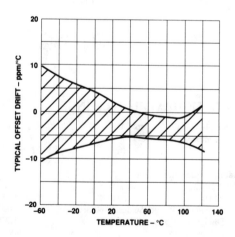

Typical Offset Drift vs. Temperature

Analog Devices AD598

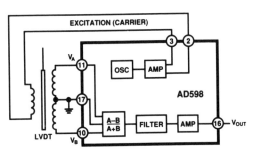

AD598 Functional Block Diagram

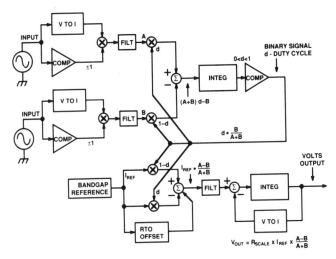

Block Diagram of Decoder

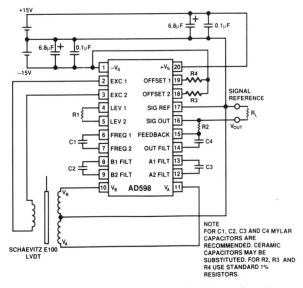

Interconnection Diagram for Dual Supply Operation

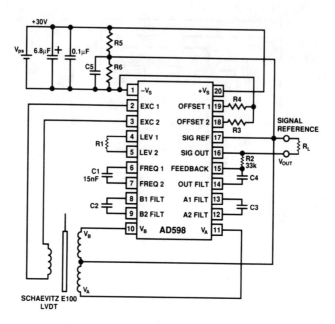

Interconnection Diagram for Single Supply Operation

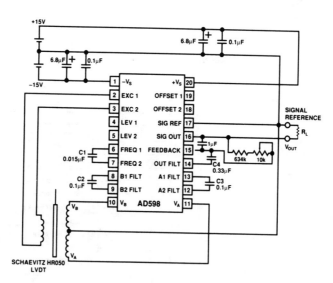

Proving Ring-Weigh Scale Circuit

Analog Devices AD598

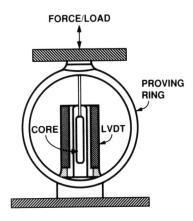

Proving Ring-Weigh Scale Cross Section

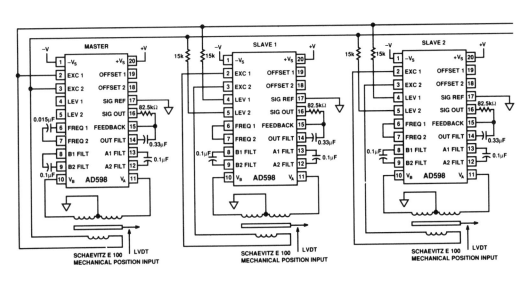

Multiple LVDTs – Synchronous Operation

Analog Devices AD598

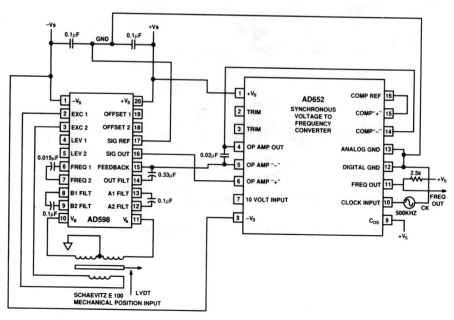

High Resolution Position-to-Frequency Converter

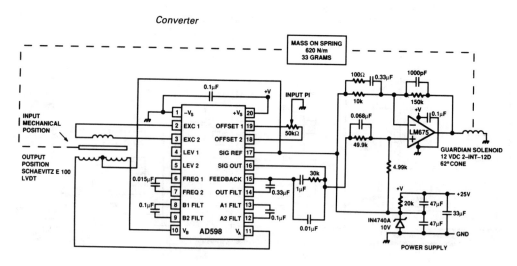

Low Cost Set-Point Controller

Analog Devices AD598   129

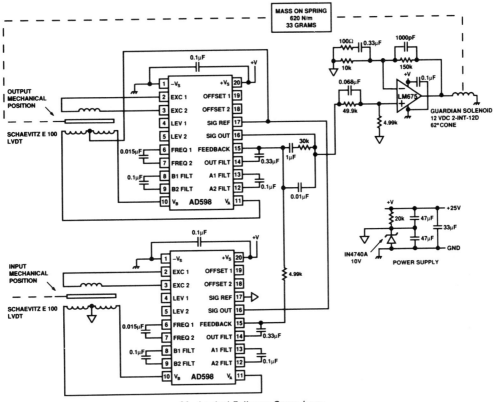

Mechanical Follower Servo-Loop

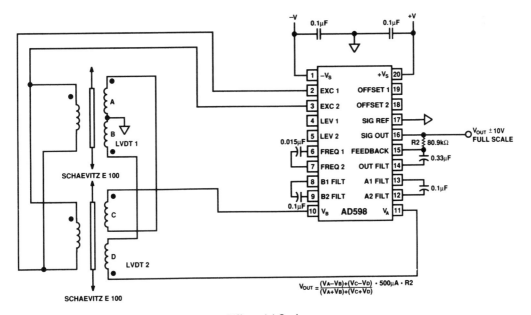

$$V_{OUT} = \frac{(V_A - V_B) + (V_C - V_D)}{(V_A + V_B) + (V_C + V_D)} \cdot 500\mu A \cdot R2$$

Differential Gaging

Analog Devices AD598

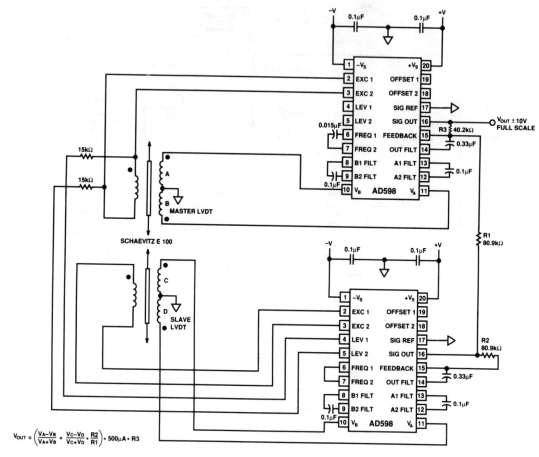

Precision Differential Gaging

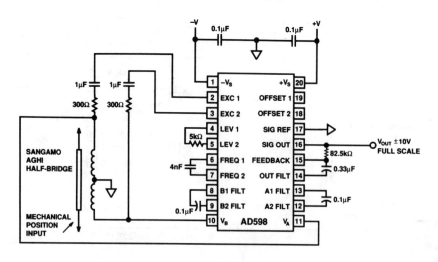

Half-Bridge Operation

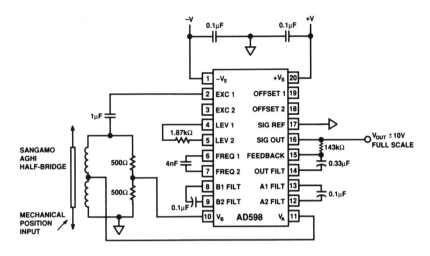

Alternate Half-Bridge Circuit

Analog Devices
AD596*/AD597*
Thermocouple Conditioners and Set-Point Controllers

*Protected by U.S. Patent No. 4,029,974

FEATURES
- Low cost
- Operates with type J (AD596) or type K (AD597) thermocouples
- Built-in ice-point compensation
- Temperature proportional operation: 10 mV/°C
- Temperature set-point operation: on/off
- Programmable switching hysteresis
- High-impedance differential input

PRODUCT HIGHLIGHTS
1. The AD596/AD597 provides cold-junction compensation and a high-gain amplifier, which can be used as a set-point comparator.
2. The input stage of the AD596/AD597 is a high quality instrumentation amplifier that allows the thermocouple to float over most of the supply voltage range.
3. Linearization not required for thermocouple temperatures close to 175°C (+100°C to +540°C for AD596).
4. Cold-junction compensation is optimized for ambient temperatures ranging from +25°C to +100°C.
5. In the stand-alone mode, the AD596/AD597 produces an output voltage that indicates its own temperature.

Analog Devices AD596/AD597

SPECIFICATIONS (@ +60°C and V_S = 10 V, Type J (AD596), Type K (AD597) Thermocouple (unless otherwise noted))

Model	AD596AH			AD597AH			Units
	Min	Typ	Max	Min	Typ	Max	
ABSOLUTE MAXIMUM RATINGS							
+V_S to −V_S			36			36	Volts
Common-Mode Input Voltage	(−V_S − 0.15)		+V_S	(−V_S − 0.15)		+V_S	Volts
Differential Input Voltage	−V_S		+V_S	−V_S		+V_S	Volts
Alarm Voltages							
+ALM	−V_S		(−V_S + 36)	−V_S		(−V_S + 36)	Volts
−ALM	−V_S		+V_S	−V_S		+V_S	Volts
Operating Temperature Range	−55		+125	−55		+125	°C
Output Short Circuit to Common	Indefinite			Indefinite			
TEMPERATURE MEASUREMENT							
(Specified Temperature Range +25°C to +100°C)							
Calibration Error[1]	−4		+4	−4		+4	°C
Stability vs. Temperature[2]		±0.02	±0.05		±0.02	±0.05	°C/°C
Gain Error	−1.5		+1.5	−1.5		+1.5	%
Nominal Transfer Function		10			10		mV/°C
AMPLIFIER CHARACTERISTICS							
Closed Loop Gain[3]		180.6			245.5		V/V
Input Offset Voltage		°C × 53.21 + 235			°C × 41.27 − 37		μV
Input Bias Current		0.1			0.1		μA
Differential Input Range	−10		+50	−10		+50	mV
Common-Mode Range	(−V_S − 0.15)		(+V_S − 4)	(−V_S − 0.15)		(+V_S − 4)	Volts
Common-Mode Sensitivity − RTO			10			10	mV/V
Power Supply Sensitivity − RTO		1	10		1	10	mV/V
Output Voltage Range							
Dual Supplies	(−V_S + 2.5)		(+V_S − 2)	(−V_S + 2.5)		(+V_S − 2)	Volts
Single Supply	0		(+V_S − 2)	0		(+V_S − 2)	Volts
Usable Output Current[4]	±5			±5			mA
3dB Bandwidth		15			15		kHz
ALARM CHARACTERISTICS							
$V_{CE(SAT)}$ at 2mA		0.3			0.3		Volts
Leakage Current			±1			±1	μA
Operating Voltage at −ALM			(+V_S − 4)			(+V_S − 4)	Volts
Short Circuit Current		20			20		mA
POWER REQUIREMENTS							
Operating[5]	(+V_S to −V_S) ≤ 30			(+V_S to −V_S) ≤ 30			Volts
Quiescent Current							
+V_S		160	300		160	300	μA
−V_S		100	200		100	200	μA
PACKAGE OPTION							
TO-100 (H-10A)	AD596AH			AD597AH			

NOTES
[1] This is a measure of the deviation from ideal with a measuring thermocouple junction of 175°C and a chip temperature of 60°C. The ideal transfer function is given by:
AD596: $V_{OUT} = 180.57 \times (V_m − V_a + \text{(ambient in °C)} \times 53.21 \mu V/°C + 235 \mu V)$
AD597: $V_{OUT} = 245.46 \times (V_m − V_a + \text{(ambient in °C)} \times 41.27 \mu V/°C − 37 \mu V)$
Where V_m and V_a represent the measuring and ambient temperatures and are taken from the appropriate J or K thermocouple table. The ideal transfer function minimizes the error over the ambient temperature range of 25°C to 100°C with a thermocouple temperature of approximately 175°C.
[2] Defined as the slope of the line connecting the AD596/AD597 CJC errors measured at 25°C and 100°C ambient temperature.
[3] Pin 6 shorted to pin 7.
[4] Current Sink Capability in single supply configuration is limited to current drawn to ground through a 50kΩ resistor at output voltages below 2.5V.
[5] −V_S must not exceed −16.5V.

Specifications subject to change without notice.

Specifications shown in **boldface** are tested on all production units at final electrical test. Results from those tests are used to calculate outgoing quality levels. All min and max specifications are guaranteed, although only those shown in boldface are tested on all production units.

Analog Devices AD596/AD597

Thermocouple Temperature °C	Type J Voltage mV	AD596 Output mV	Type K Voltage mV	AD597 Output mV	Thermocouple Temperature °C	Type J Voltage mV	AD596 Output mV	Type K Voltage mV	AD597 Output mV
−200	−7.890	−1370	−5.891	−1446	500	27.388	5000	20.640	5066
−180	−7.402	−1282	−5.550	−1362	520	28.511	5203	21.493	5276
−160	−6.821	−1177	−5.141	−1262	540	29.642	5407	22.346	5485
−140	−6.159	−1058	−4.669	−1146	560	30.782	5613	23.198	5694
−120	−5.426	−925	−4.138	−1016	580	31.933	5821	24.050	5903
−100	−4.632	−782	−3.553	−872	600	33.096	6031	24.902	6112
−80	−3.785	−629	−2.920	−717	620	34.273	6243	25.751	6321
−60	−2.892	−468	−2.243	−551	640	35.464	6458	26.599	6529
−40	−1.960	−299	−1.527	−375	660	36.671	6676	27.445	6737
−20	−.995	−125	−.777	−191	680	37.893	6897	28.288	6944
−10	−.501	−36	−.392	−96	700	39.130	7120	29.128	7150
0	0	54	0	0	720	40.382	7346	29.965	7355
10	.507	146	.397	97	740	41.647	7575	30.799	7560
20	1.019	238	.798	196	750	42.283	7689	31.214	7662
25	1.277	285	1.000	245	760	−	−	31.629	7764
30	1.536	332	1.203	295	780	−	−	32.455	7966
40	2.058	426	1.611	395	800	−	−	33.277	8168
50	2.585	521	2.022	496	820	−	−	34.095	8369
60	3.115	617	2.436	598	840	−	−	34.909	8569
80	4.186	810	3.266	802	860	−	−	35.718	8767
100	5.268	1006	4.095	1005	880	−	−	36.524	8965
120	6.359	1203	4.919	1207	900	−	−	37.325	9162
140	7.457	1401	5.733	1407	920	−	−	38.122	9357
160	8.560	1600	6.539	1605	940	−	−	38.915	9552
180	9.667	1800	7.338	1801	960	−	−	39.703	9745
200	10.777	2000	8.137	1997	980	−	−	40.488	9938
220	11.887	2201	8.938	2194	1000	−	−	41.269	10130
240	12.998	2401	9.745	2392	1020	−	−	42.045	10320
260	14.108	2602	10.560	2592	1040	−	−	42.817	10510
280	15.217	2802	11.381	2794	1060	−	−	43.585	10698
300	16.325	3002	12.207	2996	1080	−	−	44.439	10908
320	17.432	3202	13.039	3201	1100	−	−	45.108	11072
340	18.537	3402	13.874	3406	1120	−	−	45.863	11258
360	19.640	3601	14.712	3611	1140	−	−	46.612	11441
380	20.743	3800	15.552	3817	1160	−	−	47.356	11624
400	21.846	3999	16.395	4024	1180	−	−	48.095	11805
420	22.949	4198	17.241	4232	1200	−	−	48.828	11985
440	24.054	4398	18.088	4440	1220	−	−	49.555	12164
460	25.161	4598	18.938	4649	1240	−	−	50.276	12341
480	26.272	4798	19.788	4857	1250	−	−	50.633	12428

Output Voltage vs. Thermocouple Temperature (Ambient +60°C, $V_S = -5V, +15V$)

AD596/AD597 FUNCTIONAL BLOCK DIAGRAM

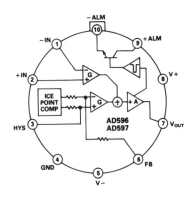

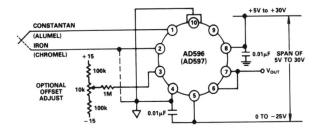

Temperature Proportional Output Connection

Analog Devices AD596/AD597

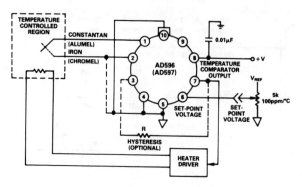

Set-Point Control Mode

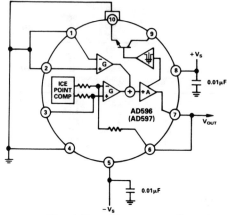

*Stand-Alone Temperature Transducer
Temperature Proportional Output Connection*

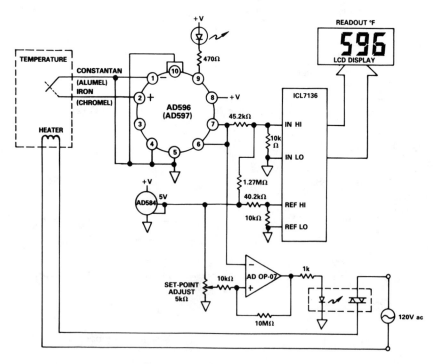

Temperature Measurement and Control

The information included herein is believed to be accurate and reliable. However, LSI Computer Systems, Inc. assumes no responsibilities for inaccuracies, nor for any infringements of patent rights of others which may result from it use.

LSI
LS7237
Touch-Control Step-Dimmer
Light Switch and ac Motor-Speed Controller

FEATURES
- Phase-locked loop synchronization allows use in wall switch applications and produces pure ac waveform across output load (no dc offset)
- Produces on/off or brightness control of incandescent lamps and on/off control of fluorescent lamps (mode "0" only) without the use of mechanical switches
- Provides speed control of ac motors, such as shaded pole and universal series motors
- Controls the "duty cycle" from 25% to 88% (on time angles for ac half cycles between 45° and 159° respectively)
- Operates on 50-Hz/60Hz line frequency
- Provides control through transformers for low-voltage lighting applications
- Input for extensions or remote sensors
- 12- to 18-V supply voltage
- 8-pin plastic DIP

ABSOLUTE MAXIMUM RATINGS

PARAMETER	SYMBOL	VALUE	UNITS
DC supply voltage	V_{SS}	+20	Volt
Any input voltage	V_{IN}	V_{SS} -20 to V_{SS} + .5	Volt
Operating temperature	T_A	0 to +80	°C
Storage temperature	T_{stg}	-65 to +150	°C

CONNECTION DIAGRAM

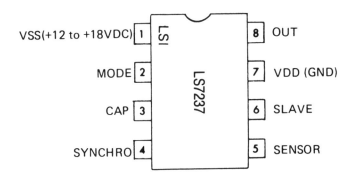

BLOCK DIAGRAM

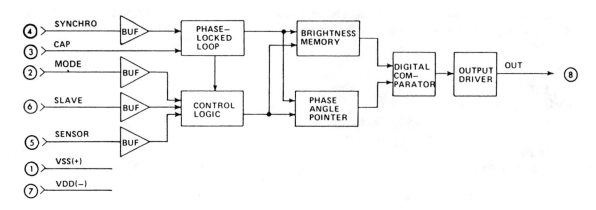

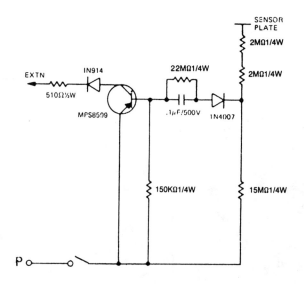

ELECTRONIC SWITCH EXTENSION

All switching and dimming functions can also be implemented by utilizing the SLAVE input. This can be done by either a mechanical switch or the electronic switch in conjunction with sensor plate as shown in Fig. 6. When the plate is touched, a logical high level is generated at the EXTENSION terminal for both half cycles of the line frequency.

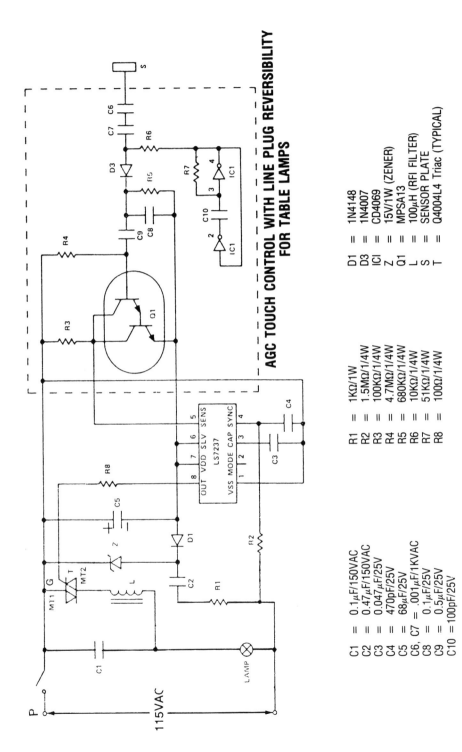

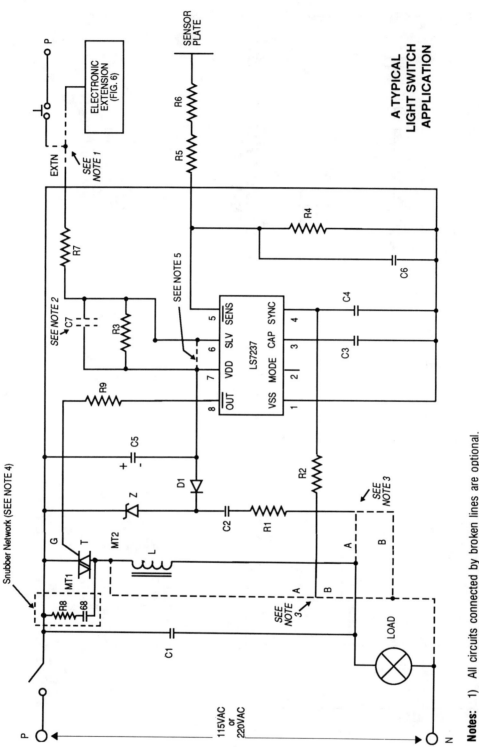

LSI
LS7260/LS7261, LS7262
Brushless dc Motor Commutator/Controller

FEATURES
- Direct drive of p-channel and n-channel FETs (LS7260)
- Direct drive of pnp and npn transistors (LS7261/LS7262)
- Open- or closed-loop motor-speed control
- 5- to 28-V operation
- Externally selectable input to output code for 60°, 120°, 240°, or 300° electrical sensor spacing
- Three- or four-phase operation
- Analog speed-control input
- Forward/reverse control
- Output enable control
- Positive static braking
- Overcurrent sensing
- Six outputs drive switching bridge directly

MAXIMUM RATINGS

PARAMETER	SYMBOL	VALUE	UNITS
Storage temperature	T_{stg}	−65 to +150	°C
Operating temperature			
1. Plastic	T_{ap}	−25 to +70	°C
2. Ceramic	T_{ac}	−55 to +125	°C
Voltage (any pin to V_{SS})	V_{max}	−30 to +.5	VOLTS

CONNECTION DIAGRAM
TOP VIEW STANDARD 20 PIN PLASTIC DIP

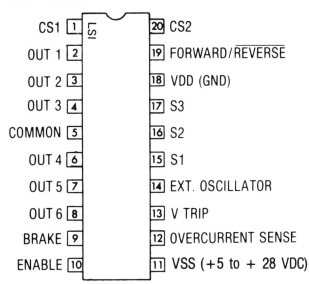

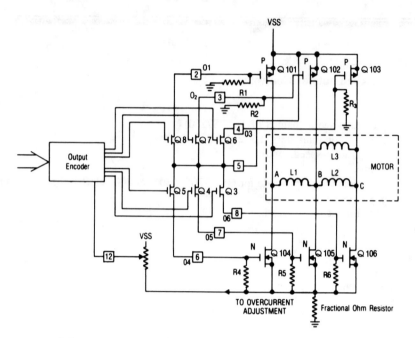

LS7260 THREE PHASE OUTPUT DRIVER CIRCUITRY

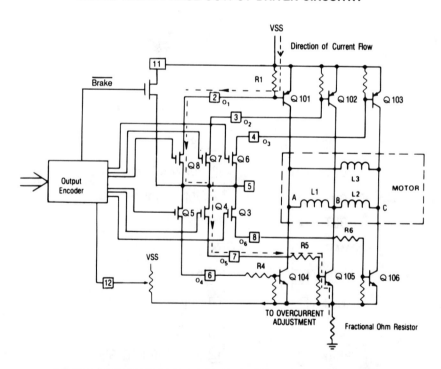

LS7261/LS7262 THREE PHASE OUTPUT DRIVER CIRCUITRY

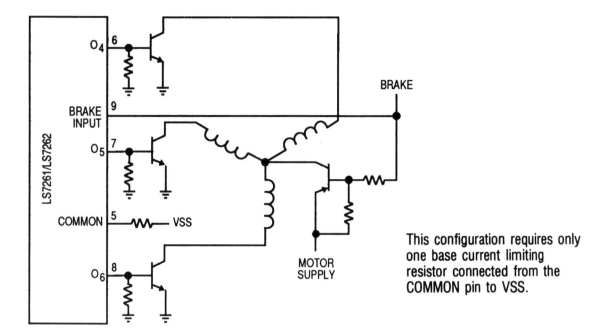

SINGLE ENDED DRIVER CIRCUIT

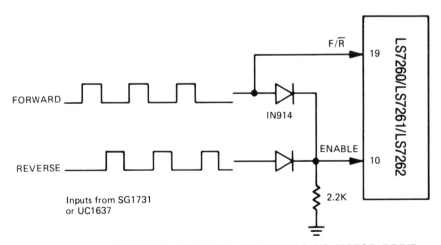

PRECISION CONTROL BRUSHLESS DC MOTOR DRIVE

FOUR PHASE OUTPUT DRIVER CIRCUITRY

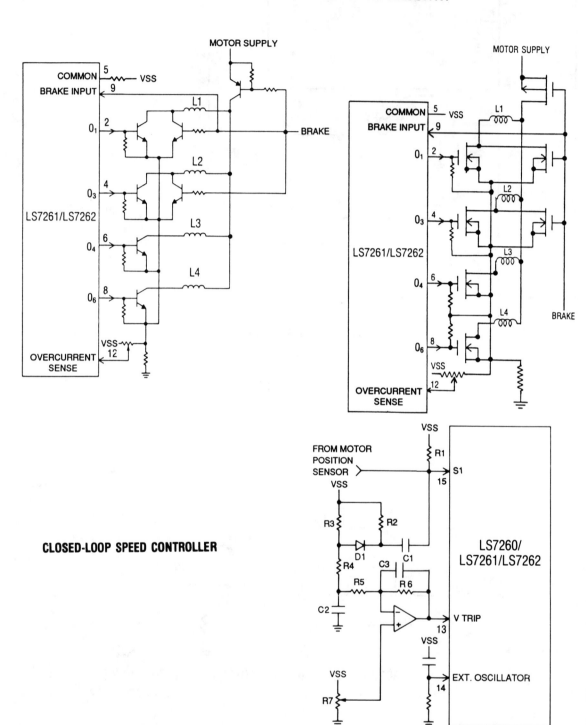

CLOSED-LOOP SPEED CONTROLLER

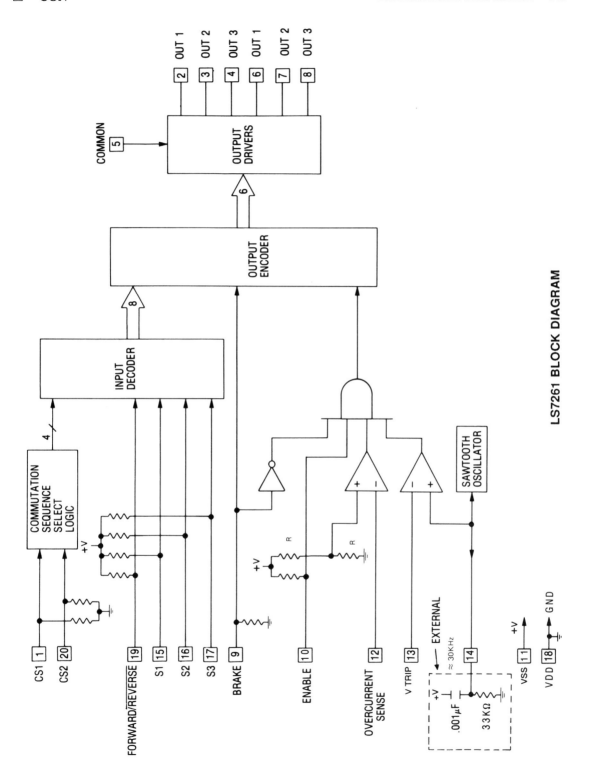

LS7261 BLOCK DIAGRAM

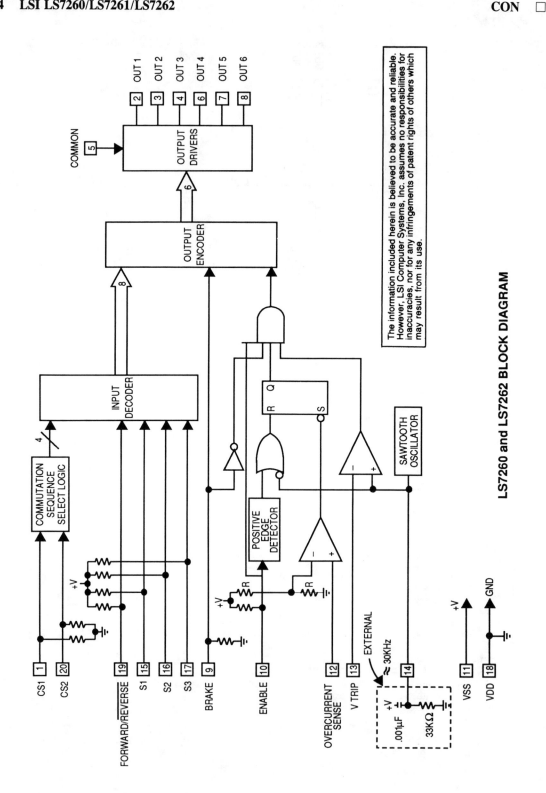

LS7260 and LS7262 BLOCK DIAGRAM

LS7263
Brushless dc Motor-Speed Controller

FEATURES
- Highly accurate speed regulation (±1% derived from XTL controlled time base.)
- Rapid acceleration to speed with little overshoot
- Static braking
- 10- to 28-V supply range
- Low speed detection output
- Over-current detection logic
- Power on reset
- Six outputs drive power switching bridge directly
- 18-pin dual-in-line package

MAXIMUM RATINGS

PARAMETER	SYMBOL	VALUE	UNITS
Storage temperature	T_{stg}	−65 to +150	°C
Operating temperature			
1. Plastic	T_{ap}	−25 to +70	°C
2. Ceramic	T_{ac}	−55 to +125	°C
Voltage (any pin to V_{SS})	V_{max}	−30 to +0.5	VOLTS

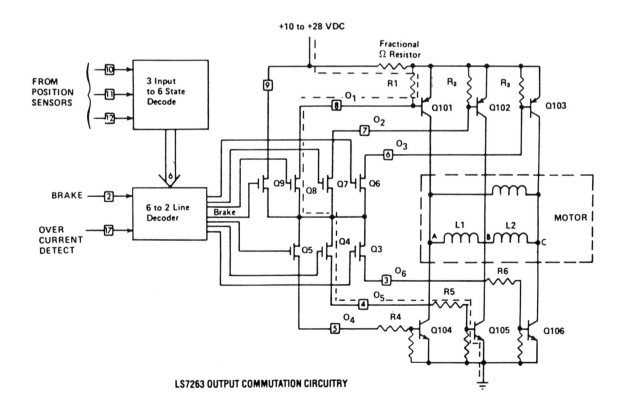

LS7263 OUTPUT COMMUTATION CIRCUITRY

LSI LS7263

LS DETECTS OUTPUT	1	18	TACHOMETER INPUT
BRAKE INPUT	2	17	OVERCURRENT DETECT
OUTPUT O_6	3	16	VDD (GND)
OUTPUT O_5	4	15	FREQUENCY TEST POINT
OUTPUT O_4	5	14	OSC OUT
OUTPUT O_3	6	13	OSC IN
OUTPUT O_2	7	12	INPUT C
OUTPUT O_1	8	11	INPUT B
VSS (+10 to +28 VDC)	9	10	INPUT A

TOP VIEW

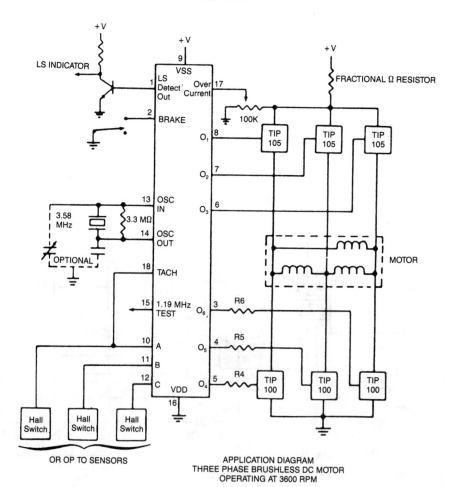

APPLICATION DIAGRAM
THREE PHASE BRUSHLESS DC MOTOR
OPERATING AT 3600 RPM

LSI LS7263

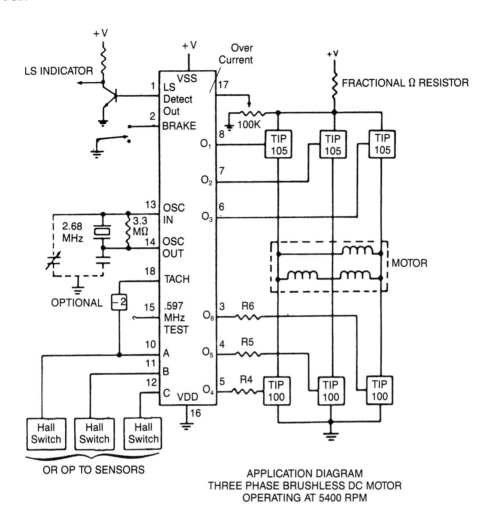

APPLICATION DIAGRAM
THREE PHASE BRUSHLESS DC MOTOR
OPERATING AT 5400 RPM

Tach input using ÷2 with output data rate doubled to achieve 5400 RPM operation.

LSI LS7263

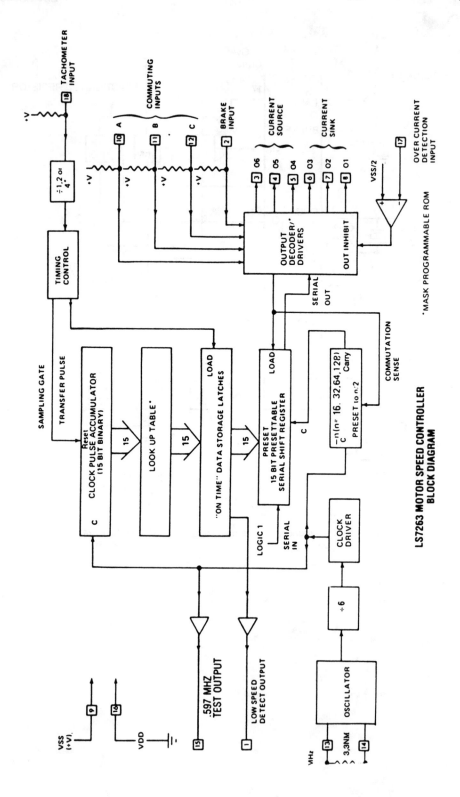

LS7263 MOTOR SPEED CONTROLLER BLOCK DIAGRAM

LSI
LS7270
Programmable Integrated Controller/Sequencer

FEATURES
- Hardware-oriented simple instruction set
- 4 on-chip 12-bit programmable down-counters
- 4 priority interrupt (JAM) inputs
- 12 discrete inputs
- 12 latched outputs
- 12 discrete memory-bit registers
- Anti-bounce circuits on DI, CNT, and JAM inputs for direct interface with mechanical switches, keyboards, etc.
- Simple serial interface to external program memory (PROM or ROM)
- External program memory up to 2048 instructions
- On-chip clock generator
- Inputs TTL, NMOS, and CMOS compatible
- Outputs TTL, NMOS, and CMOS compatible
- Single power-supply operation. +4.75 Vdc to +12 Vdc
- 40-pin plastic DIP

MAXIMUM RATINGS

PARAMETER	SYMBOL	VALUE	UNITS
Storage temperature	T_{STG}	−55 to +150	°C
Operating temperature	T_A	0 to 70	°C
Voltage (any pin to V_{SS})	V_{max}	+15 to −0.3	V

TOP VIEW

```
DI7   [1]            [40]  DI8
DI6   [2]            [39]  DI9
DI5   [3]            [38]  DI10
DI4   [4]            [37]  DI11
DI3   [5]            [36]  DI12
DI2   [6]            [35]  VDD (+5V)
DI1   [7]            [34]  CNT4
LO12  [8]            [33]  CNT3
LO11  [9]            [32]  CNT2
LO10  [10]           [31]  CNT1
LO9   [11]           [30]  SHIFT CLOCK
LO8   [12]           [29]  SHIFT/LOAD
LO7   [13]           [28]  MCLR
LO6   [14]           [27]  OSC
LO5   [15]           [26]  VSS (GND)
LO4   [16]           [25]  JAM1
LO3   [17]           [24]  JAM2
LO2   [18]           [23]  JAM3
LO1   [19]           [22]  JAM4
ADDRESS [20]         [21]  INST
```

150 LSI LS7270

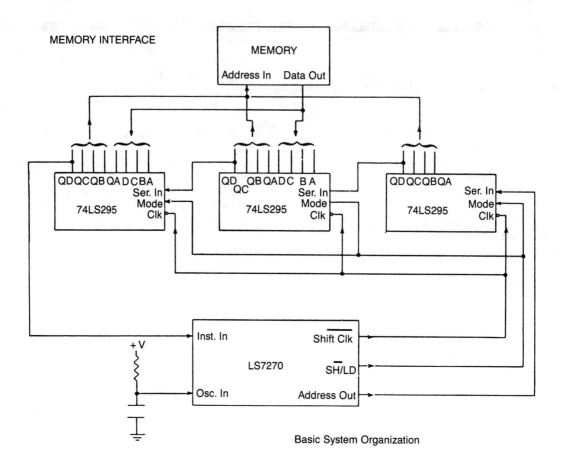

Basic System Organization

PROGRAM EXAMPLE

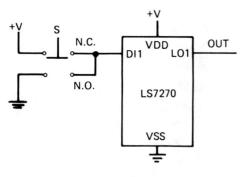

FIGURE 7

A simple example is given below to illustrate how the codes are constructed.

A momentary push-button switch, S is connected to the DI1 input of the ICS as shown in Fig. 7. It is required that every time S is pushed, the output LO1 will toggle (change state). Note that only the transition from the nondepressed to the depressed state should cause LO1 to toggle; if S is held depressed, it will have no further effect on the output.

Let us assign ICS internal register M1 to store the status of S and M2 to store the status of the output latch LO1 during each sample cycle. A flow chart to describe the program steps is given in Fig 8. The program is in mnemonic code and its binary equivalent is given below.

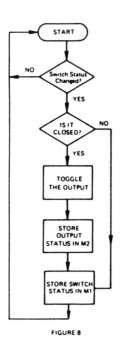

FIGURE 8

Mem. Address (Decimal)	Mnemonic	Binary	Comment
0	STRT: LD DI1, T2	0 0 1 0 0 0 0 0	Read status of S and
1		1 0 0 0 0 0 0 0	save it in T2.
2	XOR M2, −2	0 1 0 0 1 1 1 0	If S changed go to next step,
3		1 0 1 1 0 1 0 0	otherwise back to start.
4	LD T2	0 1 0 0 0 0 0 0	S changed; get ready to
5		1 0 0 0 0 1 1 1	test for open/close.
6	XOR T1, +1	0 0 0 0 0 0 0 1	Closed? If so, skip next step
7		1 0 1 1 0 1 1 1	
8	J UPDT	0 0 0 1 0 0 0 0	Not closed. Go to UPDT
9		0 1 0 1 0 0 0 0	routine to update M2
10	LD M1	0 0 0 0 0 0 0 0	S is closed; so get ready
11		1 0 0 0 1 0 0	to toggle.
12	STRC LO1	0 0 0 0 0 0 0 0	Toggled.
13		1 1 1 0 1 0 0 0	
14	STRC M1	0 0 0 0 0 0 0 0	Save current output
15		1 1 1 0 1 1 0 0	status in M1
16	UPDT: LD T2	0 1 0 0 0 0 0 0	Current status of S
17		1 0 0 0 0 1 1 1	
18	STR M2	0 1 0 0 0 0 0 0	Save status in M2
19		1 1 1 0 0 1 0 0	
20	J STRT	0 0 0 0 0 0 0 0	Start new sample cycle
21		0 1 0 1 0 0 0 0	

Note here that memory addresses for successive instructions have incremental value of 2. This is because each memory location can only hold a single byte (8 bits), whereas, an instruction consists of 2 bytes. The low byte of the first instruction is stored at address 0 and the high byte at address 1. The low byte of the second instruction at address 2 and the high byte at address 3 and so on.

Silicon Systems
SSI 32B451 SCSI Controller

FEATURES

- Supports asynchronous data transfer up to 1.5 Mbytes/sec
- Supports target role in SCSI applications
- Includes high-current drivers and Schmitt trigger receivers for direct connection to the SCSI bus
- Full hardware compliance to ANSI X3T9.2 Rev. 17B specification as a target peripheral adapter
- Contains circuitry to support SCSI arbitration, (re)selection and parity features
- Complements the SSI 32C453 buffer controller
- Plug compatible with AIC 500L
- Available in 44-pin PLCC
- Single +5-V supply

DC OPERATING CHARACTERISTICS
(Ta = 0 to 70 °C, VCC = +5 V±5%, VSS = 0 V)

PARAMETER		CONDITION	MIN	MAX	UNITS
IIL	Input Leakage (BREQ, LO, $\overline{BOE}$, $\overline{BIE}$, $\overline{ET}$ $\overline{SELOUT}$, $\overline{BSYOUT}$, CDIN, I/OIN, MSGIN, PAR/$\overline{RST}$, CLK, $\overline{ACK}$)	0 < Vin < VCC	-10	+10	µA
IOL	SCSI Output Leakage ($\overline{SEL}$, $\overline{BSY}$, $\overline{DB0}$-$\overline{DB7}$, $\overline{DBP}$, $\overline{MSG}$, $\overline{C/D}$, $\overline{I/O}$)	0.5 < Vout < VCC	-50	+50	µA
IOL	D0-D7	0.45 < Vout < VCC	-10	+10	µA
VIL	Input Low Voltage		0	0.8	V
VIH	Input High Voltage		2.0		V
VOH	Output High Voltage	IOH = -400 µA	2.4		V
VOL	SCSI Output Low Voltage	IOL = 48 mA		0.5	V
VOL	All others	IOL = 2 mA		0.4	V
	Power Dissipation			500	mW
Vhsy	Hysteresis Voltage (all SCSI signals)		200		mV
Iccs	Standby Current	Ta = 70°C		600	µA
Icc	Supply Current	Ta = 70°C		30	mA
Cin	Input Capacitance			15	pF

ORDERING INFORMATION

PART DESCRIPTION	ORDER NO.	PKG. MARK
SSI 32B451 44-pin PLCC	SSI 32B451-CH	32B451-CH

No responsibility is assumed by Silicon Systems for use of this product nor for any infringements of patents and trademarks or other rights of third parties resulting from its use. No license is granted under any patents, patent rights or trademarks of Silicon Systems. Silicon Systems reserves the right to make changes in specifications at any time without notice. Accordingly, the reader is cautioned to verify that the data sheet is current before placing orders.

ABSOLUTE MAXIMUM RATINGS

(Maximum limits indicate where permanent device damage occurs. Continuous operation at these limits is not intended and should be limited to those conditions specified in the DC Operating Characteristics.)

PARAMETER	RATING	UNIT
VCC with respect to VSS (GND)	+7	V
Max. voltage on any pin with respect to VSS	-0.5 to +7	V
Operating temperature	0 to 70	°C
Storage temperature	-55 to +125	°C

BLOCK DIAGRAM

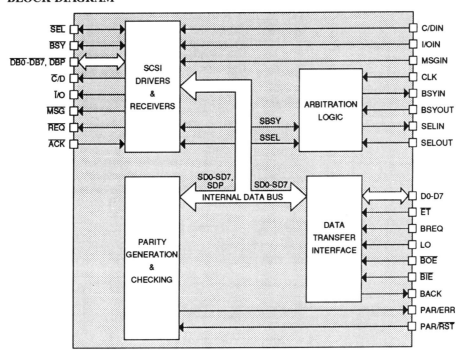

PACKAGE PIN DESIGNATIONS
(TOP VIEW)

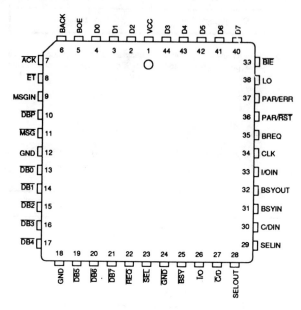

44-pin PLCC

CAUTION: Use handling procedures necessary for a static sensitive component.

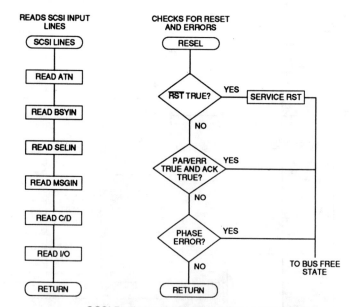

SCSI Background Routines Using SSI 32B451

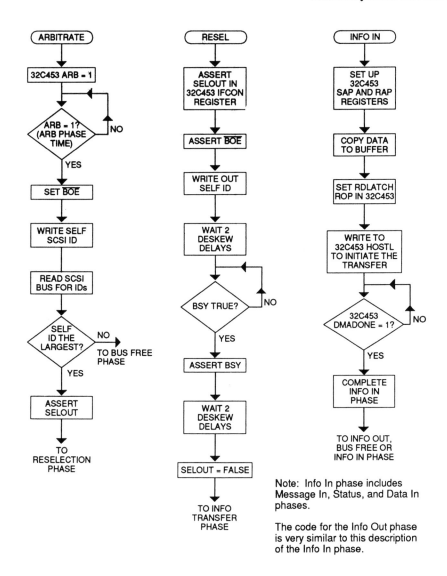

Flow Charts for Various SSI 32B451 Routines

Partial Schematic for SCSI Implementation with Arbitration Support Using SSi Devices

Silicon Systems
SSI 32C260 PC AT/XT Combo Controller
Advance Information

FEATURES
- PC AT/XT bus interface
  - ~ Single-chip PC AT/XT controller
  - ~ Supports ST506/412, ST412HP, ESDI, and SMD disk interfaces
  - ~ Direct bus interface logic with on-chip 24-mA drivers
  - ~ Logic for daisy chaining 2 embedded controller drives on a PC AT
  - ~ Supports 15 Mbit/s concurrent disk transfer on a 12-MHz PC AT without wait states
- Buffer manager
  - ~ Supports buffer memory throughput to 6 MB/s
  - ~ Direct buffer memory addressing up to 64 kB static RAM
  - ~ Dual port circular buffer control
- Storage controller
  - ~ NRZ Data rate up to 15 Mbit/s
  - ~ Selectable 16-bit CRC or 56-bit ECC polynomial with fast hardware correction circuitry
  - ~ Support sector level defect management
  - ~ Support 1:1 interleaved operation

- Microprocessor interface
  - Supports both Intel 8051, and Motorola 68HC11 family of microprocessors
  - Interrupt or polled microprocessor interface
- Others
  - Low power CMOS technology
  - Plug and play compatible with Cirrus CL-SH 260 chip
  - Available in 84-pin PLCC or 100-pin QFP

Advance Information: Indicates that a product is still in the design cycle, and any specifications are based on design goals only. Do not use it for final design.

No responsibility is assumed by Silicon Systems for use of this product nor for any infringements of patents and trademarks or other rights of third parties resulting from its use. No license is granted under any patents, patent rights or trademarks of Silicon Systems. Silicon Systems reserves the right to make changes in specifications at any time without notice. Accordingly, the reader is cautioned to verify that the data sheet is current before placing orders.

ABSOLUTE MAXIMUM RATINGS

Maximum limits indicate where a permanent device damage occurs. Continuous operation at these limits is not intended and should be limited to those conditions specified in the DC operating characteristics

PARAMETER	RATING	UNITS
Power Supply Voltage, VCC	7	V
Ambient Temperature	0 to 70	°C
Storage Temperature	-65 to 150	°C
Power Dissipation	750	mW
Input, Output pins	-0.5 to VCC+0.5	V

ELECTRICAL CHARACTERISTICS

PARAMETER	CONDITION	MIN	NOM	MAX	UNITS
VCC Power Supply Voltage		4.75		5.25	V
VIL Input Low Voltage		-0.5		0.8	V
VIH Input High Voltage		2.0		VCC+0.5	V
VOL Output Low Voltage	All pins except PC interface, IOL = 2 mA			0.4	
VOL Output Low Voltage	PC interface pins, IOL = 24 mA			0.5	V
VOH Output High Voltage	IOH = -400 µA			2.4	V
ICC Supply Current				50	mA
ICCS Supply Current Standby	All Inputs at GND or VCC	250			µA
IL Input Leakage Current	0 < VIN < VCC	-10		10	µA
CIN Input Capacitance				10	pF
COUT Output Capacitance				10	pF

BLOCK DIAGRAM

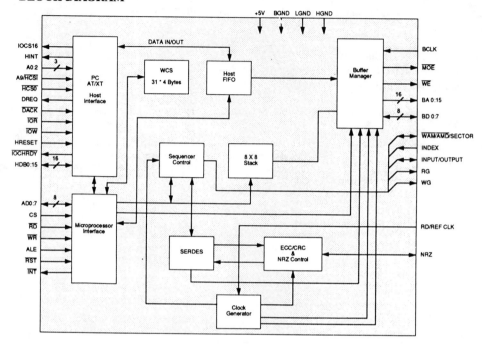

PACKAGE PIN DESIGNATIONS
(TOP VIEW)

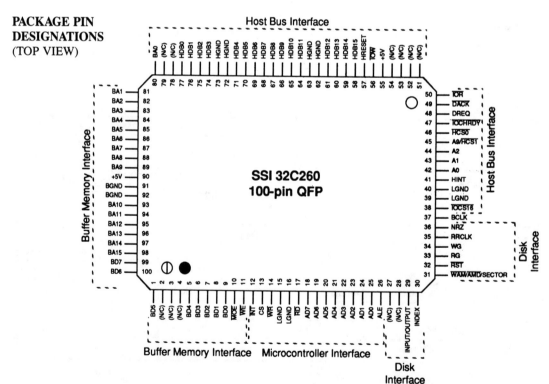

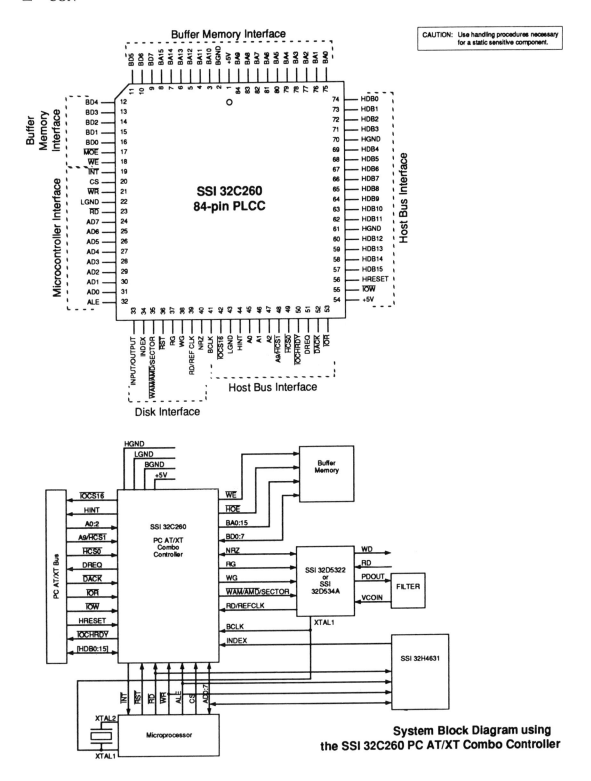

Silicon Systems
SSI 32C452
Storage Controller
Preliminary Data

FEATURES

- Supports ST506/412, ST412HP, SA100, SMD, ESDI and custom interfaces
- Operates with 16-MHz microprocessors
- Internal RAM-based control sequencer
- Internal user-programmable ECC to 32 bits
- Noninterleaved data transfer to 20 Mbit/s
- Hard- or soft-sector formats
- Programmable sector lengths up to a full track
- High-performance low-power CMOS device
- Plug- and software-compatible with AIC-010F storage controller
- Single 5-V supply
- Available in 44-pin PLCC or 40-pin DIP package

ORDERING INFORMATION

PART DESCRIPTION		ORDER NO.	PKG. MARK
SSI 32C452 Storage Controller	40 Pin DIP	SSI 32C452-CP	32C452-CP
	44 Pin PLCC	SSI 32C452-CH	32C452-CH

Preliminary Data: Indicates a product not completely released to production. The specifications are based on preliminary evaluations and are not guaranteed. Small quantities are available, and Silicon Systems should be consulted for current information.

No responsibility is assumed by Silicon Systems for use of this product nor for any infringements of patents and trademarks or other rights of third parties resulting from its use. No license is granted under any patents, patent rights or trademarks of Silicon Systems. Silicon Systems reserves the right to make changes in specifications at any time without notice. Accordingly, the reader is cautioned to verify that the data sheet is current before placing orders.

RECOMMENDED OPERATING CONDITIONS

PARAMETER	CONDITIONS	MIN	NOM	MAX	UNIT
VCC Supply Voltage		4.75		5.25	V
TA Operating Free Air Temp.		0		70	°C
Input Low Voltage		0		0.4	V
Input High Voltage		2.4		VCC	V

ABSOLUTE MAXIMUM RATINGS

PARAMETER	RATING	UNIT
Ambient Temperature Under Bias	0 to 70	°C
Storage Temperature	-65 to 150	°C
Voltage On Any Pin With Respect To Ground	GND -0.5 or VCC + 0.5	V
Power Supply Voltage	7.0	V
Max Current Injection	25	mA

NOTE: Stress above those listed under Absolute Maximum Ratings may cause permanent damage to the device. This is a stress rating only and functional operation of the device at these or any conditions above those indicated in the operational sections of this specification is not implied. Exposure to absolute maximum rating conditions for extended periods may affect device reliability.

BLOCK DIAGRAM

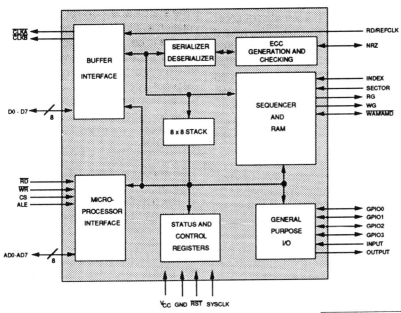

CAUTION: Use handling procedures necessary for a static sensitive components

Silicon Systems SSI 32C452

PACKAGE PIN DESIGNATIONS
(TOP VIEW)

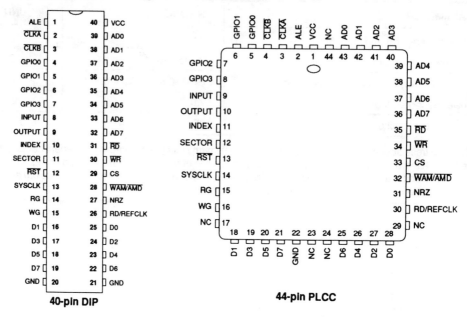

40-pin DIP

44-pin PLCC

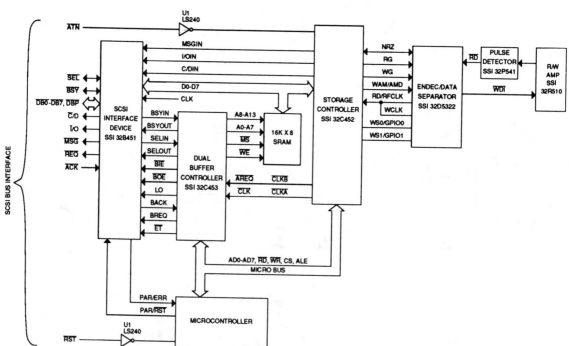

Partial Schematic for SCSI Implementation with Arbitration Support using Silicon Systems Microperipheral Devices

Silicon Systems
SSI 32C453
Dual-Port Buffer Controller

FEATURES

- Dual-port circular FIFO buffer controller
- SCSI bus arbitration control
- DMA handshake control
- Multiplexed-mode buffer addressing up to 64 Kbytes
- Direct-mode buffer addressing up to 1 Kbyte (DIP) or 16 Kbytes (PLCC)
- High speed CMOS device has 16-MHz microprocessor interface
- Compatible with SSI 32C452 storage controller
- Plug and software compatible with AIC-300 buffer controller
- Single 5-V supply
- Available in 44-pin PLCC or 40-pin DIP package

ORDERING INFORMATION

PART DESCRIPTION	ORDER NO.	PKG. MARK
SSI 32C453 Dual Port Buffer Controller		
40 Pin DIP	SSI 32C453-CP	SSI 32C453-CP
44-Pin PLCC	SSI 32C453-CH	SSI 32C453-CH

No responsibility is assumed by Silicon Systems for use of this product nor for any infringements of patents and trademarks or other rights of third parties resulting from its use. No license is granted under any patents, patent rights or trademarks of Silicon Systems. Silicon Systems reserves the right to make changes in specifications at any time without notice. Accordingly, the reader is cautioned to verify that the data sheet is current before placing orders.

BLOCK DIAGRAM

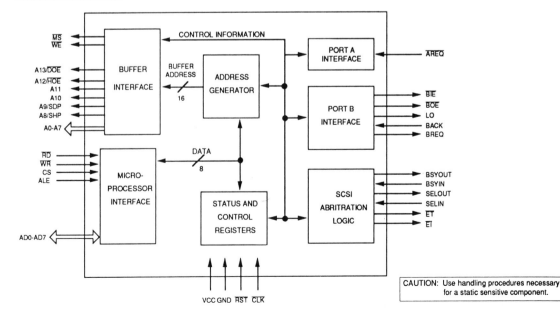

CAUTION: Use handling procedures necessary for a static sensitive component.

PACKAGE PIN DESIGNATIONS
(TOP VIEW)

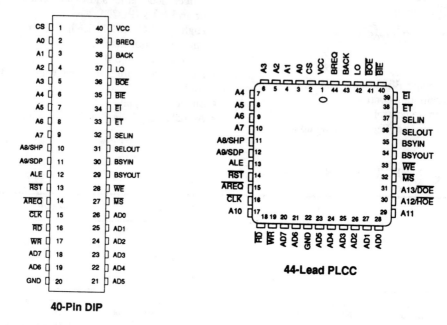

40-Pin DIP

44-Lead PLCC

ABSOLUTE MAXIMUM RATINGS

PARAMETER	RATING	UNIT
Ambient Temperature Under Bias	0 to 70	°C
Storage Temperature	-65 to 150	°C
Voltage on any Pin with respect to Ground	-0.5 to 7	V
Power Dissipation	0.475	W
Maximum Current Injection	±20	mA

Note: Stresses above those listed under Absolute Maximum Ratings may cause permanent damage to the device. This is a stress rating only and functional operation of the device at these or any other conditions above those indicated in the operational sections of this specification is not implied. Exposure to absolute maximum rating conditions for extended periods may affect device reliability.

RECOMMENDED OPERATING CONDITIONS

PARAMETERS	CONDITIONS	MIN	NOM	MAX	UNIT
VCC, Supply Voltage		4.75		5.25	V
TA, Operating Free Air Temperature		0		70	°C
Input Low Voltage		0		0.4	V
Input High Voltage		2.4		VCC	V

dc CHARACTERISTICS (TA = 0°C to 70°C, VCC = recommended range unless otherwise specified.)

PARAMETERS		CONDITIONS	MIN	NOM	MAX	UNIT
VIL	Input Low Voltage		-0.5		0.8	V
VIH	Input High Voltage		2.0		VCC+0.5	V
VOL	Output Low Voltage	IOL = 2 mA			0.4	V
VOH	Output High Voltage	IOH = 400 µA	2.4			V
ICC	Supply Current				85	mA
IL	Input Leakage	0V<VIN<VCC	-10		10	µA
IOL	Output Leakage	0.45V<VOUT<VCC	-10		10	µA
CIN	Input Capacitance				10	pF
COUT	Output Capacitance				10	pF

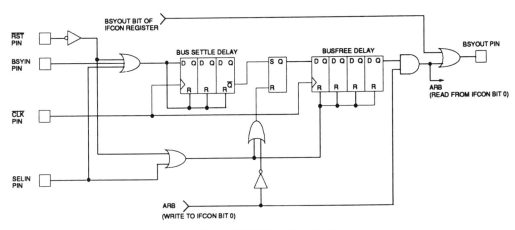

SSI 32C453 SCSI Arbitration Logic

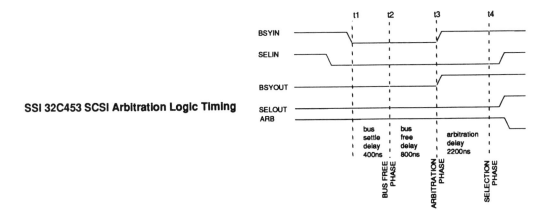

SSI 32C453 SCSI Arbitration Logic Timing

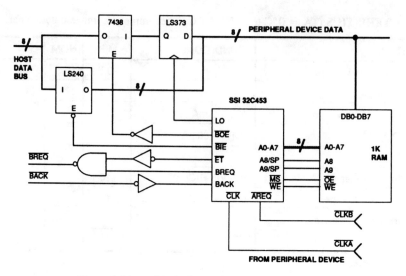

Direct Address Mode Example - 10 Address Lines

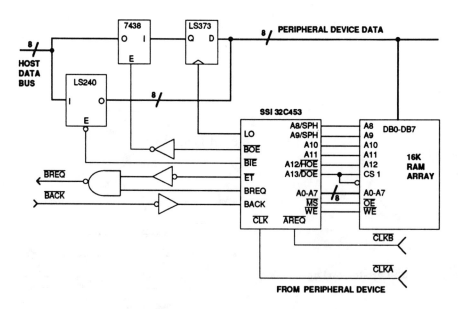

**Direct Addressing Mode Example - 14 Address Lines
(SSI 32C453 PLCC Version Only)**

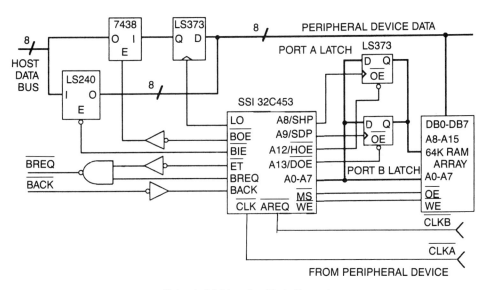

Extended Addressing Mode Example

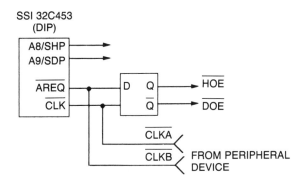

Extended Mode Address Strobes for DIP Package

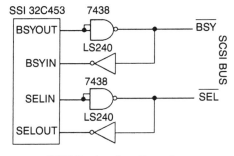

SCSI Bus Interface Example

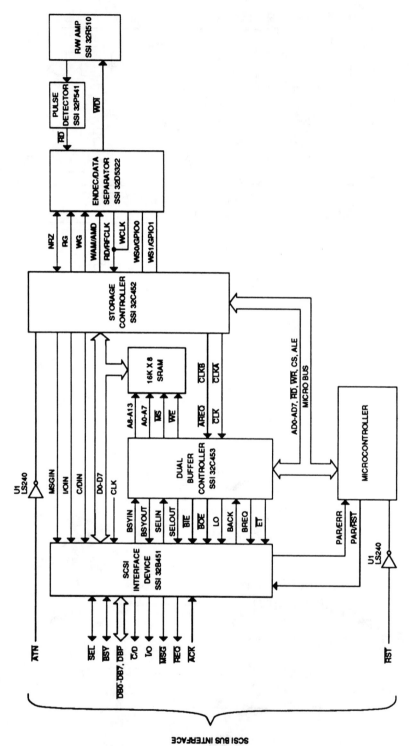

Partial Schematic for SCSI Implementation with Arbitration Support using Silicon Systems microperipheral devices

CHAPTER 4

DATA-CONVERSION AND PROCESSING CIRCUITS

Analog Devices
AD630
Balanced Modulator/Demodulator

FEATURES
- Recovers signal from +100-dB noise
- 2-MHz channel bandwidth
- 45-V/μs slew rate
- −120-dB crosstalk @ 1 kHz
- Pin-programmable closed-loop gains of ±1 and ±2
- 0.05% closed-loop gain accuracy and match
- 100-μV channel offset voltage (AD630BD)
- 350-kHz full-power bandwidth
- Chips available

PRODUCT HIGHLIGHTS
1. The configuration of the AD630 makes it ideal for signal processing applications such as: balanced modulation and demodulation, lock-in amplification, phase detection, and square-wave multiplication.
2. The application flexibility of the AD630 makes it the best choice for many applications requiring precisely fixed gain, switched gain, multiplexing, integrating-switching functions, and high-speed precision amplification.
3. The 100-dB dynamic range of the AD630 exceeds that of any hybrid or IC balanced modulator/demodulator and is comparable to that of costly signal-processing instruments.
4. The op-amp format of the AD630 ensures easy implementation of high-gain or complex switched-feedback functions. The application resistors facilitate the implementation of most common applications with no additional parts.
5. The AD630 can be used as a two-channel multiplexer with gains of +1, +2, +3 or +4. The channel separation of 100 dB @ 10 kHz approaches the limit which is achievable with an empty IC package.
6. The AD630 has pin-strappable frequency compensation (no external capacitor required) for stable operation at unity gain without sacrificing dynamic performance at higher gains.
7. Laser trimming of the comparator and amplifying channel offsets eliminates the need for external nulling in most cases.

Analog Devices AD630

SPECIFICATIONS (@ +25°C and $\pm V_S = \pm 15$ V unless otherwise specified)

Model	AD630J/A Min	AD630J/A Typ	AD630J/A Max	AD630K/B Min	AD630K/B Typ	AD630K/B Max	AD630S Min	AD630S Typ	AD630S Max	Units
GAIN										
Open Loop Gain	90	110		100	120		90	110		dB
$\pm 1, \pm 2$ Closed Loop Gain Error		0.1				0.05		0.1		%
Closed Loop Gain Match		0.1				0.05		0.1		%
Closed Loop Gain Drift		2			2			2		ppm/°C
CHANNEL INPUTS										
V_{IN} Operational Limit[1]	$(-V_S+4V)$ to $(+V_S-1V)$			$(-V_S+4V)$ to $(+V_S-1V)$			$(-V_S+4V)$ to $(+V_S-1V)$			Volts
Input Offset Voltage			500			100			500	μV
Input Offset Voltage T_{min} to T_{max}			800			160			1000	μV
Input Bias Current		100	300		100	300		100	300	nA
Input Offset Current		10	50		10	50		10	50	nA
Channel Separation @ 10kHz		100			100			100		dB
COMPARATOR										
V_{IN} Operational Limit[1]	$(-V_S+3V)$ to $(+V_S-1.5V)$			$(-V_S+3V)$ to $(+V_S-1.5V)$			$(-V_S+3V)$ to $(+V_S-1.3V)$			Volts
Switching Window			± 1.5			± 1.5			± 1.5	mV
Switching Window T_{min} to T_{max}[2]			± 2.0			± 2.0			± 2.5	mV
Input Bias Current		100	300		100	300		100	300	nA
Response Time (−5mV to +5mV step)		200			200			200		ns
Channel Status										
$I_{SINK} @ V_{OL} = -V_S + 0.4V$[3]	1.6			1.6			1.6			mA
Pull-Up Voltage			$(-V_S+33V)$			$(-V_S+33V)$			$(-V_S+33V)$	Volts
DYNAMIC PERFORMANCE										
Unity Gain Bandwidth		2			2			2		MHz
Slew Rate[4]		45			45			45		V/μs
Settling Time to 0.1% (20V step)		3			3			3		μs
OPERATING CHARACTERISTICS										
Common-Mode Rejection	85	105		90	110		90	110		dB
Power Supply Rejection	90	110		90	110		90	110		dB
Supply Voltage Range	± 5		± 16.5	± 5		± 16.5	± 5		± 16.5	Volts
Supply Current		4	5		4	5		4	5	mA
OUTPUT VOLTAGE, @ $R_L = 2k\Omega$										
T_{min} to T_{max}	± 10			± 10			± 10			Volts
Output Short Circuit Current		25			25			25		mA
TEMPERATURE RANGES										
Rated Performance – N Package	0		+70	0		+70	N/A			°C
D Package	−25		+85	−25		+85	−55		+125	°C

NOTES
[1] If one terminal of each differential channel or comparator input is kept within these limits the other terminal may be taken to the positive supply.
[2] This parameter guaranteed but not tested.
[3] $I_{SINK} @ V_{OL} = (-V_S + 1)$ volt is typically 4mA.
[4] Pin 12 Open. Slew rate with Pins 12 & 13 shorted is typically 35V/μs.

Specifications subject to change without notice.
Specifications shown in boldface are tested on all production units at final electrical test. Results from those tests are used to calculate outgoing quality levels. All min and max specifications are guaranteed, although only those shown in boldface are tested on all production units.

ABSOLUTE MAXIMUM RATINGS

Supply voltage	± 18 V
Internal power dissipation	600 mW
Output short circuit to ground	Indefinite
Storage temperature, ceramic package	−65°C to +150°C
Storage temperature, plastic package	−55°C to +125°C
Lead temperature, 10 sec. soldering	+300°C
Max junction temperature	+150°C

THERMAL CHARACTERISTICS

	θ_{JC}	θ_{JA}
20-Pin Plastic DIP (N)	24°C/W	61°C/W
20-Pin Ceramic DIP (D)	35°C/W	120°C/W
20-Pin Leadless Chip Carrier (E)	35°C/W	120°C/W

ORDERING GUIDE

Temperature Range	Size Brazed DIP (D-20)	Plastic DIP (N-20)	Leadless Chip Carrier (E-20A)	Chips
0 to +70°C		AD630JN AD630KN		AD630 J Chips
−25°C to +85°C	AD630AD AD630BD			
−55°C to +125°C	AD630SD AD630SD/883B		AD630SE AD630SE/883B	AD630S Chips

TYPICAL PERFORMANCE CHARACTERISTICS—AD630

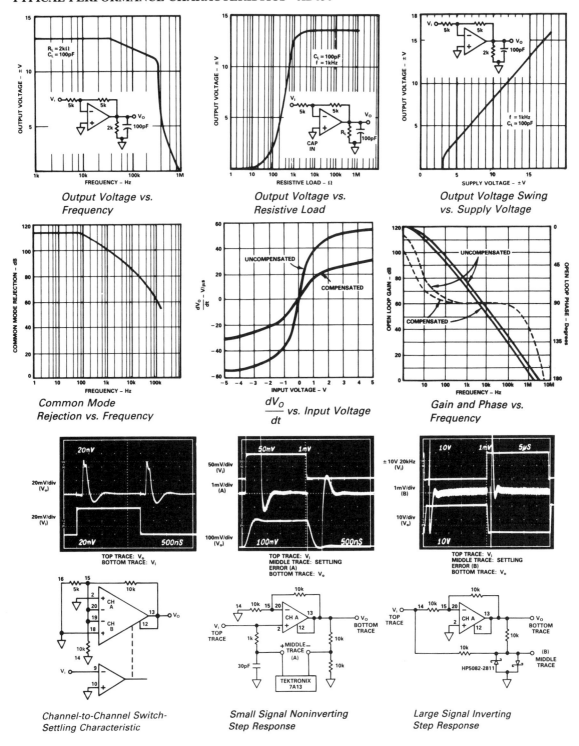

Analog Devices AD630

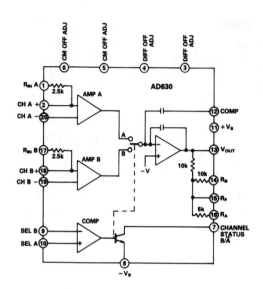

Functional Block Diagram

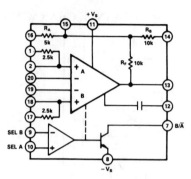

Architectural Block Diagram

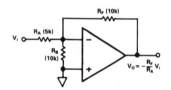

Inverting Gain Configuration

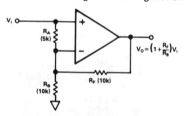

Noninverting Gain Configuration

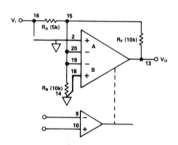

AD630 Symmetric Gain (±2)

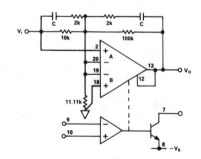

AD630 with External Feedback

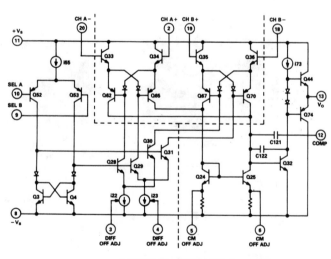

AD630 Simplified Schematic

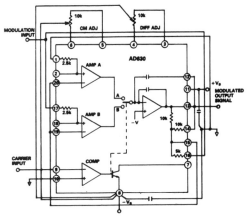

AD630 Configured as a Gain-of-One Balanced Modulator

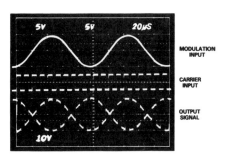

Gain-of-Two Balanced Modulator Sample Waveforms

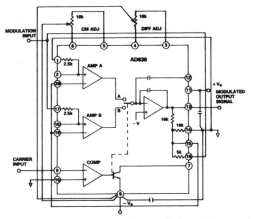

AD630 Configured as a Gain-of-Two Balanced Modulator

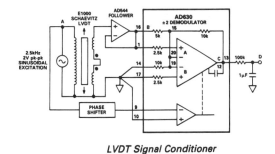

LVDT Signal Conditioner

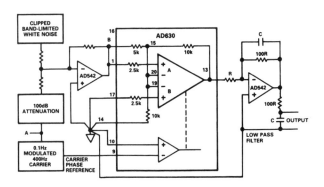

Lock-In Amplifier

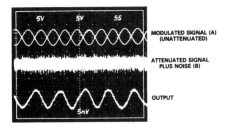

Lock-In Amplifier Waveforms

174 Analog Devices AD632　　　　　　　　　　　　　　　DCP □

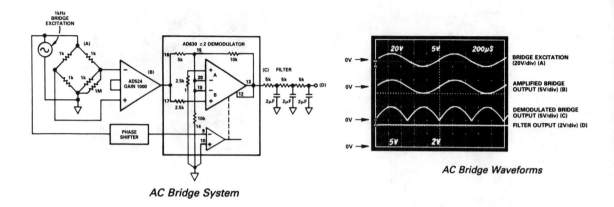

AC Bridge System

AC Bridge Waveforms

Analog Devices
AD632
Internally Trimmed Precision IC Multiplier

FEATURES

- Pretrimmed to ±0.5% max. 4-quadrant error
- All inputs (X, Y, and Z) differential, high impedance for $[(X_1-X_2)(Y_1-Y_2)/10]+Z_2$ transfer function
- Scale-factor adjustable to provide up to ×10 gain

- Low noise design: 90 μV rms, 10 Hz to 10 kHz
- Low-cost monolithic construction
- Excellent long-term stability

APPLICATIONS

- High-quality analog-signal processing
- Differential ratio and percentage computations
- Algebraic and trigonometric function synthesis
- Accurate voltage-controlled oscillators and filters

PRODUCT HIGHLIGHTS

Guaranteed Performance Over Temperature The AD632A and AD632B are specified for maximum multiplying errors of ±1.0% and ±0.5% of full scale, respectively at +25°C and are rated for operation from −25°C to +85°C. Maximum multiplying errors of ±2.0% (AD632S) and ±1.0% (AD632T) are guaranteed over the extended temperature range of −55°C to +125°C.

High Reliability The AD632S and AD632T series are also available with MIL-STD-883 Level B screening and all devices are available in either the hermetically-sealed TO-100 metal can or TO-116 ceramic DIP package.

SPECIFICATIONS (@ +25°C, $V_S = \pm 15$ V, $R \geq 2$ kΩ unless otherwise noted)

Model	AD632A			AD632B			AD632S			AD632T			Units
	Min	Typ	Max	Min	Typ	Max	Min	Typ	Max	Min	Typ	Max	
MULTIPLIER PERFORMANCE													
Transfer Function		$\frac{(X_1-X_2)(Y_1-Y_2)}{10V}+Z_2$			$\frac{(X_1-X_2)(Y_1-Y_2)}{10V}+Z_2$			$\frac{(X_1-X_2)(Y_1-Y_2)}{10V}+Z_2$			$\frac{(X_1-X_2)(Y_1-Y_2)}{10V}+Z_2$		
Total Error[1] (−10V≤X, Y≤+10V)			±1.0			±0.5			±1.0			±0.5	%
T_A = min to max		±1.5			±1.0			±2.0			±1.0		%
Total Error vs Temperature		±0.022			±0.015			±0.02			±0.01		%/°C
Scale Factor Error (SF − 10.000V Nominal)[2]		±0.25			±0.1			±0.25			±0.1		%
Temperature-Coefficient of Scaling-Voltage		±0.02			±0.01			±0.2			±0.005		%/°C
Supply Rejection (±15V ±1V)		±0.01			±0.01			±0.01			±0.01		%
Nonlinearity, X (X = 20V pk-pk, Y = 10V)		±0.4			±0.2	±0.3		±0.4			±0.2	±0.3	%
Nonlinearity, Y (Y = 20V pk-pk, X = 10V)		±0.2			±0.1	±0.1		±0.2			±0.1	±0.1	%
Feedthrough[3], X (Y Nulled, X 20V pk-pk 50Hz)		±0.3			±0.15	±0.3		±0.3			±0.15	±0.3	%
Feedthrough[3], Y (X Nulled,													

Model	AD632A Min	AD632A Typ	AD632A Max	AD632B Min	AD632B Typ	AD632B Max	AD632S Min	AD632S Typ	AD632S Max	AD632T Min	AD632T Typ	AD632T Max	Units
MULTIPLIER PERFORMANCE													
Y 20V pk-pk 50Hz		±0.01			±0.01	±0.1		±0.01			±0.01	±0.1	%
Output Offset Voltage		±5	±30		±2	±15		±5	±30		±2	±15	mV
Output Offset Voltage Drift		200			100				500			300	μV/°C
DYNAMICS													
Small Signal BW, (V_{OUT} = 0.1rms)		1			1			1			1		MHz
1% Amplitude Error (C_{LOAD} = 1000pF)		50			50			50			50		kHz
Slew Rate (V_{OUT} 20 pk-pk)		20			20			20			20		V/μs
Settling Time (to 1%, ΔV_{OUT} = 20V)		2			2			2			2		μs
NOISE													
Noise Spectral-Density SF = 10V		0.8			0.8			0.8			0.8		μV/√Hz
SF = 3V[4]		0.4			0.4			0.4			0.4		μV/√Hz
Wideband Noise A 10Hz to 5MHz		1.0			1.0			1.0			1.0		mV rms
P 10Hz to 10kHz		90			90			90			90		μV/rms
OUTPUT													
Output Voltage Swing	±11			±11			±11			±11			V
Output Impedance (f ≤ 1kHz)		0.1			0.1			0.1			0.1		Ω
Output Short Circuit Current (R_L = 0, T_A = min to max)		30			30			30			30		mA
Amplifier Open Loop Gain (f = 50Hz)		70			70			70			70		dB
INPUT AMPLIFIERS (X, Y and Z)[5]													
Signal Voltage Range (Diff. or CM)		±10			±10			±10			±10		V
Operating Diff.)		±12			±12			±12			±12		V
Offset Voltage X, Y		±5	±20		±2	±10		±5	±20		±2	±10	mV
Offset Voltage Drift X, Y		100			50			100			150		μV/°C
Offset Voltage Z		±5	±30		±2	±15		±5	±30		±2	±15	mV
Offset Voltage Drift Z		200			100				500			300	μV/°C
CMRR	60	80		70	90		60	80		70	90		dB
Bias Current		0.8	2.0		0.8	2.0		0.8	2.0		0.8	2.0	μA
Offset Current		0.1			0.1			0.1			0.1		μA
Differential Resistance		10			10			10			10		MΩ
DIVIDER PERFORMANCE													
Transfer Function ($X_1 > X_2$)		$10V \frac{(Z_2 - Z_1)}{(X_1 - X_2)} + Y_1$			$10V \frac{(Z_2 - Z_1)}{(X_1 - X_2)} + Y_1$			$10V \frac{(Z_2 - Z_1)}{(X_1 - X_2)} + Y_1$			$10V \frac{(Z_2 - Z_1)}{(X_1 - X_2)} + Y_1$		
Total Error[1] (X = 10V, -10V ≤ Z ≤ +10V)		±0.75			±0.35			±0.75			±0.35		%
(X = 1V, -1V ≤ Z ≤ +1V)		±2.0			±1.0			±2.0			±1.0		%
(0.1V ≤ X ≤ 10V, -10V ≤ Z ≤ 10V)		±2.5			±1.0			±2.5			±1.0		%
SQUARER PERFORMANCE													
Transfer Function		$\frac{(X_1 - X_2)^2}{10V} + Z_2$			$\frac{(X_1 - X_2)^2}{10V} + Z_2$			$\frac{(X_1 - X_2)^2}{10V} + Z_2$			$\frac{(X_1 - X_2)^2}{10V} + Z_2$		
Total Error (-10V ≤ X ≤ 10V)		±0.6			±0.3			±0.6			±0.3		%
SQUARE-ROOTER PERFORMANCE													
Transfer Function, ($Z_1 \le Z_2$)		$\sqrt{10V(Z_2 - Z_1)} + X_2$			$\sqrt{10V(Z_2 - Z_1)} + X_2$			$\sqrt{10V(Z_2 - Z_1)} + X_2$			$\sqrt{10V(Z_2 - Z_1)} + X_2$		
Total Error[1] (1V ≤ Z ≤ 10V)		±1.0			±0.5			±1.0			±0.5		%
POWER SUPPLY SPECIFICATIONS													
Supply Voltage Rated Performance		±15			±15			±15			±15		V
Operating	±8		±18	±8		±18	±8		±22	±8		±22	V
Supply Current Quiescent		4	6		4	6		4	6		4	6	mA
PACKAGE OPTIONS													
TO-100 (H-10A)		AD632AH			AD632BH			AD632SH			AD632TH		
TO-116 (D-14)		AD632AD			AD632BD			AD632SD			AD632TD		

NOTES
[1] Figures given are percent of full-scale, ±10V (i.e., 0.01% = 1mV).
[2] May be reduced down to 3V using external resistor between $-V_S$ and SF.
[3] Irreducible component due to nonlinearity: excludes effect of offsets.
[4] Using external resistor adjusted to give SF = 3V.
[5] See functional block diagram for definition of sections.

Specifications subject to change without notice.
All min and max specifications are guaranteed.
Specifications shown in **boldface** are tested on all production units at final electrical test. Results from those tests are used to calculate outgoing quality levels.

Analog Devices AD632

TYPICAL PERFORMANCE CURVES (typical at +25°C with ±V_S=15 V)

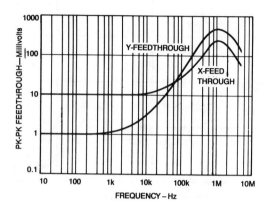

AC Feedthrough vs. Frequency

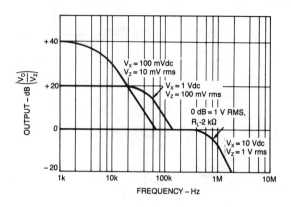

Frequency Response vs. Divider Denominator Input Voltage

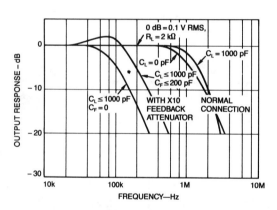

Frequency Response as a Multiplier

CHIP DIMENSIONS & PAD LAYOUT
Contact factory for latest dimensions
Dimensions shown in inches and (mm).

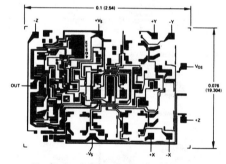

For further information, consult factory

ORDERING GUIDE

Temperature Range	Header (H)	Side Brazed DIP (D)
−25°C to +85°C	AD632AH	AD632AD
	AD632BH	AD632BD
−55°C to +125°C	AD632SH	AD632SD
	AD632SH/883B	AD632SD/883B
	AD632TH	AD632TD
	AD632TH/883B	AD632TD/883B

Thermal Characteristics

Thermal Resistance θ_{JC} = 25°C/W for H-10A
θ_{JA} = 150°C/W for H-10A
θ_{JC} = 25°C/W for D-14
θ_{JA} = 95°C/W for D-14

Analog Devices AD632

AD632 PIN CONFIGURATIONS

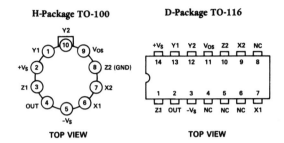

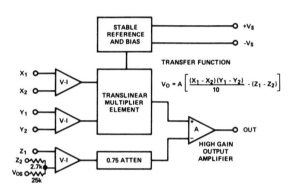

AD632 Functional Block Diagram

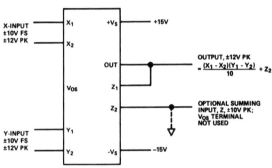

Basic Multiplier Connection

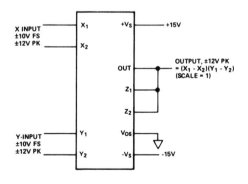

Connections for Scale-Factor of Unity

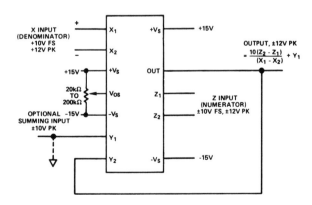

Basic Divider Connection

GEC Plessey
MS2014
Digital Filter and Detector (FAD)
Advance Information

FEATURES
- Linear 16-bit data
- 13-bit coefficient
- 2-MHz operating-clock frequency
- Serial operation
- 448 bits of on-chip shift-register data storage for 8th-order multiplex
- Nth-order multiplexing ($N \leq 8$)
- TTL compatible
- Single +5-V supply

APPLICATIONS
- Low-cost digital filtering
- Level detection
- Spectral analysis
- Tone detectors (multi-frequency receivers)
- Speech synthesis and analysis
- Data modems
- Group delay equalizers (all-pass networks)

ABSOLUTE MAXIMUM RATINGS

Supply voltage (V_{DD})	−0.5 V to +7 V
Input voltage	−0.5 V to +7 V
Maximum output voltage	+7 V
Temperature: Storage	−65 °C to 125 °C
Operating	0 °C to 70 °C

NOTE
All voltages with respect to V_{SS}.

RECOMMENDED OPERATING CONDITIONS

Characteristic	Symbol	Min.	Max.	Units	Conditions
Supply voltage	V_{DD}	4.75	5.25	V	
Input voltage (high state) except clock	V_{IH}	2.2	-	V	
Input voltage (low state) except clock	V_{IL}	-	0.7	V	
Input voltage (high state) clock	V_{IHC}	4.5	-	V	
Input voltage (low state) clock	V_{ILC}	-	0.5	V	
Clock rise and fall time	t_{cl}		30	ns	10% - 90% (Note 1)
Clock frequency	f_{cl}	0.5	2.048	MHz	
Operating temperature	T_{amb}	0	70	°C	

ELECTRICAL CHARACTERISTICS Test conditions (unless otherwise stated):
$V_{DD} = +5$ V $T_{amb} = 25$ °C

Characteristic	Symbol	Min.	Typ.	Max.	Units	Conditions
Supply current	I_{DD}		90	120	mA	
Output voltage, low	V_{OL}	-	-	0.5	V	$I_{OL} = 0.4$ mA (Note 2)
Output voltage, high	V_{OH}	2.7	3.4	-	V	$I_{OH} = -40\mu$A (Note 2)
Input capacitance (except clock)	C_{in}		5	7	pF	
Input capacitance (clock)	C_{inc}		25		pF	
Input data set up time	t_{is}	50	-	-	ns	Fig.7
Input data hold time	t_{ih}	150	-	-	ns	Fig.7
Output data delay time	t_{os}	-	-	200	ns	Fig.7

NOTES
1. An operating clock frequency of 2.048MHz is guaranteed over the supply voltage range and the full operating temperature range.
2. The output stage is designed to drive a standard TTL LS gate (74LS series).

Plessey MS2014

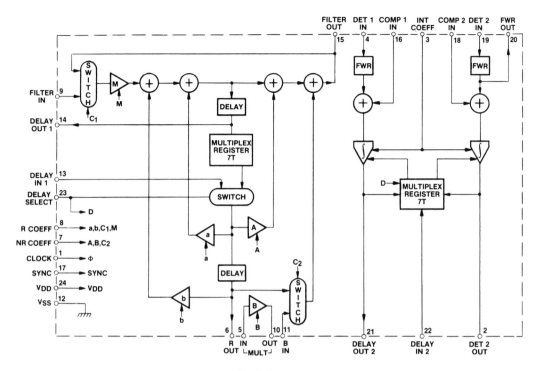

Block diagram

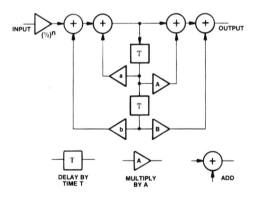

Basic 2nd order filter section

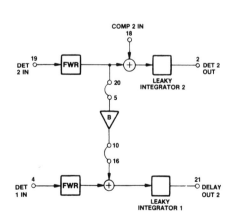

Relative level detection

180 Plessey MS2014

Leak factor	Rise time (0 to 90%)
1/2	3 Ts + T
3/4	8 Ts + T
7/8	17 Ts + T
15/16	35 Ts + T
31/32	72 Ts + T
63/64	146 Ts + T
127/128	293 Ts + T
255/256	588 Ts + T

Integrator rise times

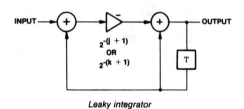

Leaky integrator

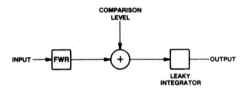

2nd order 32kHz bandwidth filter

Simple level detector

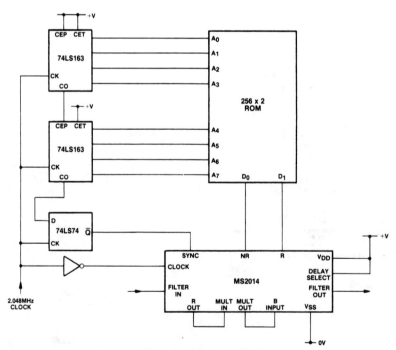

A 16th order 4kHz bandwidth filter

GEC Plessey
MV1441
2-MBIT PCM Signaling Circuit
HDB3 Encoder/Decoder/Clock Regenerator

FEATURES
- On-chip digital clock regenerator
- HDB3 encoding and decoding to CCITT rec. G703
- Asynchronous operation
- Simultaneous encoding and decoding
- Clock recovery signal allows off-chip clock regeneration
- Loop-back control
- HDB3 error monitor
- "All-ones" error monitor
- Loss-of-input alarm (all-zeros detector)
- Decode data in NRZ form
- Low-power operation
- 2.048-MHz or 1.544-MHz operation

ABSOLUTE MAXIMUM RATINGS
The absolute maximum ratings are limiting values above which operating life might be shortened or specified parameters might be degraded.

ELECTRICAL RATINGS
$+V_{CC}$	-0.5 V to $+7$ V
Inputs	$V_{CC} +0.5$ V Gnd -0.3 V
Outputs	V_{CC}, Gnd -0.3 V

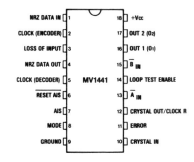

Pin connections - top view

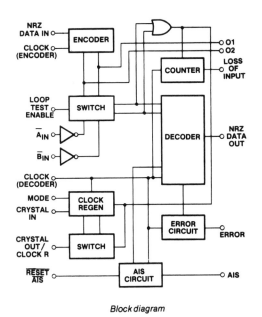

Block diagram

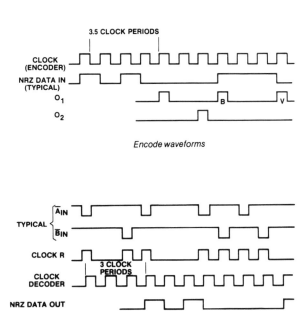

Encode waveforms

Decode waveforms

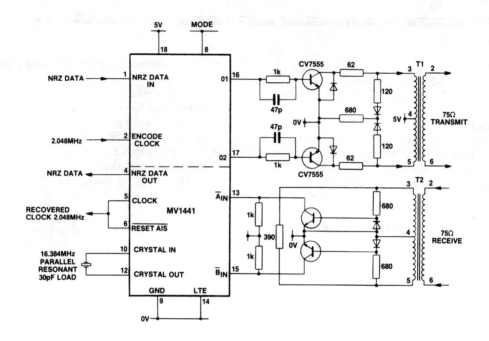

A typical application of the MV1441 with the interfacing to the transmission lines included

GEC Plessey
MV3506 A-LAW Filter/Codec
MV3507 μ-LAW Filter/Codec
MV3507A μ-LAW Filter/Codec with A/B Signaling
MV3508 A-LAW Filter/Codec with Optional Squelch
MV3509 μ-LAW Filter/Codec with Optional Squelch

FEATURES
- Low-power CMOS 80 mW (operating) 10 mW (standby)
- Meets or exceeds AT and T3, and CCITT G.711, G.712, and G.733 specifications
- Input analog filter eliminates need for external anti-aliasing prefilter
- Uncommitted input and output op amps for programming gain
- Output op amp provides ±3.1 V into a 1200 Ω load or can be switched off for reduced power (70 mW)
- Encoder has dual-speed auto-zero loop for fast acquisition on power-up
- Low absolute group delay = 410 ms at 1 kHz

ABSOLUTE MAXIMUM RATINGS
Exceeding these ratings might cause permanent damage. Functional operation under these conditions is not implied.

Positive supply voltage V_{DD}	-0.5 V to $+6.0$ V
Analog ground V_{AGND}	-0.1 V to $+0.1$ V
Negative supply voltage V_{SS}	-6.0 V to $+0.5$ V
Storage temperature T_S	$-65\,°\text{C}$ to $+150\,°\text{C}$
Voltage at digital or analog pins V_P	$V_{SS}-0.3$ V to $V_{DD}+0.3$ V
Package power dissipation P	1000 mW

DCP

Plessey MV3506/MV3507/MV3507A/MV3508/MV3509

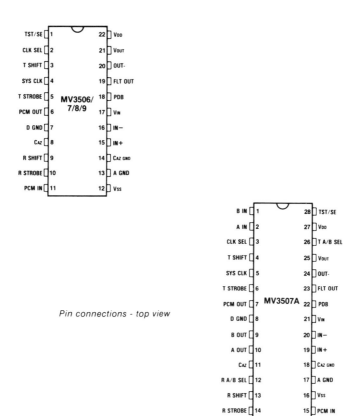

Pin connections - top view

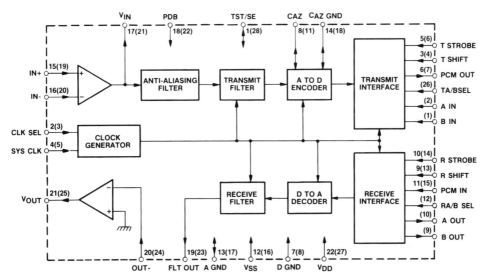

Functional block diagram (pin numbers for the MV3507A are in brackets)

ELECTRICAL CHARACTERISTICS
Test conditions – Voltages are with respect to digital ground (V_{DGND})

Characteristic	Symbol	Min.	Typ.(1)	Max.	Units
Digital supply voltage	V_{DD}	4.75	5	5.25	V
Negative supply voltage	V_{SS}	-5.25	-5	-4.75	V
Analog ground voltage	V_{AGND}	-0.1	0	0.1	V
Ambient temperature	V_{AMB}	0		70	°C
Input low voltage - digital inputs	V_{IL}	0	0.4	0.8	V
Input high voltage - digital inputs	V_{IH}	2.0	2.4	V_{DD}	V
System clock frequency					
CLK SEL tied to V_{DD}	f_s	2047.90	2048	2048.10	kHz
CLK SEL tied to D GND		255.99	256	256.01	
CLK SEL tied to V_{SS}		1549.92	1544	1544.08	
Capacitive loading - digital outputs	C_{LD}	0		100	pF
Pull-up resistance for PCM OUT pin	R_{PU}	510			Ω
Analog input voltage	V_{IA}	$V_{AGND}-3.1$		$V_{AGND}+3.1$	V
Capacitive loading - analog outputs	C_{LA}			50	pF
Resistive loading - V_{OUT} pin	R_{VOUT}	1200			Ω
Resistive loading - V_{IN} pin	R_{VIN}	10			kΩ
Resistive loading - FLT OUT pin	$R_{RLT\ OUT}$	20			kΩ

Analog Channel Characteristics - μ-Law

Characteristic	Symbol	Min.	Typ.(1)	Max.	Units	Conditions
0dBm0 level (see Note 2)	0dBm0	5.3	5.8	6.3	dBm	±5V, 25°C
Variation in 0dBm0 level	Δ_{dBm0}	-0.3	0	0.3	dB	Over test conditions
Weighted idle channel noise	ICN_W		5	17	dBrnc0	AT&T D3 (see Note 3)
Single frequency idle channel noise	ICN_{SF}			-60	dBm0	AT&T D3
Weighted receive idle channel noise	ICN_{WR}			15	dBrnc0	AT&T D3
Spurious out-band noise	N_{SOB}			-28	dBm0	AT&T D3
Spurious in-band noise	N_{SIB}			-40	dBm0	AT&T D3
Two tone interdemodulation	IMD_{2T}			-35	dBm0	AT&T D3
Tone + power inter-demodulation	IMD_{TP}			-49	dBm0	AT&T D3
Crosstalk attenuation between V_{IN} and V_{OUT}	A_X	75	80		dB	AT&T D3

NOTES
1. Typical figures are for design aid only. They are not guaranteed and not subject to production testing.
2. The typical 0dBm0 level of 5.8dBm corresponds to an RMS voltage of 1.51V and a maximum coding level of 3.1V.
3. The maximum value reduces to -68dBm0p without squelch (MV3508 with TST/SE pin unconnected).
4. The maximum value reduces to 22dBrnc0 without squelch (MV3509 with TST/SE pin unconnected).

Power Supply Requirements - $V_{DD} = 5V$, $V_{SS} = -5V$

Characteristic	Symbol	Value			Units	Conditions
		Min.	Typ.(1)	Max.		
Power dissipation - normal	P_N		80	110	mW	Unloaded
Power dissipation - without output amp.	P_{WA}		70		mW	Unloaded
Power dissipation - standby	P_S		10	20	mW	Unloaded

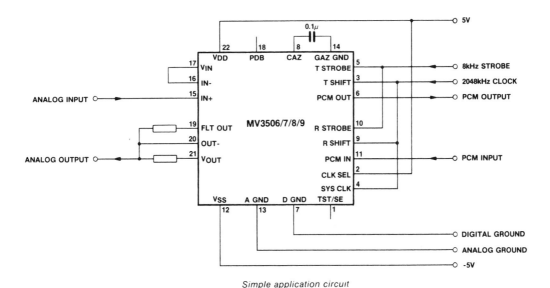

Simple application circuit

GEC Plessey
SL9009
Adaptive Balance Circuit
Preliminary Information

- Extracts the received signal
- Adapts automatically to line variations
- No microprocessor required
- Simple application circuit
- 40-dB (typ.) rejection of transmitted signal

APPLICATIONS
- Modems: extracting the received data
- Feature phones: extracting the received voice
- PBX/PABX/CO line cards: extracting the incoming signal from the telephone line

ABSOLUTE MAXIMUM RATINGS
Exceeding these ratings may cause permanent damage. Functional operation under these conditions is not implied.

ABSOLUTE MAXIMUM RATINGS
Exceeding these ratings may cause permanent damage. Functional operation under these conditions is not implied.

Positive supply voltage (pin 9), V_{CC} -0.3 V to $+10$ V
Negative supply voltage (pin 8), V_{EE} -10 V to $+0.3$ V
Input voltages (pins 2, 5, 6, 10, 11, 12, 13, 14, and 15), V_I V_{EE} to V_{CC}
Output voltages (pins 1, 3, 4, 7, 16), V_O V_{EE} to V_{CC}
Cell voltage (pins 1, .3, 4) minus Common voltage (pin 2), V_C -5 V to $+5$ V
Storage temperature, T_{ST} $-10\,°C$ to $+125\,°C$

ELECTRICAL CHARACTERISTICS
Test conditions – Voltages are with respect to digital ground (V_{DGND} [$V_{CC} - V_{EE}$]/2)

Characteristic	Symbol	Min.	Typ.(1)	Max.	Units
Positive supply voltage (pin 9)	V_{CC}	4.5	5	7.0	V
Negative supply voltage (pin 8)	V_{EE}	-7.0	-5	-4.5	V
Ambient temperature	T_{amb}	0		70	°C
Common cell pin voltage (pin 2)	V_{COM}	V_{EE} +2.7		V_{CC} -2	V
Cell input currents (pins 1,3 and 4)	I_{CELL}	-10		10	µA
X control voltage (pin 5)	V_X	V_{EE} +2.7		V_{EE} -6.0	V
Y control voltage (pin 6)	V_Y	-2.0		1.8	V
Detector input voltages (pins 12 and 13)	V_{IN}	V_{EE} +2.7		V_{CC} -2.7	V
Analysis input voltages (pins 10,11,14 and 15)	V_A	V_{EE} +2.7		V_{CC} -2.7	V
Detector output voltages (pins 7 and 16)	V_{OUT}	V_{EE} +2.5		V_{CC} -2	V

Operating Characteristics: Cells - Voltages are with respect to ground ($V_{GND} = [V_{CC} - V_{EE}]/2$)

Characteristic	Symbol	Min.	Typ.(1)	Max.	Units	Conditions
Internal resistance (pins 1, 3 and 4)	R_I	3		14	kΩ	
Control input leakage (pins 5 and 6)	I_C			0.12	µA	
Minimum cell gain	G_{MIN}		0.05			
Maximum cell gain	G_{MAX}		10			
DC bias current (pin 2)	I_{BDC}	-12	0	12	µA	
Residual impedance (pin 2)	Z_R	500			kΩ	Cell pins open circuit

Operating Characteristics: Detectors - Voltages are with respect to ground ($V_{GND} = [V_{CC} - V_{EE}]/2$)

Characteristic	Symbol	Min.	Typ.(1)	Max.	Units	Conditions
Differential input offset (pins 10 to 15)	V_{DOFF}			13	mV	
Input offset current (pins 10 to 15)	I_{OFF}	-0.15		0.15	µs	
Input bias current (pins 10 to 15)	I_{IB}		0.1	0.7	µA	
Transconductance gain	G_T	250	500	1000	µΩ	Magnitude
Output offset current (pins 7 and 16)	I_{OFF}	-1.2		1.2	µA	
Maximum output current (pins 7 and 16)	I_{MAX}		50		µA	Sink or Source
Output impedance (pins 7 and 16)	Z_{OUT}		5		MΩ	

NOTE
1. Typical figures are for design aid only. They are not guaranteed and not subject to production testing.

Operating Characteristics: General - Voltages are with respect to ground ($V_{GND} = [V_{CC} - V_{EE}]/2$)

Characteristic	Symbol	Value			Units	Conditions
		Min.	Typ.(1)	Max.		
Power dissipation	P_D			30	mW	
Supply current	I_{CC}		1.3	15	mA	
Pin capacitance	C_P		7	15	pF	Pin to supplies

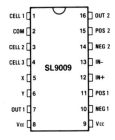

Pin connections - top view

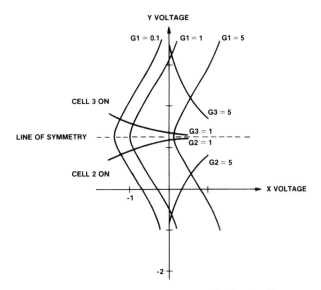

Typical gain characteristics for CELLS 1, 2 and 3
(G1, G2 and G3 respectively)

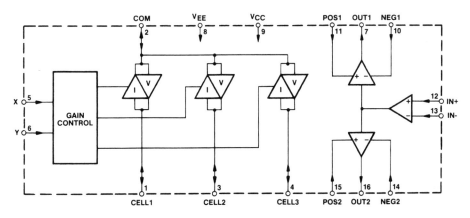

Functional block diagram

Plessey SL9009

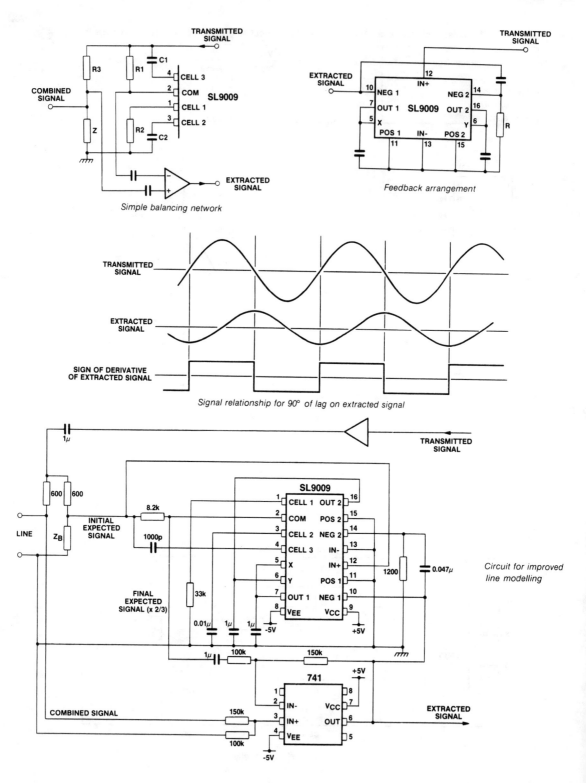

Simple balancing network

Feedback arrangement

Signal relationship for 90° of lag on extracted signal

Circuit for improved line modelling

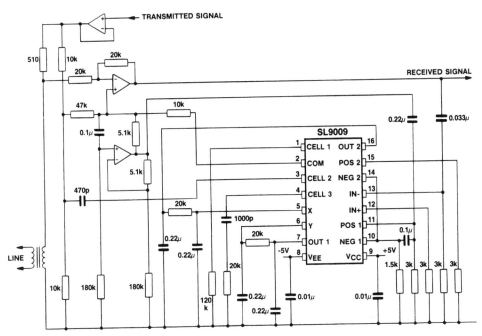

Line transformer:- Total primary + Secondary resistance = 100Ω; Inductance = 0.9H.

Complete circuit

GEC Plessey
ZN5683E/J
PCM Line Interface Circuit

FEATURES
- Operation up to 8.448 Mbit/s in both transmit and receive directions
- Supports balanced and unbalanced receiver inputs
- Single +5-V supply
- TTL compatible
- Suitable for T1, T2, 2.048, and 8.448 Mbit systems
- 18-pin ceramic or plastic DIL

ZN5683E/J

ELECTRICAL CHARACTERISTICS

Test conditions (unless otherwise stated):
T_{amb}: $-40\,°C$ to $+85\,°C$, V_{CC} $+5\,V$ $\pm 0.25\,V$

dc CHARACTERISTICS

Characteristic	Symbol	Pins	Min.	Typ.	Max.	Units	Conditions
Supply current	I_{CC}			35.0	60.0	mA	Output drivers open
Low level input current - data	I_{ILD}	12, 17	-200		-50	µA	V_{IN} = 0V
Low level input current - clock	I_{ILC}	16	-400		-100	µA	V_{IN} = 0V
Low level output voltage - clock	V_{OLC}	8		0.4	0.8	V	I_{CL} = 2mA
High level output voltage - clock	V_{OHC}	8	3	3.6		V	
Low level output voltage - data	V_{OLD}	10, 11		0.4	0.8	V	I_{CL} = 2mA
High level output voltage - data	V_{OHD}	10, 11	3	3.6		V	
Low level output voltage - line driver	V_{OLO}	13, 15	0.6	0.8	0.9	V	See Note 3
High level output current - line driver	I_{OHO}	13, 15			100	µA	
Output driver current sink	I_{OLO}	13, 15			40	mA	
Input voltage		2, 3		3	3.3	V	See Note 4

ac CHARACTERISTICS

Characteristic	Symbol	Min.	Typ.	Max.	Units	Conditions
Output driver rise time	t_{ro}		20	25	ns	See Note 5

ABSOLUTE MAXIMUM RATINGS (See Note 1)

Supply voltage, V_{CC} +20V
Input voltage, V_{IN} (See Note 2)   -0.3V to +20V
Current sink   40mA
Operating temperature range   -40°C to +85°C
Storage temperature range   -55°C to +125°C

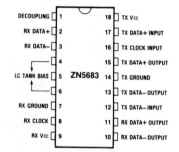

Pin connections - top view

NOTES
1. The absolute maximum ratings are limiting values above which operating life may be shortened or specified parameters may be degraded.
2. V_{IN} = input voltage relative to pins 7, 14.
3. Measured at pins 13, 15 with 300 Ohms pull up to 5.0V.
4. Per CCITT G.703 pulse mask. See Fig.3.
5. Measured at pins 13, 15 with 150 Ohms pull up to 5.0V.

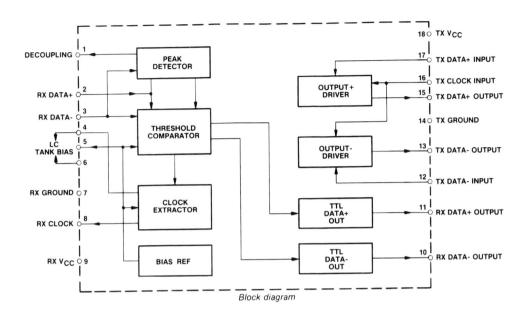

Block diagram

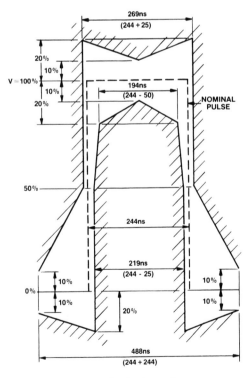

Input signal pulse mask

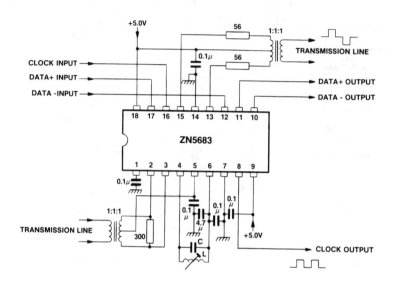

Recommended circuit

Raytheon
DAC-08
8-Bit High-Speed Multiplying D/A Converter

FEATURES

- Fast-settling output current: 85 nS
- Full scale current prematched to ±1.0 LSB
- Direct interface to TTL, CMOS, ECL, HTL, PMOS
- Nonlinearity to ±0.1% max. over temperature range
- High output impedance and compliance: −10 to +18 V
- Differential current outputs
- Wide range multiplying capability: 1.0 MHz bandwidth
- Low FS current drift: ±10 ppm/°C
- Wide power supply range: ±4.5 to ±18 V
- Low power consumption: 33 mW @ ±5.0 V
- Low cost

ABSOLUTE MAXIMUM RATINGS
($T_A = +25\,°C$ unless otherwise noted)

Supply voltage (between $+V_S$ and $-V_S$)	36 V
Logic inputs	$-V_S$ to $-V_S$ plus 36 V
V_{LC}	$-V_S$ to $+V_S$
Analog current outputs	4 mA
Reference inputs (V_{14} to V_{15})	$-V_S$ to $+V_S$
Reference input differential voltage (V_{14} to V_{15})	±18 V
Reference input current (I_{14})	5.0 mA
Operating temperature range	
DAC-08AD, D	−55 °C to +125 °C
DAC-08HN, EN, CN	0 °C to +70 °C
Storage temperature range	−65 °C to +150 °C
Lead soldering temperature (60 sec)	+300 °C

ORDERING INFORMATION

Part Number	Package	Operating Temperature Range	Non-linearity
DAC-08HN	N	0°C to +70°C	±0.1%
DAC-08EN	N	0°C to +70°C	±0.19%
DAC-08CN	N	0°C to +70°C	±0.39%
DAC-08AD	D	-55°C to +125°C	±0.1%
DAC-08D	D	-55°C to +125°C	±0.19%
DAC-08D/883B	D	-55°C to +125°C	±0.19%
DAC-08AD/883B	D	-55°C to +125°C	±0.1%

Notes:
/883B suffix denotes Mil-Std-883, Level B processing
N= 16-lead plastic DIP
D =16-lead ceramic DIP
Contact a Raytheon sales office or representative for ordering information on special package/temperature range combinations.

DCP

CONNECTION INFORMATION

16-Lead Dual-In-Line (Top View)

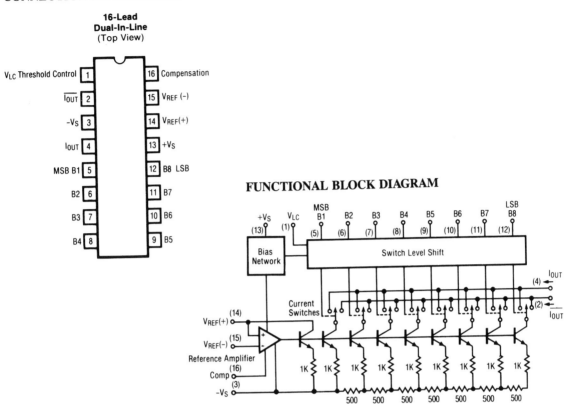

FUNCTIONAL BLOCK DIAGRAM

TYPICAL APPLICATIONS

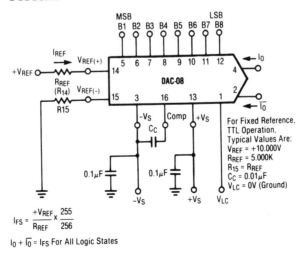

$$I_{FS} = \frac{+V_{REF}}{R_{REF}} \times \frac{255}{256}$$

$I_O + \overline{I_O} = I_{FS}$ For All Logic States

Basic Positive Reference Operation

Raytheon DAC-08

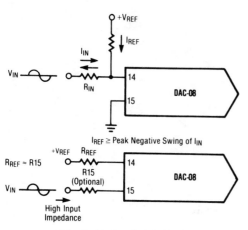

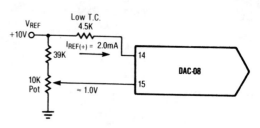

Recommended Full Scale Adjustment Circuit

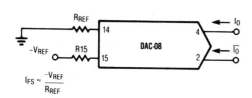

+V$_{REF}$ Must be Above Peak Positive Swing of V$_{IN}$

Accommodating Bipolar References

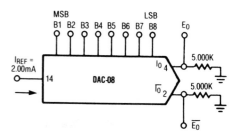

Scale	B1 B2 B3 B4 B5 B6 B7 B8	I$_0$mA	$\overline{I_0}$mA	E$_0$	$\overline{E_0}$
Full Scale	1 1 1 1 1 1 1 1	1.992	0.000	−9.960	−0.000
Half Scale +LSB	1 0 0 0 0 0 0 1	1.008	0.984	−5.040	−4.920
Half Scale	1 0 0 0 0 0 0 0	1.000	0.992	−5.000	−4.960
Half Scale −LSB	0 1 1 1 1 1 1 1	0.992	1.000	−4.960	−5.000
Zero Scale +LSB	0 0 0 0 0 0 0 1	0.008	1.984	−0.040	−9.920
Zero Scale	0 0 0 0 0 0 0 0	0.000	1.992	0.000	−9.960

Note: R$_{REF}$ sets I$_{FS}$; R15 is for bias current cancellation.

Basic Negative Reference Operation

Basic Unipolar Negative Operation

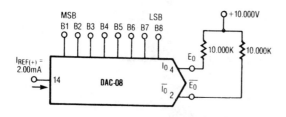

Scale	B1 B2 B3 B4 B5 B6 B7 B8	E$_0$	$\overline{E_0}$
Pos Full Scale	1 1 1 1 1 1 1 1	−9.920	+10.000
Pos Full Scale −LSB	1 1 1 1 1 1 1 0	−9.840	+9.920
Zero Scale +LSB	1 0 0 0 0 0 0 1	−0.080	+0.160
Zero Scale	1 0 0 0 0 0 0 0	0.000	+0.080
Zero Scale −LSB	0 1 1 1 1 1 1 1	+0.080	0.000
Neg Full Scale +LSB	0 0 0 0 0 0 0 1	+9.920	−9.840
Neg Full Scale	0 0 0 0 0 0 0 0	+10.000	−9.920

Basic Bipolar Output Operation

Raytheon DAC-08

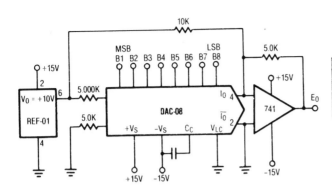

Scale	B1 B2 B3 B4 B5 B6 B7 B8	E_0
Pos Full Scale	1 1 1 1 1 1 1 1	+4.960
Zero Scale	1 0 0 0 0 0 0 0	0.00
Neg Full Scale +1 LSB	0 0 0 0 0 0 0 1	-4.960
Neg Full Scale	0 0 0 0 0 0 0 0	-5.000

Offset Binary Operation

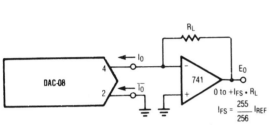

For complementary output (operation as a negative logic DAC), connect inverting input of Op-Amp to I_O (pin 2); connect $\overline{I_O}$ (pin 4) to ground.

Positive Low Impedance Output Operation

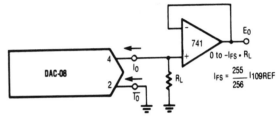

For complementary output (operation as a negative logic DAC), connect non-inverting input of Op-Amp to I_O (pin 2); connect I_O (pin 4) to ground.

Negative Low Impedance Output Operation

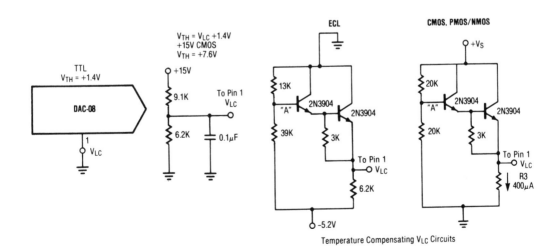

Temperature Compensating V_{LC} Circuits

Interfacing With Various Logic Families

196 Raytheon DAC-10

Settling Time Test Fixture

Raytheon
DAC-10
10-Bit High-Speed Multiplying D/A Converter

FEATURES
- Nonlinearity to 0.05% max. over temperature range
- Low full-scale drift: 10 ppm/°C
- Wide-range multiplying capability: 1.0-MHz bandwidth
- Wide power-supply range: +5.0 V/−7.5 V to ±18 V
- Two-quadrant multiplying
- High-output compliance
- High speed: 85 nS

APPLICATIONS
- A/D converters
- Servo controls
- Waveform generators
- Programmable power supplies
- High-speed modems

ABSOLUTE MAXIMUM RATINGS

Operating temperature range	
DAC-10BD, CD	−55 °C to +125 °C
DAC-10FD, GD	0 °C to +70 °C
Storage temperature range	−65 °C to +150 °C
Lead soldering temperature (60 sec)	+300 °C
Supply voltage ($+V_S$ to $-V_S$)	+36 V
Logic inputs	$-V_S$ to $-V_S$ plus 36 V
V_{LC}	$-V_S$ to $+V_S$
Analog current outputs	$-V_S$ to $+V_S$
Reference inputs (V_{16} to V_{17})	$-V_S$ to $+V_S$
Reference input differential voltage (V_{16} to V_{17})	±18 V
Reference input current (I_{16})	2.5 mA

Raytheon DAC-10

ORDERING INFORMATION

Part Number	Package	Operating Temperature Range	Non-linearity
DAC-10FD	D	0°C to +70°C	±0.05%
DAC-10GD	D	0°C to +70°C	±0.01%
DAC-10BD	D	-55°C to +125°C	±0.05%
DAC-08BD/883B	D	-55°C to +125°C	±0.05%
DAC-08CD	D	-55°C to +125°C	±0.05%
DAC-08CD/883B	D	-55°C to +125°C	±0.05%

Notes:
/883B suffix denotes Mil-Std-883, Level B processing
D = 18-lead ceramic DIP
Contact a Raytheon sales office or representative for ordering information on special package/temperature range combinations.

CONNECTION INFORMATION

18-Lead Hermetic Dual In-Line
(Top View)

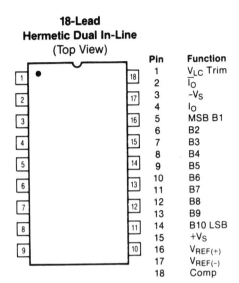

Pin	Function
1	V_{LC} Trim
2	$\overline{I_O}$
3	$-V_S$
4	I_O
5	MSB B1
6	B2
7	B3
8	B4
9	B5
10	B6
11	B7
12	B8
13	B9
14	B10 LSB
15	$+V_S$
16	$V_{REF(+)}$
17	$V_{REF(-)}$
18	Comp

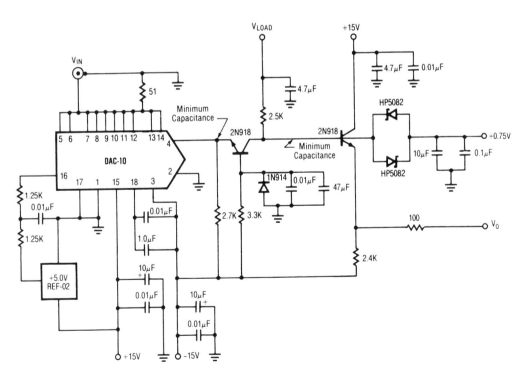

Settling Time Test Fixture

Raytheon DAC-10

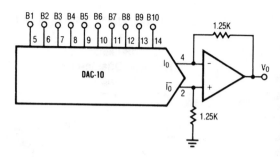

Bipolar Operation

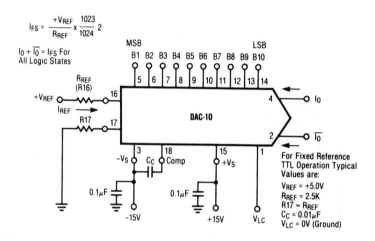

Positive Reference Operation

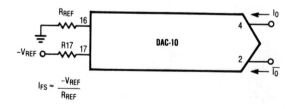

Note: R_{REF} Sets I_{FS}, R17 is for Bias Current Cancellation, so R17 may be 5% Tolerance.

Negative Reference Operation

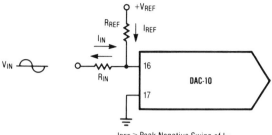

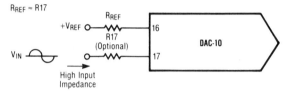

Providing Offsets to Accommodate Bipolar References

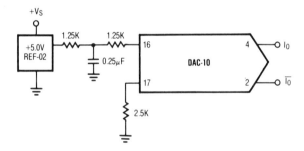

Input Reference Noise Limiting Filter

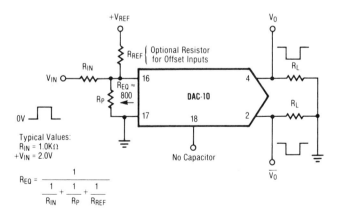

Pulsed Reference Operation

Raytheon DAC-10

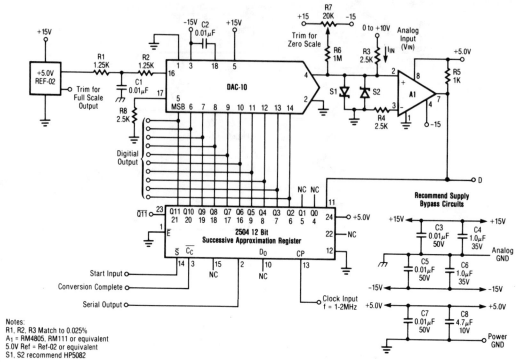

10-Bit Successive Approximation A/D Converter

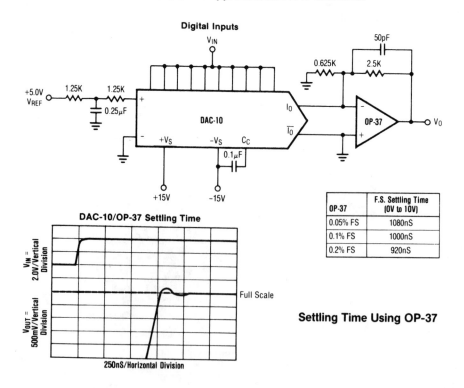

DAC-10/OP-37 Settling Time

OP-37	F.S. Settling Time (0V to 10V)
0.05% FS	1080nS
0.1% FS	1000nS
0.2% FS	920nS

Settling Time Using OP-37

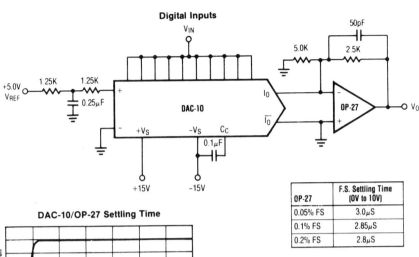

OP-27	F.S. Settling Time (0V to 10V)
0.05% FS	3.0 µS
0.1% FS	2.85 µS
0.2% FS	2.8 µS

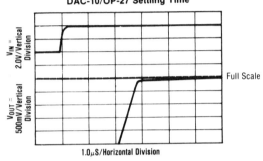

Settling Time Using OP-27

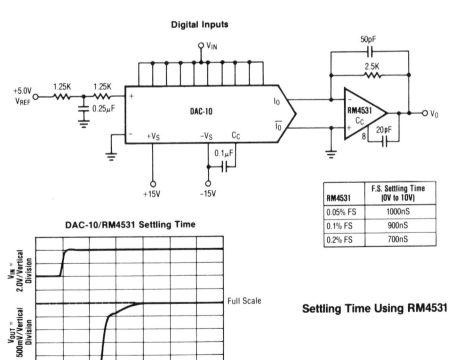

RM4531	F.S. Settling Time (0V to 10V)
0.05% FS	1000 nS
0.1% FS	900 nS
0.2% FS	700 nS

Settling Time Using RM4531

Raytheon DAC-10

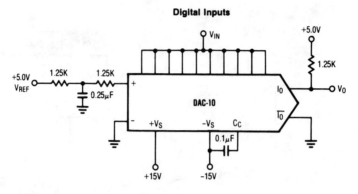

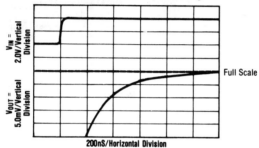

1.25KΩ Resistor	F.S. Settling Time (5.0V to 5.0mV)
0.05% FS	450nS
0.1% FS	320nS
0.2% FS	240nS

Settling Time Using 1.25kΩ Resistor Output

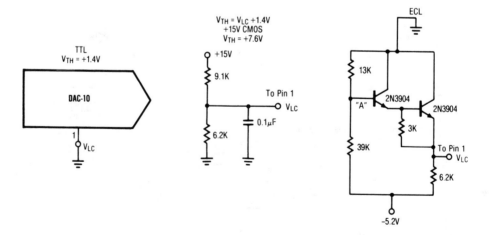

Interfacing With Various Logic Families

Raytheon DAC-10

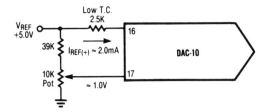

Recommended Full Scale Adjustment Circuit

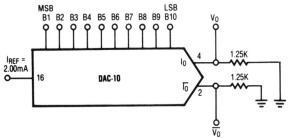

	B1 B2 B3 B4 B5 B6 B7 B8 B9 B10	I_O mA	$\overline{I_O}$ mA	V_O	$\overline{V_O}$
Full Scale	1 1 1 1 1 1 1 1 1 1	3.996	0.000	-4.995	-0.000
Half Scale +LSB	1 0 0 0 0 0 0 0 0 1	2.004	1.992	-2.505	-2.490
Half Scale	1 0 0 0 0 0 0 0 0 0	2.000	1.996	-2.500	-2.495
Half Scale -LSB	0 1 1 1 1 1 1 1 1 1	1.996	2.000	-2.495	-2.500
Zero Scale +LSB	0 0 0 0 0 0 0 0 0 1	0.004	3.992	-0.005	-4.990
Zero Scale	0 0 0 0 0 0 0 0 0 0	0.000	3.996	-0.000	-4.995

Basic Unipolar Negative Operation

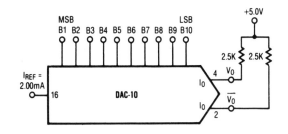

	B1 B2 B3 B4 B5 B6 B7 B8 B9 B10	V_O	$\overline{V_O}$
Pos Full Scale	1 1 1 1 1 1 1 1 1 1	-4.990	+5.000
Pos Full Scale -LSB	1 1 1 1 1 1 1 1 1 0	-4.980	+4.990
Zero Scale +LSB	1 0 0 0 0 0 0 0 0 1	-0.010	+0.020
Zero Scale	1 0 0 0 0 0 0 0 0 0	0.000	+0.010
Zero Scale -LSB	0 1 1 1 1 1 1 1 1 1	+0.010	0.000
Neg Full Scale +LSB	0 0 0 0 0 0 0 0 0 1	+4.990	-4.980
Neg Full Scale	0 0 0 0 0 0 0 0 0 0	+5.000	-4.990

Basic Bipolar Output Operation

Raytheon RC4151/4152

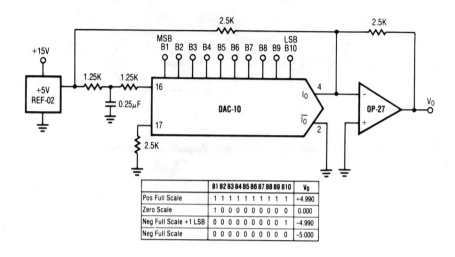

	B1 B2 B3 B4 B5 B6 B7 B8 B9 B10	V_O
Pos Full Scale	1 1 1 1 1 1 1 1 1 1	+4.990
Zero Scale	1 0 0 0 0 0 0 0 0 0	0.000
Neg Full Scale +1 LSB	0 0 0 0 0 0 0 0 0 1	-4.990
Neg Full Scale	0 0 0 0 0 0 0 0 0 0	-5.000

Offset Binary Operation

Simplified Schematic Diagram

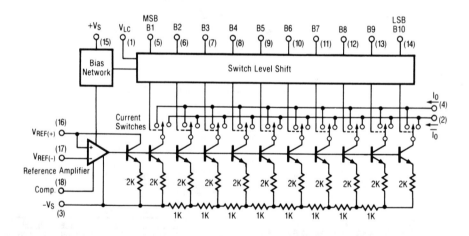

Raytheon
RC4151, 4152
Voltage-to-Frequency Converters

FEATURES

- Single-supply operation
- Pulse-output DTL/TTL/CMOS compatible
- Programmable scale factor (K)
- High noise rejection
- Inherent monotonicity
- Easily transmittable output
- Simple full-scale trim
- Single-ended input, referenced to ground
- V-F or F-V conversion
- Voltage or current input
- Wide dynamic range

APPLICATIONS

- Precision voltage-to-frequency converters
- Pulse-width modulators
- Programmable pulse generators
- Frequency-to-voltage converters
- Integrating analog-to-digital converters
- Long-term analog integrators
- Signal conversion:
  Current-to-frequency
  Temperature-to-frequency
  Pressure-to-frequency
  Capacitance-to-frequency
  Frequency-to-current
- Signal isolation:
  VFC—optoisolation—FVC
  ADC with optoisolation
- Signal encoding:
  FSK modulation/demodulation
  Pulse-width modulation
- Frequency scaling
- dc motor-speed control

CONNECTION INFORMATION

8-Lead Plastic DIP

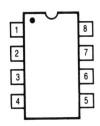

Pin	Function
1	Switched Current Source Output (I_O)
2	Switched Voltage Reference (R_S)
3	Logic Output (Open Collector) (F_O)
4	Ground (GND)
5	One-Shot R, C Timing (C_O)
6	Threshold (V_{TH})
7	Input Voltage (V_{IN})
8	$+V_S$

FUNCTIONAL BLOCK DIAGRAM

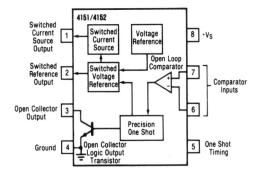

ORDERING INFORMATION

Part Number	Package	Operating Temperature Range
RC4151N	N	0°C to +70°C
RC4152N	N	0°C to +70°C

Notes:
N = 8- lead plastic DIP
Contact a Raytheon sales office or representative for ordering information on special package/temperature range combinations.

ABSOLUTE MAXIMUM RATINGS

Supply voltage	+22 V
Internal power dissipation	500 mW
Input voltage	−0.2 V to $+V_s$
Output sink current (frequency output)	20 mA
Output short circuit to ground	Continuous
Storage temperature range	−65°C to +150°C
Operating temperature range	0°C to +70°C

Raytheon RC4151/4152

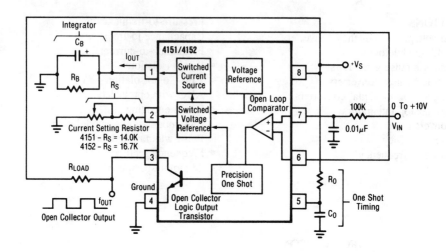

Single Supply VFC

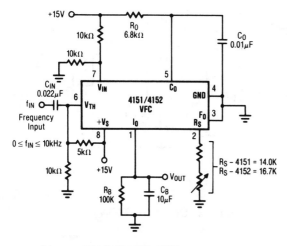

Single Supply FVC

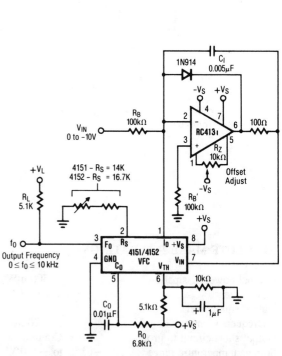

Precision Current — Sourced VFC

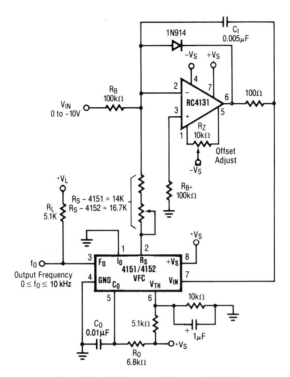

Precision Voltage — Sourced VFC

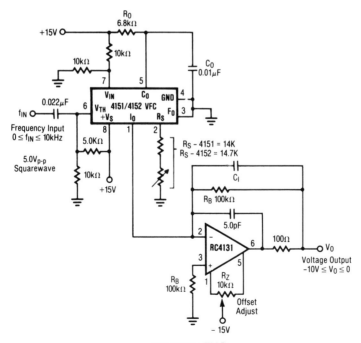

Precision FVC

Raytheon RC4153
Voltage-to-Frequency Converter

FEATURES
- 0.1-Hz to 250-kHz dynamic range
- 0.01% F.S. maximum nonlinearity error: 0.1 Hz to 10 kHz
- 50 ppm/°C maximum-gain temperature coefficient (external reference)
- Few external components required

APPLICATIONS
- Precision voltage-to-frequency converters
- Serial transmission of analog information
- Pulse-width modulators
- Frequency-to-voltage converters
- A/D converters and long-term integrators
- Signal isolation
- FSK modulation/demodulation
- Frequency scaling
- Motor-speed controls
- Phase-lock loop stabilization

ABSOLUTE MAXIMUM RATINGS

Supply voltage	±18 V
Internal power dissipation	500 mW
Input voltage range	$-V_S$ to $+V_S$
Output sink current (freq. output)	20 mA
Storage temperature range	−65°C to +150°C
Operating temperature range	
RM4153	−55°C to +125°C
RC4153	0°C to +70°C

ORDERING INFORMATION

Part Number	Package	Operating Temperature Range
RC4153D	D	0°C to +70°C
RM4153D	D	-55°C to +125°C

Notes:
D = 14- lead ceramic DIP
Contact a Raytheon sales office or representative for ordering information on special package/temperature range combinations.

FUNCTIONAL BLOCK DIAGRAM

14 Lead Dual In-Line Package
(Top View)

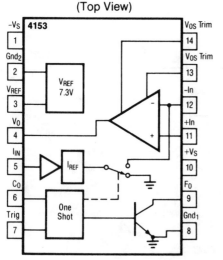

Pin	Function	Pin	Function
1	$-V_S$	8	Circuit Gnd
2	REF Gnd	9	Frequency Output (Open Collector)
3	V_{REF} Output	10	$+V_S$
4	V_{OUT} (Op Amp)	11	(+) Op Amp Input
5	I_{IN} (REF Input)	12	(−) Op Amp Input
6	C_O (Pulse Width)	13	V_{OS} Trim
7	Trigger Input	14	V_{OS} Trim

TYPICAL APPLICATION CIRCUITS

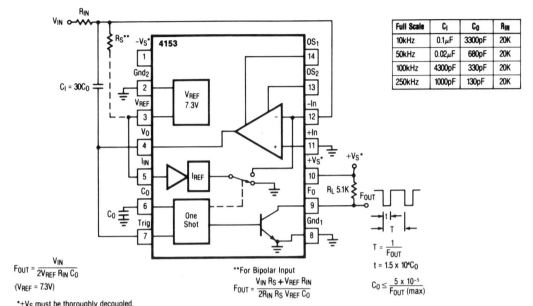

Voltage-to-Frequency Converter Minimum Circuit

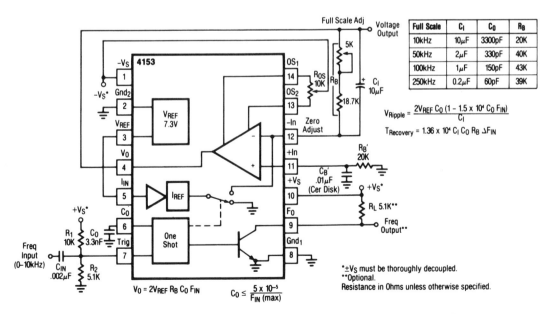

Frequency-to-Voltage Converter — V_O (Volts) = F_{IN} (kHz) — 100kHz Max

Silicon Systems SSI 32F8011

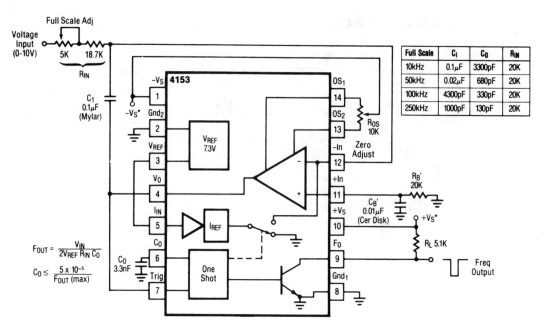

Voltage-to-Frequency Converter With Offset and Gain Adjusts

Silicon Systems
SSI 32F8011
Programmable Electronic Filter
Preliminary Data

FEATURES
- Compatible with 24-Mbit/s operation
- Ideal for constant-density recording applications
- Programmable filter cutoff frequency (f_C=5 to 13 MHz)
- Programmable pulse slimming equalization (0 to 9 dB boost at the filter cutoff frequency)
- Matched normal and differentiated low-pass outputs
- Differential filter input and outputs
- ±10% cutoff frequency accuracy
- ±0.75 ns group delay variation from $0.2\ F_c$ to f_c=13 MHz
- Total harmonic distortion less than 1%
- No external-filter components required
- +5 V only operation
- 16-pin DIP, SON, and SOL package

ORDERING INFORMATION

PART DESCRIPTION	ORDER NO.	PKG. MARK
SSI 32F8011 Programmable Electronic Filter		
16-lead SON (150 mil)	SSI 32F8011-CN	32F8011
16-lead SOL (300 mil)	SSI 32F8011-CL	32F8011
16-lead PDIP	SSI 32F8011-CP	32F8011-CP

Preliminary Data: Indicates a product not completely released to production. The specifications are based on preliminary evaluations and are not guaranteed. Small quantities are available, and Silicon Systems should be consulted for current information.

No responsibility is assumed by Silicon Systems for use of this product nor for any infringements of patents and trademarks or other rights of third parties resulting from its use. No license is granted under any patents, patent rights or trademarks of Silicon Systems. Silicon Systems reserves the right to make changes in specifications at any time without notice. Accordingly, the reader is cautioned to verify that the data sheet is current before placing orders.

BLOCK DIAGRAM

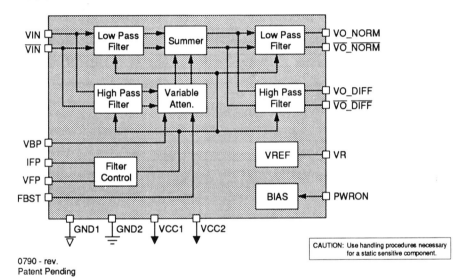

0790 - rev.
Patent Pending

CAUTION: Use handling procedures necessary for a static sensitive component.

PIN DIAGRAM

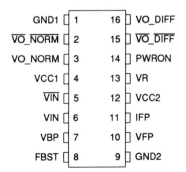

ABSOLUTE MAXIMUM RATINGS

Operation above maximum ratings may damage the device.

PARAMETER	RATINGS	UNIT
Storage Temperature	-65 to +150	°C
Junction Operating Temperature, Tj	+130	°C
Supply Voltage, VCC1, VCC2	-0.5 to 7	V
Voltage Applied to Inputs	-0.5 to VCC	V
IFP, VFP Inputs Maximum Current*	<1.5	mA
Maximum Power Dissipation, fc = 13 MHz	390	mW

* Exceeding this current may cause frequency programming lockup.

RECOMMENDED OPERATING CONDITIONS

Supply voltage, VCC1, VCC2	4.65 < VCC1,2 < 5.50	V
Junction Temperature, Tj	0 < Tj < 130	°C
Ambient Temperature	0 < Ta < 70	°C

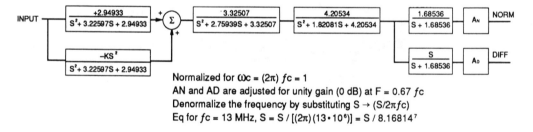

Normalized for $\omega c = (2\pi) fc = 1$
AN and AD are adjusted for unity gain (0 dB) at F = 0.67 fc
Denormalize the frequency by substituting S → (S/2πfc)
Eq for fc = 13 MHz, S = S / [(2π)(13·10⁶)] = S / 8.16814⁷

32F8011 Normalized Block Diagram

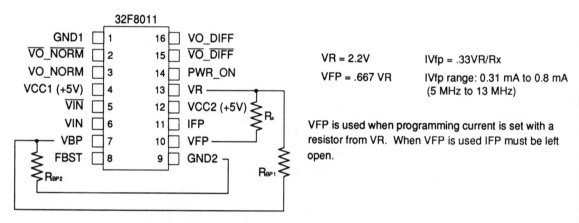

VR = 2.2V
VFP = .667 VR

IVfp = .33VR/Rx
IVfp range: 0.31 mA to 0.8 mA
(5 MHz to 13 MHz)

VFP is used when programming current is set with a resistor from VR. When VFP is used IFP must be left open.

32F8011 Applications Setup, 16-Pin SO or DIP

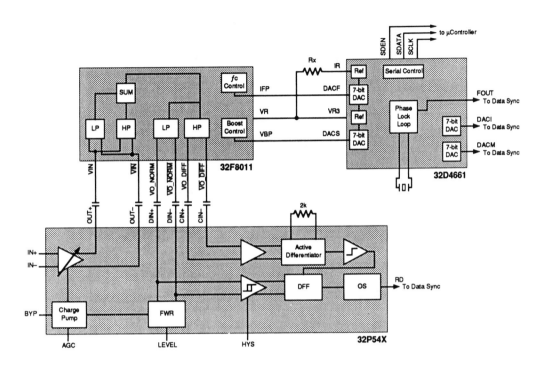

**Applications Setup, Constant Density Recording
32F8011, 32P54X, 32D4661**

IOF = DACF output current

IOF = (0.98F·VR)/127Rx

Rx = (0.98F·VR)/127IOF

Rx = current reference setting resistor

VR = Voltage Reference = 2.2V

F = DAC setting: 0-127

Full scale, F = 127

For range of Max f_c = 13 MHz then IFP = 0.8 mA

Therefore, for Max programming current range to 0.8 mA:

Rx = (0.98)(2.2/0.8) = 2.7 kΩ

Please note that in setups such as this where IFP is used for cutoff frequency programming VFP must be left open.

Silicon Systems
SSI 32F8020
Low-Power Programmable Electronic Filter
Advance Information

FEATURES
- Ideal for constant-density recording applications
- Programmable-filter cutoff frequency (f_c=1.5 to 8 MHz)
- Programmable-pulse slimming equalization (0- to 9-dB boost at the filter cutoff frequency)
- Matched normal and differentiated low-pass outputs
- Differential filter input and outputs
- ±10% cutoff frequency accuracy
- ±2% maximum group delay variation from 1.5 to 8 MHz
- Total harmonic distortion less than 1%
- No external filter components required
- +5-V only operation
- 16-pin DIP, SON, and SOL package

BLOCK DIAGRAM

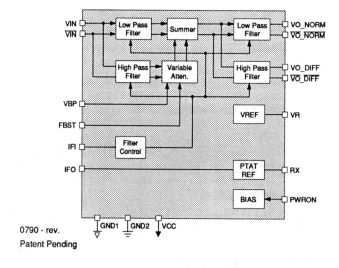

0790 - rev.
Patent Pending

PIN DIAGRAM

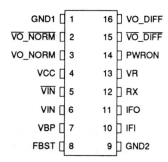

CAUTION: Use handling procedures necessary for a static sensitive component.

CHAPTER 5

LOGIC CIRCUITS

**Allegro
3611 through 3614
Dual 2-Input Peripheral/Power Drivers**

FEATURES
- Four logic types
- DTL/TTL/PMOS/CMOS-compatible inputs
- Low input current
- Standoff voltage of 80 V
- Hermetically sealed package
- High-reliability screening

ABSOLUTE MAXIMUM RATINGS

Supply voltage, V_{CC}	7.0 V
Input voltage, V_{IN}	30 V
Output off-state voltage, V_{OFF}	80 V
Output on-state sink current, I_{ON}	600 mA
Power dissipation, P_D	
(one output)	1.0 W
(total package)	See graph
Operating temperature range, T_A	$-55\,°C$ to $+125\,°C$
Storage temperature range, T_S	$-65\,°C$ to $+150\,°C$

These devices are noncompliant regarding MIL-STD-883C because of package dimensions.

Always order by complete part number:

Part Number	Description
UDS3611H883	Dual AND Driver
UDS3612H883	Dual NAND Driver
UDS3613H883	Dual OR Driver
UDS3614H883	Dual NOR Driver

UDS3612H DUAL NAND DRIVER
ELECTRICAL CHARACTERISTICS over operating temperature range (unless otherwise noted).

Characteristic	Symbol	Test Conditions				Limits			Units	Notes
		Temp.	V_{CC}	Driven Input	Other Input	Output	Min.	Typ.	Max.		
"1" Output Reverse Current	I_{OFF}	—	MIN	0.8 V	V_{CC}	80 V	—	—	100	µA	—
		—	OPEN	0.8 V	V_{CC}	80 V	—	—	100	µA	—
"0" Output Voltage	V_{ON}	—	MIN	2.0 V	2.0 V	150 mA	—	0.4	0.5	V	—
		—	MIN	2.0 V	2.0 V	300 mA	—	0.6	0.8	V	—
"1" Level Supply Current	$I_{CC(1)}$	NOM	MAX	0 V	0 V	—	—	12	15	mA	1, 2
"0" Level Supply Current	$I_{CC(0)}$	NOM	MAX	5.0 V	5.0 V	—	—	40	53	mA	1, 2

NOTES: 1. Typical values are at V_{CC} = 5.0 V, T_A = 25°C.
2. Per package.
3. Capacitance values specified include probe and test fixture capacitance.

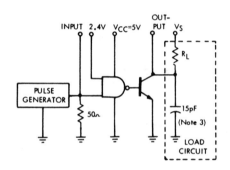

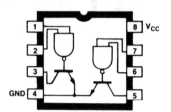

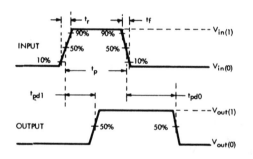

UDS3611H DUAL AND DRIVER
ELECTRICAL CHARACTERISTICS over operating temperature range (unless otherwise noted).

Characteristic	Symbol	Test Conditions				Limits			Units	Notes
		Temp.	V_{CC}	Driven Input	Other Input	Output	Min.	Typ.	Max.		
"1" Output Reverse Current	I_{OFF}	—	MIN	2.0 V	2.0 V	80 V	—	—	100	µA	—
		—	OPEN	2.0 V	2.0 V	80 V	—	—	100	µA	—
"0" Output Voltage	V_{ON}	—	MIN	0.8 V	V_{CC}	150 mA	—	0.4	0.5	V	—
		—	MIN	0.8 V	V_{CC}	300 mA	—	0.6	0.8	V	—
"1" Level Supply Current	$I_{CC(1)}$	NOM	MAX	5.0 V	5.0 V	—	—	8.0	12	mA	1, 2
"0" Level Supply Current	$I_{CC(0)}$	NOM	MAX	0 V	0 V	—	—	35	49	mA	1, 2

NOTES: 1. Typical values are at V_{CC} = 5.0 V, T_A = 25°C.
2. Per package.
3. Capacitance values specified include probe and test fixture capacitance.

☐ LOG   Allegro 3611 through 3614

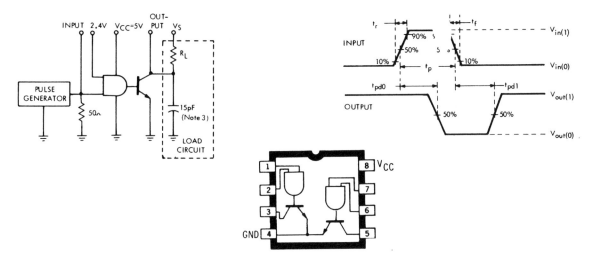

UDS3613H DUAL OR DRIVER
ELECTRICAL CHARACTERISTICS over operating temperature range (unless otherwise noted).

Characteristic	Symbol	Test Conditions					Limits			Units	Notes
		Temp.	V_{CC}	Driven Input	Other Input	Output	Min.	Typ.	Max.		
"1" Output Reverse Current	I_{OFF}	—	MIN	2.0 V	0 V	80 V	—	—	100	µA	—
		—	OPEN	2.0 V	0 V	80 V	—	—	100	µA	—
"0" Output Voltage	V_{ON}	—	MIN	0.8 V	0.8 V	150 mA	—	0.4	0.5	V	—
		—	MIN	0.8 V	0.8 V	300 mA	—	0.6	0.8	V	—
"1" Level Supply Current	$I_{CC(1)}$	NOM	MAX	5.0 V	5.0 V	—	—	8.0	13	mA	1, 2
"0" Level Supply Current	$I_{CC(0)}$	NOM	MAX	0 V	0 V	—	—	36	50	mA	1, 2

NOTES: 1. Typical values are at V_{CC} = 5.0 V, T_A = 25°C.
            2. Per package.
            3. Capacitance values specified include probe and test fixture capacitance.

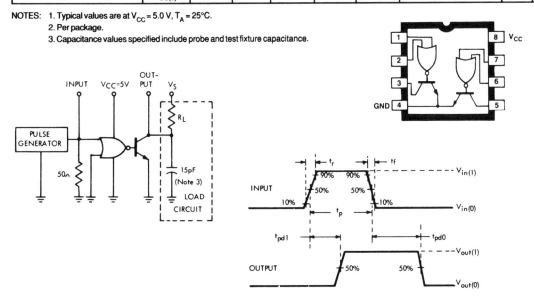

UDS3614H DUAL NOR DRIVER
ELECTRICAL CHARACTERISTICS over operating temperature range (unless otherwise noted).

Characteristic	Symbol	Test Conditions					Limits			Units	Notes
		Temp.	V_{CC}	Driven Input	Other Input	Output	Min.	Typ.	Max.		
"1" Output Reverse Current	I_{OFF}	—	MIN	0.8 V	0.8 V	80 V	—	—	100	µA	—
		—	OPEN	0.8 V	0.8 V	80 V	—	—	100	µA	—
"0" Output Voltage	V_{ON}	—	MIN	2.0 V	0 V	150 mA	—	0.4	0.5	V	—
		—	MIN	2.0 V	0 V	300 mA	—	0.6	0.8	V	—
"1" Level Supply Current	$I_{CC(1)}$	NOM	MAX	0 V	0 V	—	—	12	15	mA	1, 2
"0" Level Supply Current	$I_{CC(0)}$	NOM	MAX	5.0 V	5.0 V	—	—	40	50	mA	1, 2

NOTES: 1. Typical values are at $V_{CC} = 5.0$ V, $T_A = 25°C$.
2. Per package.
3. Capacitance values specified include probe and test fixture capacitance.

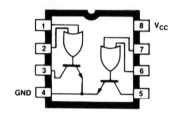

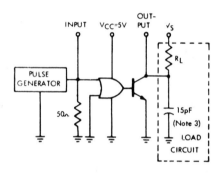

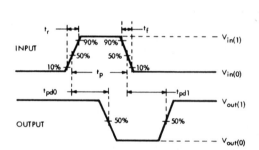

Allegro
5800 and 5801
MIL-STD-883 Compliant BiMOS II Latched Drivers

FEATURES
- 4.4-MHz minimum data input rate
- CMOS/PMOS/NMOS/TTL-compatible inputs
- Internal pull-down resistors
- Low-power CMOS control and latches
- High-voltage high-current outputs
- Transient-protected outputs
- Operating temperature −55°C to +125°C
- High-reliability screening to MIL-STD-883, class B

Always order by complete part number:

Part Number	Description
UCS5800H883	4-Bit Latched Driver
UCS5801H883	8-Bit Latched Driver

LOG

ABSOLUTE MAXIMUM RATINGS
at +25°C Free-Air Temperature

Output voltage, V_{CE}	50 V
Supply voltage, V_{DD}	15 V
Input voltage range, V_{IN}	-0.3 V to $V_{DD}+0.3$ V
Continuous collector current, I_C	500 mA
Package power dissipation, P_D	See graph
Operating ambient temperature range, T_A	$-55°C$ to $125°C$
Storage temperature range, T_S	$-65°C$ to $+150°C$

Note: Output current rating may be limited by duty cycle, ambient temperature, air flow, and number of outputs conducting. Under any set of conditions, do not exceed a maximum junction temperature of +150°C.

Caution: CMOS devices have input static protection but are susceptible to damage when exposed to extremely high static electrical charges.

UCS5800H

UCS5801H

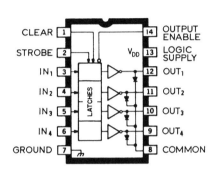

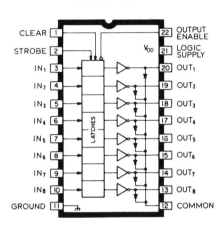

FUNCTIONAL BLOCK DIAGRAM

TYPICAL INPUT CIRCUIT

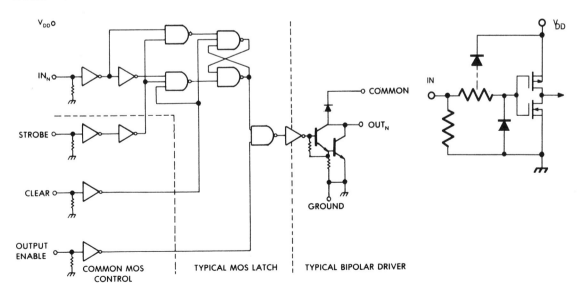

220 Allegro 5800/5801

TYPICAL APPLICATIONS

INCANDESCENT LAMP DRIVER

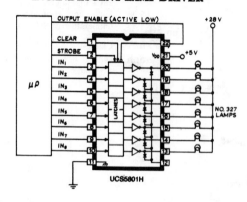

UNIPOLAR STEPPER-MOTOR DRIVE

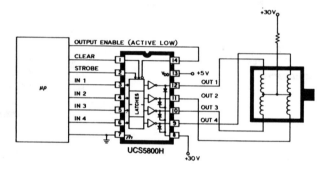

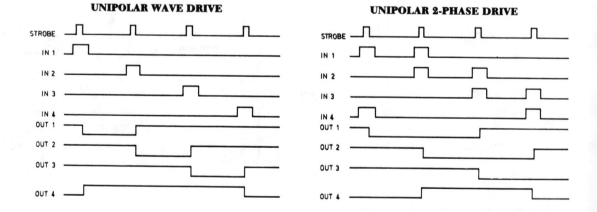

Allegro
5810 MIL-STD-883 Compliant BiMOS II 10-Bit Serial-Input, Latched Source Driver

FEATURES
- 5-MHz minimum-data input rate
- High-voltage source outputs
- CMOS/PMOS/NMOS/TTL-compatible inputs
- Low-power CMOS logic and latches
- Internal pull-down resistors
- Wide supply-voltage range
- High-reliability screening to MIL-STD-883, class B
- Operating temperature $-55\,°C$ to $+125\,°C$

ABSOLUTE MAXIMUM RATINGS
at $+25\,°C$ Free-Air Temperature

Output voltage, V_{OUT}	60 V
Logic supply voltage range, V_{DD}	4.5 V to 15 V
Driver supply voltage range, V_{BB}	5.0 V to 60 V
Input voltage range, V_{IN}	-0.3 V to $V_{DD}+0.3$ V
Continuous output current, I_{OUT}	-40 mA
Package power dissipation, P_D	1.67 W*
Operating temperature range, T_A	$-55\,°C$ to $+125\,°C$
Storage temperature range, T_S	$-65\,°C$ to $+150\,°C$

*Derate at 13.3 mW/°C above $+25\,°C$

Caution: This CMOS device has input static protection but is susceptible to damage when exposed to extremely high static electrical charges.

TYPICAL INPUT CIRCUIT

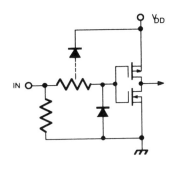

TYPICAL OUTPUT DRIVER

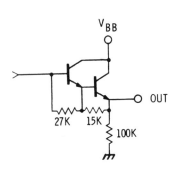

FUNCTIONAL BLOCK DIAGRAM

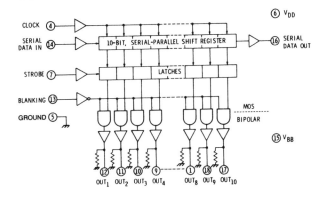

Allegro 5822
MIL-STD-883 Compliant
BiMOS II 8-Bit Serial Input, Latched Driver

FEATURES
- 3.3-MHz minimum data input rate
- High-voltage current-sink outputs
- CMOS/PMOS/NMOS/TTL-compatible
- Low-power CMOS logic and latches
- Internal pull-up/pull-down resistors
- Hermetically sealed packages to MIL-M-38510
- High-reliability screening to MIL-STD-883, class B

ABSOLUTE MAXIMUM RATINGS
at +25°C Free-Air Temperature

Output voltage, V_{OUT}	80 V
Logic supply voltage, V_{DD}	15 V
Input voltage range, V_{IN}	-0.3 V to $V_{DD}+0.3$ V
Continuous output current, I_{OUT}	500 mA
Package power dissipation, P_D	See graph
Operating temperature range, T_A	-55°C to $+125$°C
Storage temperature range, T_S	-65°C to $+150$°C

Caution: CMOS devices have input static protection but are susceptible to damage when exposed to extremely high static electrical charges.

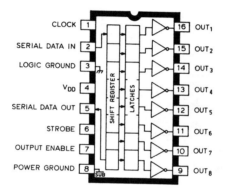

FUNCTIONAL BLOCK DIAGRAM

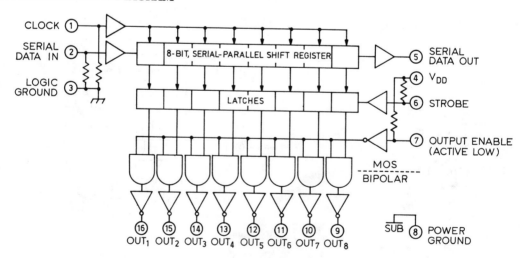

☐ LOG   Plessey PDSP16256/A   223

TYPICAL INPUT CIRCUITS

TYPICAL OUTPUT DRIVER

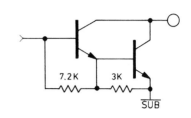

GEC Plessey
PDSP16256/A
Programmable Fir Filter

FEATURE

- Sixteen MACs in a single device
- Basic mode is 16 tap filter with 25-MHz sample rates
- 16-bit data and 32-bit accumulators
- Programmable to give up to 128 taps with sampling rates proportionally reducing to 3.13 MHz
- Can be configured as one long filter or two half-length filters
- Decimate by two option will double the filter length
- Coefficients supplied from a host system or a local EPROM
- Advanced 144-PGA package with integral ground and supply planes

APPLICATIONS

- High-performance digital filters
- Pulse compression for radar and sonar
- Matrix multiplication
- Correlation

ELECTRICAL CHARACTERISTICS
Operating Conditions (unless otherwise stated)
Commercial: $T_{AMB}=0\,°C$ to $+70\,°C$ $T_{J\,(MAX)}=100\,°C$ $V_{CC}=5.0\,V\pm5\%$ Ground$=0\,V$
Industrial: $T_{AMB}=-40\,°C$ to $+85\,°C$ $T_{J\,(MAX)}=110\,°C$ $V_{CC}=5.0\,V\pm10\%$ Ground$=0\,V$
Military: $T_{AMB}=-55\,°C$ to $+125\,°C$ $T_{J\,(MAX)}=150\,°C$ $V_{CC}=5.0\,V\pm10\%$ Ground$=0\,V$

ORDERING INFORMATION
PDSP16256A   C0   AC   25 MHz   Commercial
PDSP16256   B0   AC   20 MHz   Industrial
PDSP16256   A0   AC   20 MHz   Military
Call for availability on high-reliability parts and MIL-STD-883C screening

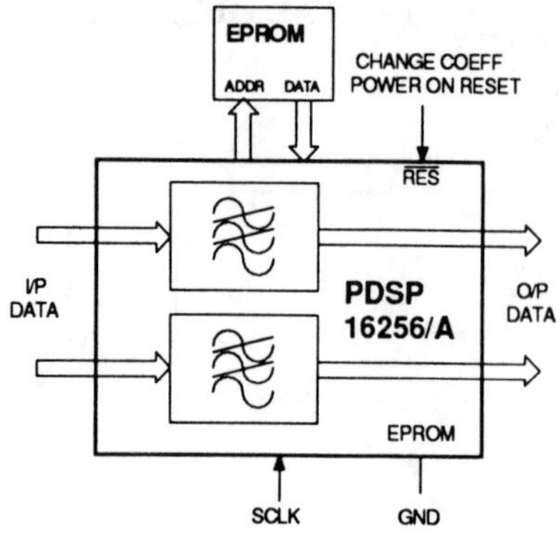

Plessey PDSP16256/A

LOG

Dual Filter

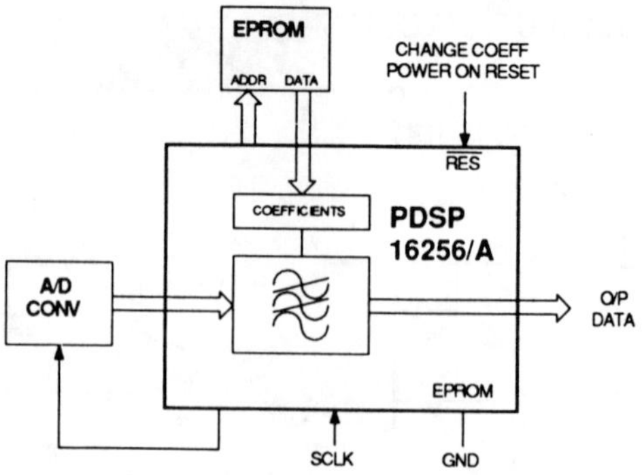

Typical System Application

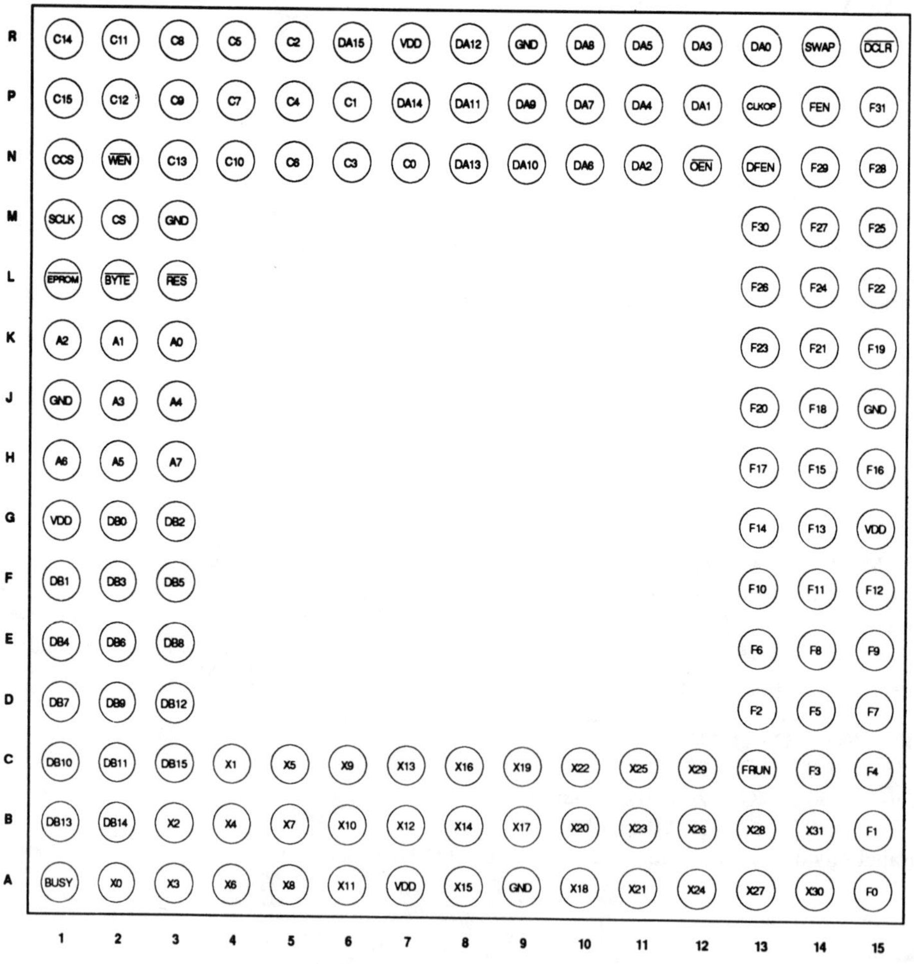

Device Pinout - Bottom view

LOG

Plessey PDSP16256/A

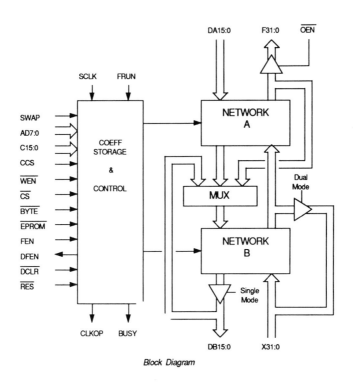

Block Diagram

Filter Network Diagram

CR 14 13 12	Input Rate	Output Rate	Filter Length	Setup Latency
0 0 0	SCLK	SCLK	16 Taps	16
0 0 1	SCLK	SCLK/2	32 Taps	17
0 1 0	SCLK/2	SCLK/2	32 Taps	16
0 1 1	SCLK/2	SCLK/4	64 Taps	18
1 0 0	SCLK/4	SCLK/4	64 Taps	20
1 0 1	SCLK/4	SCLK/8	128 Taps	24
1 1 0	SCLK/8	SCLK/8	128 Taps	24

Single Filter Options

CR 14 13 12	Input Rate	Output Rate	Filter Length	Setup Latency Ind	Setup Latency Cas
0 0 0	SCLK	SCLK	8 Taps	16	27
0 0 1	SCLK	SCLK/2	16 Taps	17	-
0 1 0	SCLK/2	SCLK/2	16 Taps	16	28
0 1 1	SCLK/2	SCLK/4	32 Taps	18	-
1 0 0	SCLK/4	SCLK/4	32 Taps	20	36
1 0 1	SCLK/4	SCLK/8	64 Taps	24	-
1 1 0	SCLK/8	SCLK/8	64 Taps	24	40

Dual Filter Options

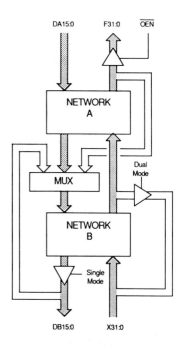

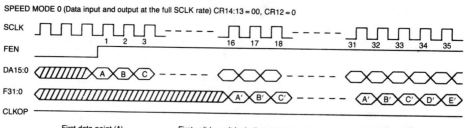

SPEED MODE 0 (Data input and output at the full SCLK rate) CR14:13 = 00, CR12 = 0

First data point (A) read edge cycle 1

First valid result including data point A available after edge 16

Valid result contain the first 16 data points available after edge 31

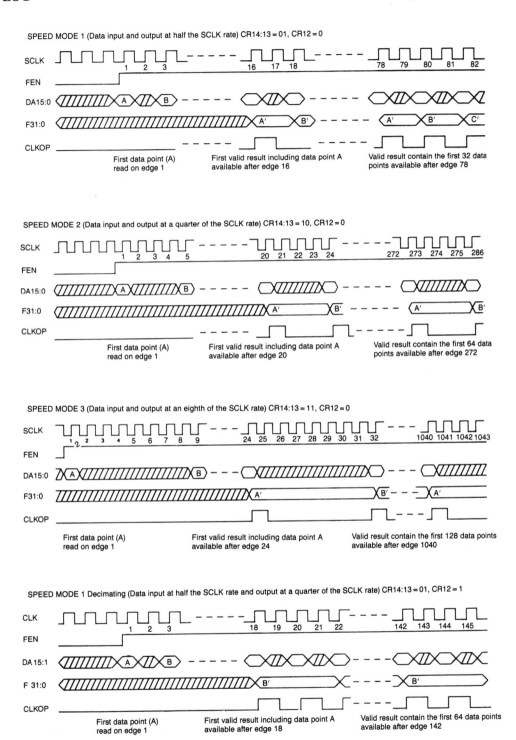

Single Filter Timing Diagrams

228 Plessey PDSP16256/A

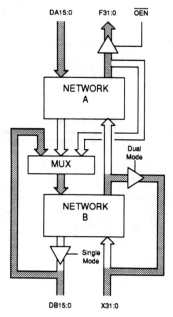

Dual Independent Filter Bus Utilisation

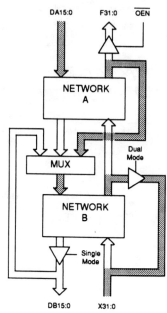

Dual Cascaded Filter Bus Utilisation

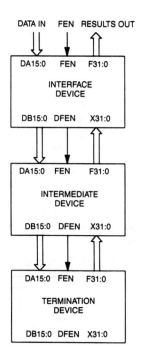

Three Device Cascaded System

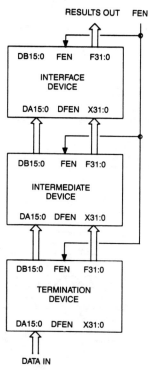

Full Speed Cascaded System

LOG

Plessey PDSP16256/A

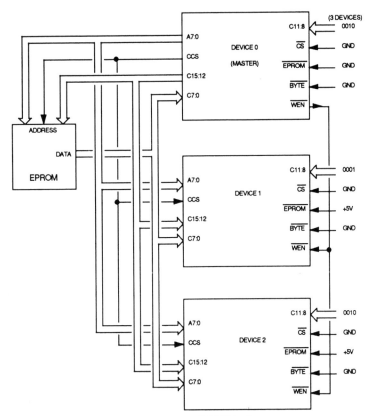

Three device auto EPROM load

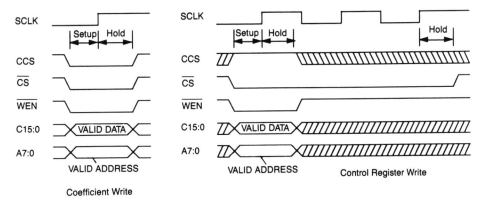

Remote Master Setup & Hold Timings

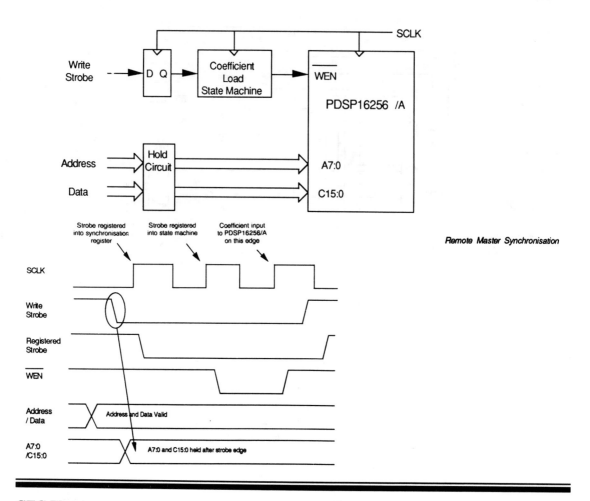

Remote Master Synchronisation

GEC Plessey
PDSP16330/A/B
Pythagoras Processor

FEATURES
- 25-MHz cartesian to polar conversion
- 16-bit cartesian inputs
- 16-bit magnitude output
- 12-bit phase output
- 2s' complement or sign-magnitude input formats
- Three-state outputs and independent data enables simplify system interfacing
- Magnitude-scaling facility with overflow flag
- Less than 400-mW power dissipation at 10 MHz
- 84-pin LCC/PGA package

APPLICATIONS
- Digital signal processing
- Digital radio
- Radar processing
- Sonar processing
- Robotics

ABSOLUTE MAXIMUM RATINGS

Supply voltage, V_{CC}	-0.5 to $+7.0$ V
Input voltage, V_{IN}	-0.5 V to $V_{CC}+0.5$ V
Output voltage, V_{OUT}	-0.5 V to $V_{CC}+0.5$ V
Clamp diode current per pin, I_X (see note 2)	± 18 mA
Static discharge voltage (HMB), V_{STAT}	500 V
Storage temperature, T_{stg}	$-65\,°C$ to $+150\,°C$
Ambient temperature with power applied, T_{AMB}	
Commercial	$0\,°C$ to $+70\,°C$

☐ **LOG**

Industrial	−40 °C to +85 °C
Military	−55 °C to +125 °C
Package power dissipation P_{TOT}	1200 mW
Junction temperature	150 °C

ORDERING INFORMATION
Commercial (0 °C to +70 °C)
PDSP16330 C0 LC (10 MHz - LCC package)
PDSP16330 C0 AC (10 MHz - PGA package)
PDSP16330A C0 LC (20 MHz - LCC package)
PDSP16330A C0 AC (20 MHz - PGA package)
PDSP16330B C0 AC (25 MHz - PGA package)
Industrial (−40 °C to +85 °C)
PDSP16330 B0 LC (10 MHz - LCC package)
PDSP16330 B0 AC (10 MHz - PGA package)
PDSP16330A B0 LC (20 MHz - LCC package)
PDSP16330A B0 AC (20 MHz - PGA package)
PDSP16330B B0 AC (25 MHz - PGA package)

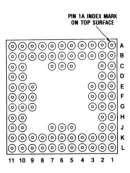

Pin connections - bottom view

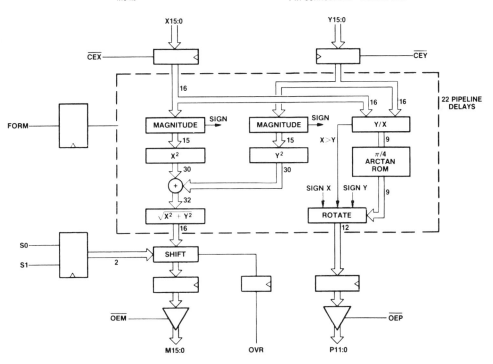

Block diagram

Plessey PDSP16330/A/B

Test	Waveform - measurement level
Delay from output high to output high impedance	V_H falling, 0.5V
Delay from output low to output high impedance	V_L rising, 0.5V
Delay from output high impedance to output low	1.5V falling, 0.5V
Delay from output high impedance to output high	1.5V rising, 0.5V

NOTES
1. V_H - Voltage reached when output driven high.
2. V_L - Voltage reached when output driven low.

Three state delay measurement load

GEC Plessey
PDSP16340
Polar to Cartesian Converter

FEATURES
- Provides R cos(θ) and R sin(θ) in 16-bit streams using a CORDIC processor
- Look-up table equivalent to 64k by 32-bit ROM
- 20-MHz clock rate
- Tri-state outputs and independent data enables
- 84-pin PGA package

APPLICATIONS
- Digital signal processing
- Radar systems
- Sonar systems
- Robotics
- Medical imaging

ABSOLUTE MAXIMUM RATINGS (Note 1)

Supply voltage V_{CC}	-0.5 V to 7.0 V
Input voltage V_{IN}	-0.5 V to $V_{CC}+0.5$ V
Output voltage V_{OUT}	-0.5 V to $V_{CC}+0.5$ V
Clamp diode current per pin I_K (see note 2)	18 mA
Static discharge voltage (HMB)	500 V
Storage temperature T_S	$-65\,°C$ to $150\,°C$
Ambient temperature with power applied T_{AMB}	
Military	$-55\,°C$ to $+125\,°C$
Industrial	$-40\,°C$ to $85\,°C$
Junction temperature	$150\,°C$
Package power dissipation	3500 mW
Thermal resistances	
Junction to case θ_{JC}	$5\,°C/W$

NOTES ON MAXIMUM RATINGS
1. Exceeding these ratings might cause permanent damage. Functional operation under these conditions is not implied.
2. Maximum dissipation or 1 second should not be exceeded, only one output to be tested at any one time.
3. Exposure to absolute maximum ratings for extended periods might affect device reliability.
4. V_{CC}=Max. outputs unloaded, Clock freq=max.
5. CMOS levels are defined as
   $V_{IH} = V_{CC} - 0.5$ V
   $V_{IL} = +0.5$ V
6. Current is defined as positive into the device.
7. θ_{JC} data assumes that heat is extracted from the top face of the package.

ORDERING INFORMATION
PDSP16340 B0 AC (Industrial - PGA package)
PDSP16340 A0 AC (Military - PGA package)
Call for availability of high-reliability parts and MIL-STD-883C screening.

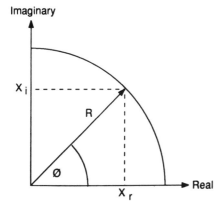

Cartesian to Polar Coordinates

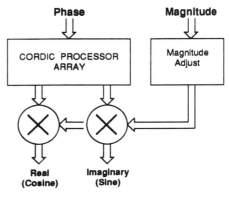

Simplified Block Diagram

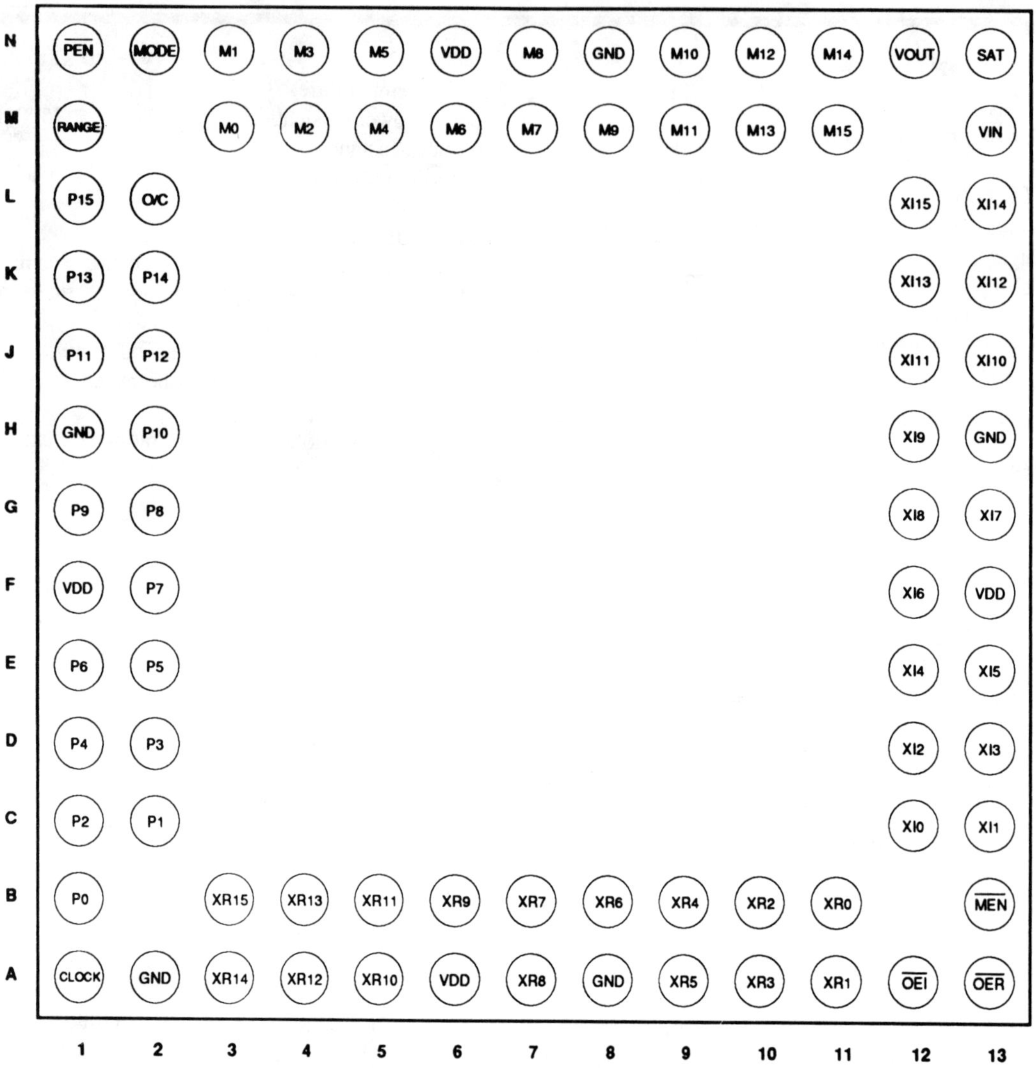

Plessey PDSP16340

LOG

	1	2	3	4	5	6	7	8	9	10	11	12	13
N	PEN	MODE	M1	M3	M5	VDD	M8	GND	M10	M12	M14	VOUT	SAT
M	RANGE		M0	M2	M4	M6	M7	M9	M11	M13	M15		VIN
L	P15	O/C										XI15	XI14
K	P13	P14										XI13	XI12
J	P11	P12										XI11	XI10
H	GND	P10										XI9	GND
G	P9	P8										XI8	XI7
F	VDD	P7										XI6	VDD
E	P6	P5										XI4	XI5
D	P4	P3										XI2	XI3
C	P2	P1										XI0	XI1
B	P0		XR15	XR13	XR11	XR9	XR7	XR6	XR4	XR2	XR0		MEN
A	CLOCK	GND	XR14	XR12	XR10	VDD	XR8	GND	XR5	XR3	XR1	OEI	OER

Device Pinout - Bottom view

LOG

Plessey PDSP16340

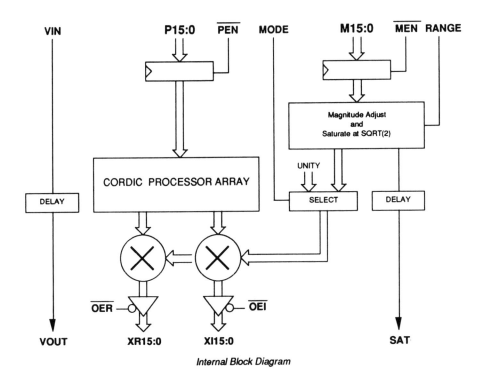

Internal Block Diagram

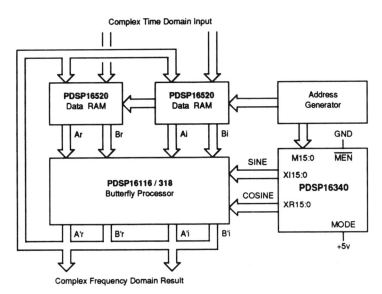

Sin / Cos generator for 20 MHz FFT System

Test	Waveform - measurement level
Delay from output high to output high impedance	V_H ⊢ 0.5 V
Delay from output low to output high impedance	V_L ⊢ 0.5 V
Delay from output high impedance to output low	1.5 V ⊢ 0.5 V
Delay from output high impedance to output high	1.5 V ⊢ 0.5 V

V_H - Voltage reached when output driven high
V_L - Voltage reached when output driven low

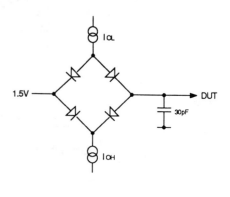

Tri-state delay measurement load.

GEC Plessey
PDSP16350
I/Q Splitter/NCO

FEATURES
- Direct digital synthesizer producing simultaneous sine and cosine values
- 16-bit phase and amplitude accuracy, giving spur levels down to -90 dB
- Synthesized outputs from dc to 10 MHz with accuracies better than 0.001 Hz
- Amplitude and phase modulation modes
- 84-pin PGA package

APPLICATIONS
- Numerically controlled oscillator (NCO)
- Quadrature signal generator
- FM, PM, or AM signal modulator
- Sweep oscillator
- High-density signal constellation applications with simultaneous amplitude and phase modulation
- VHF reference for UHF generators
- Signal demodulator

ABSOLUTE MAXIMUM RATINGS (Note 1)
Supply voltage V_{CC}	-0.5 V to 7.0 V
Input voltage V_{IN}	-0.5 to $V_{CC}+0.5$ V
Output voltage V_{OUT}	-0.5 V to $V_{CC}+0.5$ V
Clamp diode current per pin I_K (see note 2)	18 mA
Static discharge voltage (HMB)	500 V
Storage temperature T_S	-65°C to 150°C
Ambient temperature with power applied T_{AMB}	
Military	-55°C to $+125$°C
Industrial	-40°C to 85°C
Junction temperature	150°C
Package power dissipation	3500 mW
Thermal resistances	
Junction to case θ_{JC}	5 °C/W

NOTES
1. Exceeding these ratings may cause permanent damage. Functional operation under these conditions is not implied.
2. Maximum dissipation or 1 second should not be exceeded, only one output to be tested at any one time.
3. Exposure to absolute maximum ratings for extended periods may affect device reliability.
4. V_{CC}=Max, outputs unloaded, Clock freq=max.
5. CMOS levels are defined as
   $V_{IH} = V_{DD} - 0.5$ V
   $V_{IL} = +0.5$ V
6. Current is defined as positive into the device.
7. The θ_{JC} data assumes that heat is extracted from the top face of the package.

ORDERING INFORMATION
PDSP16350 B0 AC (Industrial - PGA package)
PDSP16350 A0 AC (Military - PGA package)
Call for availability of high-reliability parts and MIL-STD-883C screening.

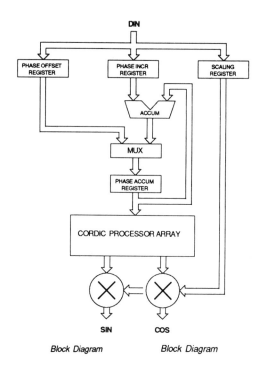

Block Diagram

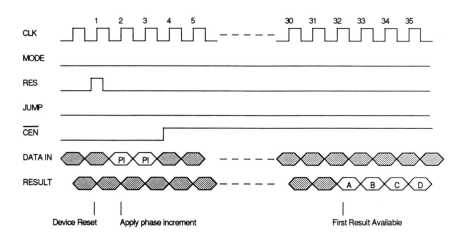

Fixed Frequency Timing Diagram

Plessey PDSP16350

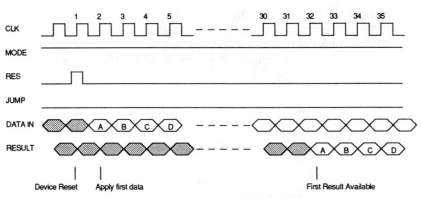

Amplitude Modulation (18bit frequency accuracy)

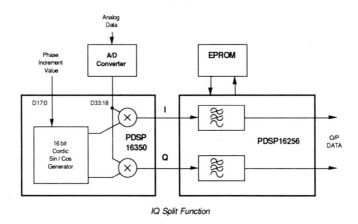

IQ Split Function

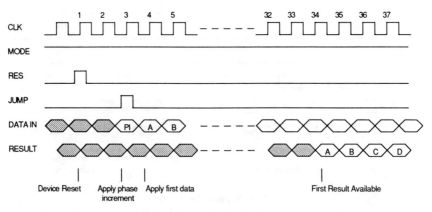

Amplitude Modulation (34bit frequency accuracy)

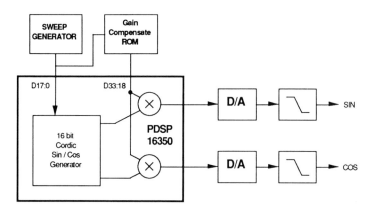

Quadrature Chirp Generator

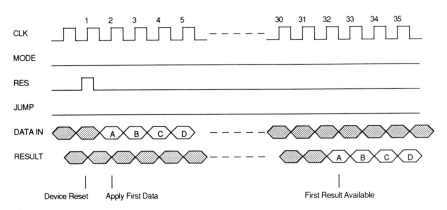

Frequency Modulation Timing Diagram

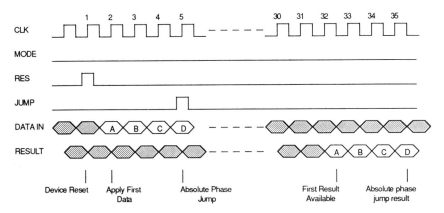

Phase Modulation Timing Diagram

Plessey PDSP16488
LOG

Pin Out Diagram Bottom View

GEC Plessey
PDSP16488
Single-Chip 2D Convolver with Integral Line Delays

FEATURES
- 8- or 16-bit pixels with rates up to 40 MHz
- Window sizes up to 8×8 with a single device
- Eight internal line delays
- Supports interlace and frame-to-frame operations
- Coefficients supplied from an EPROM or remote host
- Expandable in both X and Y for larger windows
- Gain-control and pixel-output manipulation
- 84-pin PGA package

ABSOLUTE MAXIMUM RATINGS (See Notes)
Supply voltage V_{CC} −0.5 V to 7.0 V

LOG

Input voltage V_{IN}	-0.5 V to $V_{CC}+0.5$ V
Output voltage V_{OUT}	-0.5 V to $V_{CC}+0.5$ V
Clamp diode current I_K (see note 2)	18 mA
Static discharge voltage (HMB)	500 V
Storage temperature T_S	$-65\,°C$ to $+150\,°C$
Max. junction temperature,	
Commercial	$100\,°C$
Industrial	$110\,°C$
Package power dissipation	3000 mW
Thermal resistance, junction-to-case θ_{JC}	$5\,°C/W$

NOTES ON MAXIMUM RATINGS

1. Exceeding these ratings might cause permanent damage. Functional operation under these conditions is not implied.
2. Maximum dissipation or 1 second should not be exceeded, only one output to be tested at any one time.
3. Exposure to absolute maximum ratings for extended periods might affect device reliability.
4. Current is defined as positive into the device.

ORDERING INFORMATION

PDSP16488 C0 AC (Commercial - PGA package)
PDSP16340 B0 AC (Industrial - PGA package)
Call for availability of high-reliability parts and MIL-STD-883C screening.

Data Size	Window Size Width X Depth	Max Pixel Rate	Line Delays
8	4	4	40MHz	4x1024
8	8	4	20MHz	4x1024
8	8	8	10MHz	8x512
16	4	4	20MHz	4x512
16	8	4	10MHz	4x512

Single Device Configurations

Max Pixel Rate	Pixel Size	Window size						
		3x3	5x5	7x7	9x9	11x11	15x15	23x23
10MHz	8	1	1	1	4	4	4	9
10MHz	16	1	2	2	-	-	-	-
20MHz	8	1	2	2	6	6	8	-
20MHz	16	1	4	4	-	-	-	-
40MHz	8	1	4*	4*	-	-	-	-
40MHz	16	2	-	-	-	-	-	-

* Maximum rate is limited to 30 MHz by line store expansion delays

Devices needed to implement typical window sizes

PIN NO AC PACKAGE	FUNCTION	PIN NO AC PACKAGE	FUNCTION	PIN NO AC PACKAGE	FUNCTION	PIN NO AC PACKAGE	FUNCTION
A1	L0	M3	X15	K12	$\overline{RES}$	B9	D7
B1	F1	N3	X14	K13	$\overline{CS0}$	A9	D8
C2	L1	M4	X13	J12	$\overline{CS1}$	B8	CLK
C1	L2	N4	SPARE	J13	$\overline{CS2}$	B7	SPARE
D2	L3	M5	$\overline{SINGLE}$	H12	$\overline{CS3}$	A7	D9
D1	SPARE	N5	X12	G12	$\overline{PROG}$	B6	D10
E2	L4	M6	X11	G13	$\overline{DS}$	A5	D11
E1	L5	M7	$\overline{MASTER}$	F12	$\overline{CE}$	B5	SPARE
F2	L6	N7	X10	E13	$R/\overline{W}$	A4	D12
G2	L7	M8	X9	E12	HRES	B4	D13
G1	IP7	N9	X8	D13	OV	A3	D14
H2	SPARE	M9	X7	D12	$\overline{PC1}$	B3	D15
J1	IP6	N10	X6	C13	BIN	A2	F0
J2	IP5	M10	X5	C12	$\overline{OEN}$	F1	VDD
K1	IP4	N11	X4	B13	D0	N6	VDD
K2	SPARE	M11	X3	A13	D1	F13	VDD
L1	IP3	N12	X2	A12	D2	A6	VDD
L2	IP2	N13	X1	B11	D3	H1	GND
M1	IP1	M13	X0	A11	D4	N8	GND
N1	IP0	L12	DELOP	B10	D5	H13	GND
N2	BYPASS	L13	$\overline{PC0}$	A10	D6	A8	GND

Pin out Table

242 Plessey PDSP16488

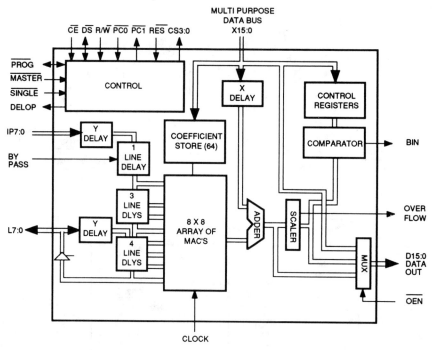

Typical, Stand Alone, Real Time System

Functional Block Diagram

LOG

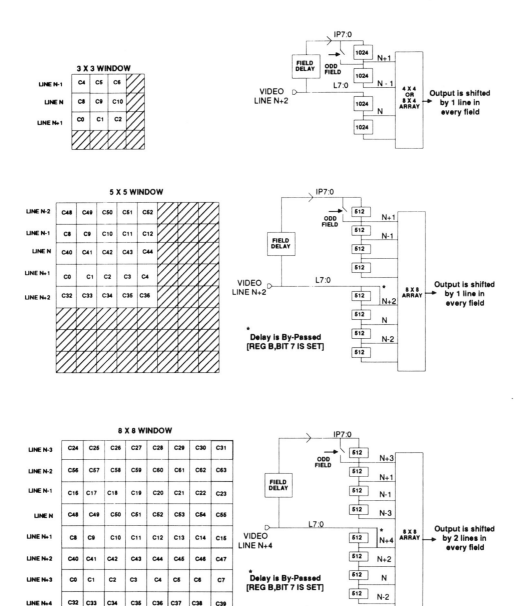

Line Delay Allocations in Single Device Interlaced Systems

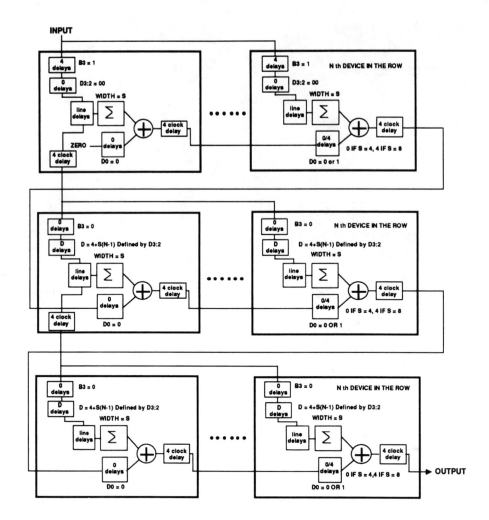

Multi-Device Delay Paths

LOG

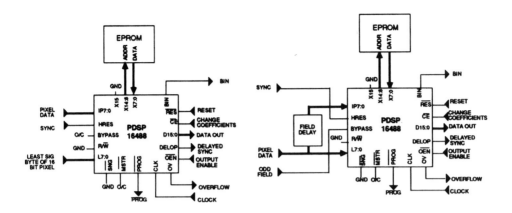

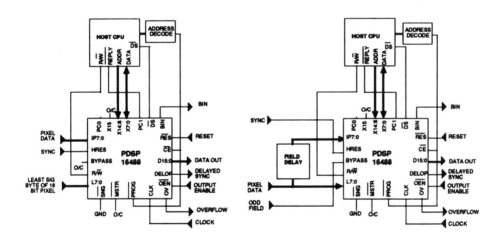

NON INTERLACED VIDEO   **INTERLACED VIDEO**

Single Device Systems

246 Plessey PDSP16488

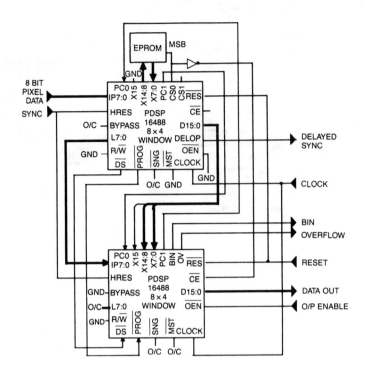

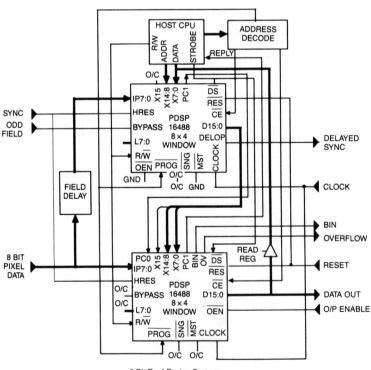

8 Bit Dual Device Systems

LOG

Plessey PDSP16488

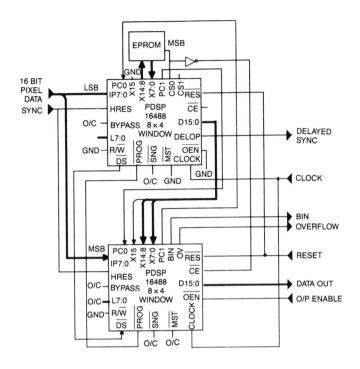

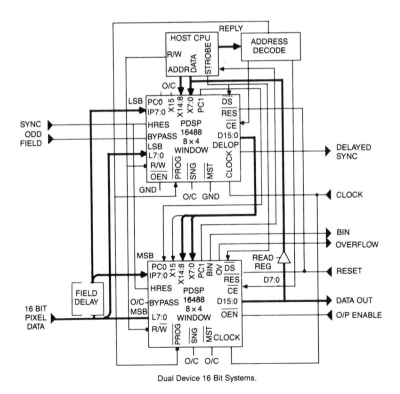

Dual Device 16 Bit Systems.

Plessey PDSP16488

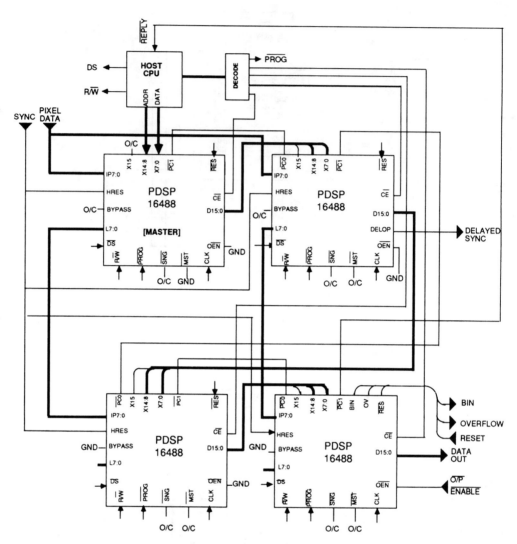

Four Device Non Interlaced System.

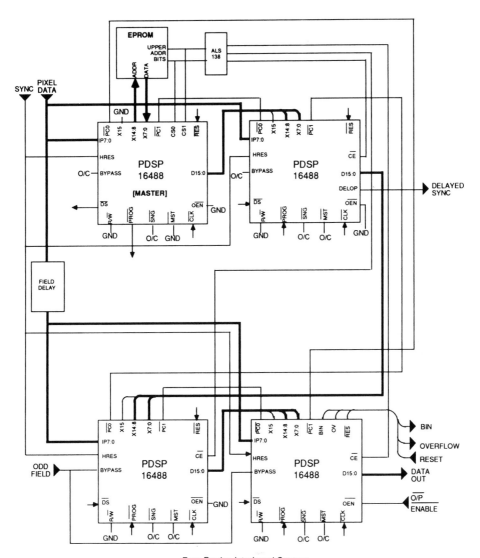

Four Device Interlaced System.

Plessey PDSP16488

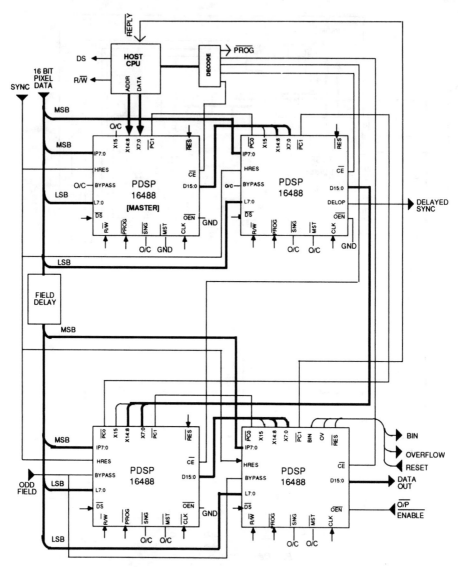

Four Device System with 16 Bit Pixels

LOG

Plessey PDSP16488

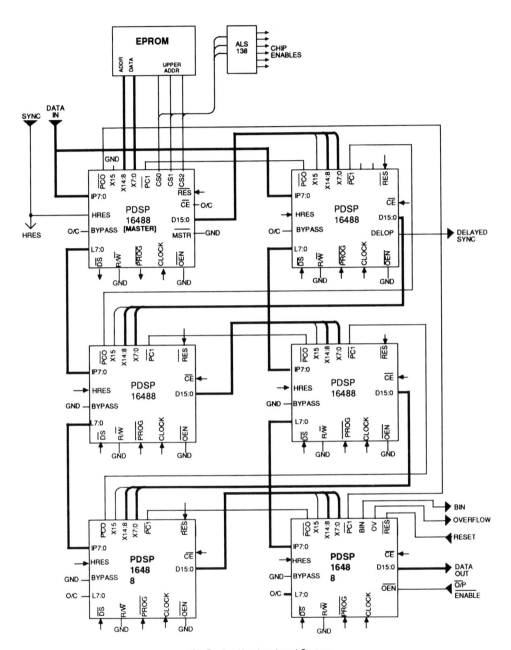

Six Device Non Interlaced System.

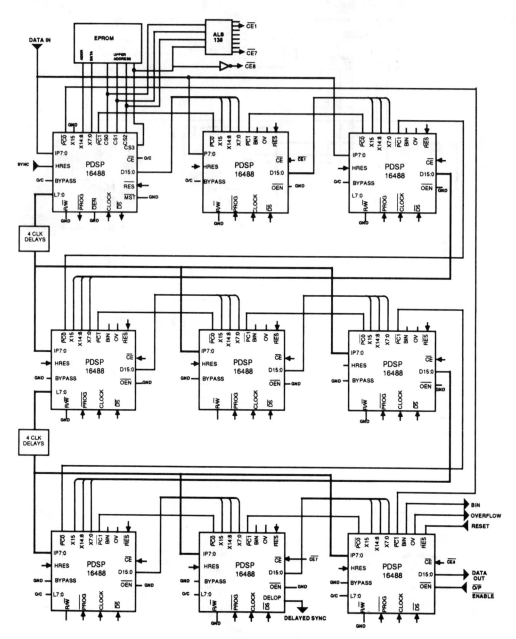

Nine Device Non Interlaced System.

The information included herein is believed to be accurate and reliable. However, LSI Computer Systems, Inc. assumes no responsibilities for inaccuracies, nor for any infringements of patent rights of others which may result from its use.

LSI
LS7220
Digital Lock Circuit

FEATURES:
- Stand-alone lock logic
- 5040, 4-digit combinations
- Out-of-sequence detection
- Direct LED and lock-relay drive
- Chip enable (for automotive applications)
- Externally controlled convenience delay
- Save memory (for valet parking, etc.)
- Internal pull-down resistors on all inputs
- High noise immunity
- Low current consumption (40 μA max.@12 Vdc)
- Single power supply operation (+5 V to +18 V)
- Momentary or static lock-control output
- All inputs protected

MAXIMUM RATINGS: (voltages referenced to V_{DD})

RATING	SYMBOL	VALUE	UNITS
dc supply voltage	V_{SS}	+5 to +18	Vdc
Operating temperature range	T_A	−25 to +70	°C
Storage temperature range	T_{STG}	−65 to +150	°C

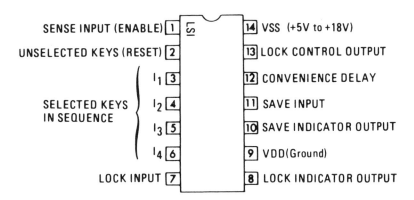

TOP VIEW
STANDARD 14 PIN DIP

254 LSI LS7220

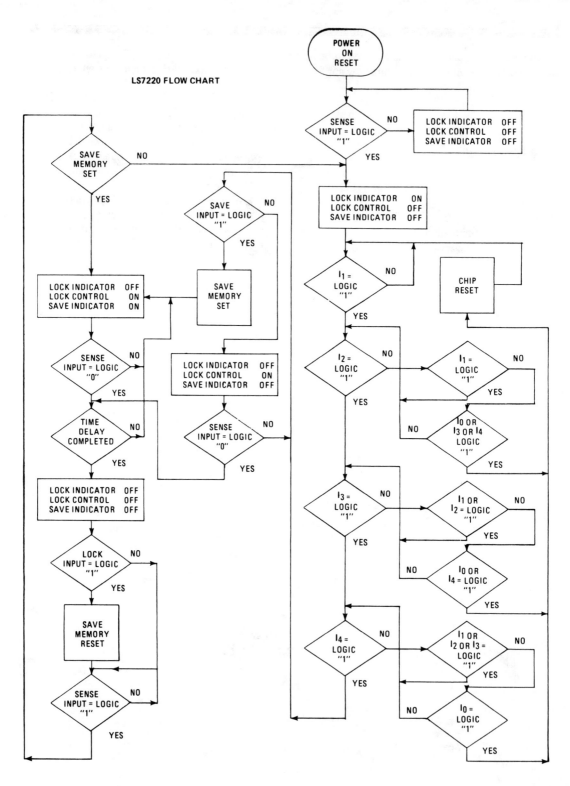

LS7220 FLOW CHART

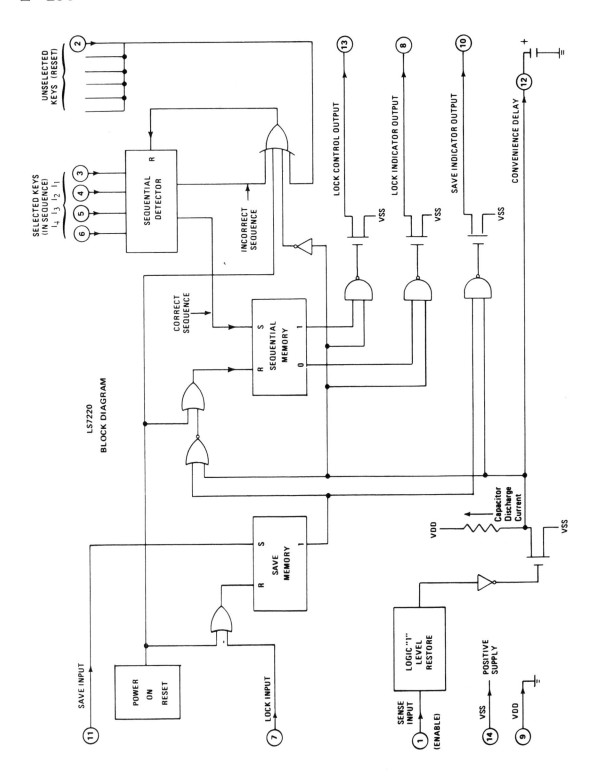

256 LSI LS7220

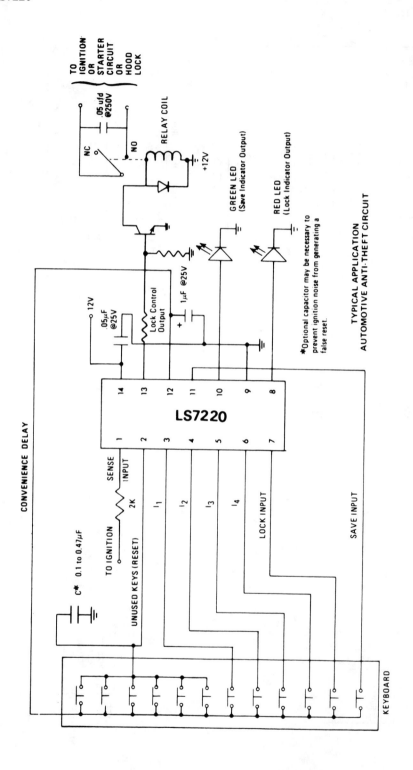

TYPICAL APPLICATION
AUTOMOTIVE ANTI-THEFT CIRCUIT

LOG

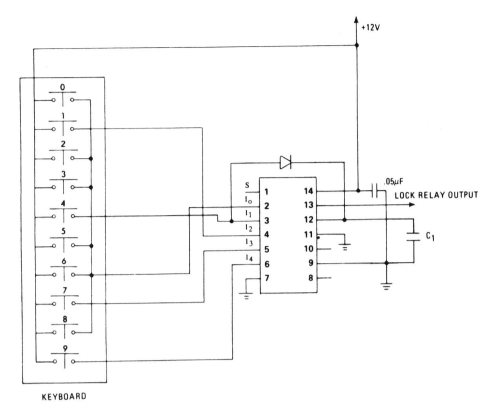

Typical 4 key code having MOMENTARY output.
The size of C_1 determines the length of entry time
(See Figure 2). The specific code shown is 4179.

LSI
LS7222
Keyboard-Programmable Digital-Lock Circuit

FEATURES
- Stand-alone lock logic
- 38416, 4-digit combinations
- 3 different user-programmable codes
- Momentary and static lock-control outputs
- Internal keyboard-debounce circuit
- Tamper-detection output
- Circuit status outputs
- Low current consumption (30 μA max.@12 Vdc)
- Single power supply operation (+4 to +15 Vdc)
- All inputs protected
- High noise immunity

MAXIMUM RATINGS (voltages reference to V_{SS})

RATING	SYMBOL	VALUE	UNIT
dc supply voltage	V_{DD}	+4 to +18	Vdc
Operating temperature range	T_A	−25 to +70	°C
Storage temperature range	T_{STG}	−65 to +150	°C

258 LSI LS7222

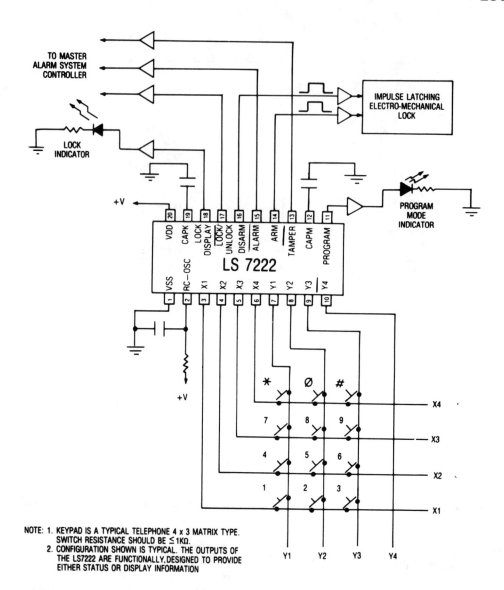

NOTE: 1. KEYPAD IS A TYPICAL TELEPHONE 4 x 3 MATRIX TYPE. SWITCH RESISTANCE SHOULD BE ≤1KΩ.
2. CONFIGURATION SHOWN IS TYPICAL. THE OUTPUTS OF THE LS7222 ARE FUNCTIONALLY, DESIGNED TO PROVIDE EITHER STATUS OR DISPLAY INFORMATION

CONNECTION DIAGRAM—TOP VIEW
STANDARD 20 PIN PLASTIC DIP

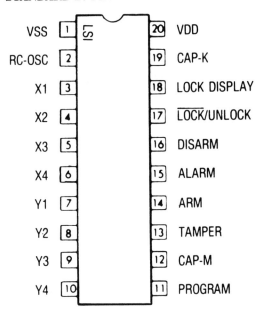

LSI
LS7223
Keyboard-Programmable Digital-Lock Circuit

FEATURES
- Stand-alone lock logic
- 38416, 4-digit combinations
- 3 different user-programmable codes
- Momentary and static lock-control outputs
- Internal keyboard-debounce circuit
- Tamper-detection output
- Circuit status outputs
- Low current consumption (30 μA max.@12 Vdc)
- Single power-supply operation (+4 to +15 Vdc)
- All inputs protected
- High noise immunity

LSI LS7223

Maximum Ratings: (Voltages references to VSS)

RATING	SYMBOL	VALUE	UNIT
DC supply voltage	VDD	+4 to +18	Vdc
Operating temperature range	TA	-25 to +70	°C
Storage temperature range	TSTG	-65 to +150	°C

DC Electrical Characteristics:
(VSS = 0V, VDD = +4 to +15V, -25°C ≤ TA ≤ +70°C unless otherwise specified)

PARAMETER	CONDITIONS	VDD	MIN	TYP	MAX	UNITS
Output source current Lock Display, Momentary $\overline{LOCK/UNLOCK\ 2}$ Alarm, $\overline{LOCK}$/UNLOCK1 Program Mode Outputs	Logic "1" Output $V_{OUT} \geq V_{DD} - 2V$	5VDC 12VDC 15VDC	1.50 5.60 7.25	2.50 8.25 10.7	— — —	ma ma ma
Output Sink Current Lock Display, Momentary, Alarm, $\overline{LOCK}$/UNLOCK1, $\overline{LOCK/UNLOCK2}$ Program Mode Outputs	Logic "0" Output $V_{OUT} \leq V_{SS} + .4V$	5VDC 12VDC 15VDC	.400 1.20 1.50	.60 1.70 2.25	— — —	ma ma ma
Output Source Current Tamper Output	Logic "1" Output $V_{OUT} \geq V_{DD} - 2V$	5VDC 12VDC 15VDC	.25 .90 1.10	.400 1.30 1.70	— — —	ma ma ma
Output Sink Current Tamper Output	Logic "0" Output $V_{OUT} \leq V_{SS} + .4V$	5VDC 12VDC 15VDC	.060 .200 .250	.100 .290 .370	— — —	ma ma ma
Input Level Detection All Inputs	V_{IH} = Logic "1"	5VDC 12VDC 15VDC	3.5 8.0 10.0	— — —	VDD VDD VDD	Vdc Vdc Vdc
	V_{IL} = Logic "0"	5VDC 12VDC 15VDC	VSS VSS VSS	— — —	1.60 4.0 5.0	Vdc Vdc Vdc

CONNECTION DIAGRAM—TOP VIEW
STANDARD 20 PIN PLASTIC DIP

```
VSS      [1]        [20] VDD
RC-OSC   [2]        [19] CAP-K
X1       [3]        [18] LOCK DISPLAY
X2       [4]        [17] LOCK/UNLOCK 1
X3       [5]        [16] LOCK/UNLOCK 2
X4       [6]        [15] ALARM
Y1       [7]        [14] MOM
Y2       [8]        [13] TAMPER
Y3       [9]        [12] CAP-M
Y4       [10]       [11] PROGRAM
```

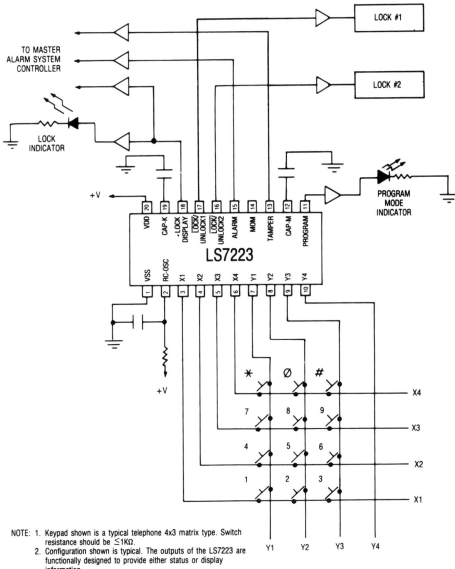

NOTE: 1. Keypad shown is a typical telephone 4x3 matrix type. Switch resistance should be ≤1KΩ.
2. Configuration shown is typical. The outputs of the LS7223 are functionally designed to provide either status or display information.

LSI
LS7222/LS7223
Program-Mode Lockout

The accompanying figure indicates how the addition of one CD4001 CMOS NOR package, two resistors, and one switch can be used to prevent entering the Program Mode. With the switch in the open position, the program mode can be entered as usual. In the closed position, the negative pulse output of pin 7 is prevented from being applied to pin 6, which inhibits the * selection from entering the circuit. Under these conditions, the circuit cannot be reprogrammed, but will otherwise function normally. Applications include any situation where the number of persons authorized to reprogram the code is far less than the number of persons who will be using the code.

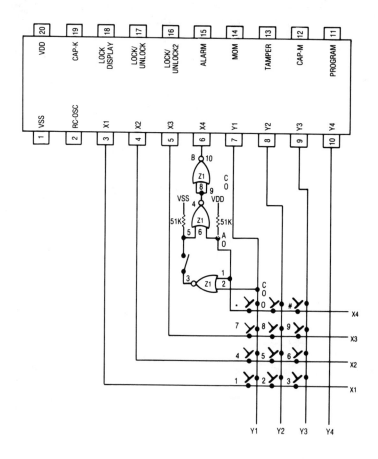

LSI
LS7225/LS7226
Digital-Lock Circuit with Tamper Output

FEATURES
- Stand-alone lock logic
- 5040, 4-digit combination with a 10-number keyboard
- Out-of-sequence detection
- Tamper output, sequence-enable input
- Direct LED and lock relay drive
- Externally controlled combination delay
- Internal pull-down resistors on all inputs
- High noise immunity
- Low current consumption (40 μA max. @12 Vdc)
- Single power-supply operation (+4 V to 18 V)
- Momentary or static lock-control output
- Auxiliary delay circuitry included

MAXIMUM RATINGS (voltages referenced to V_{DD})

RATING	SYMBOL	VALUE	UNITS
dc supply voltage	V_{SS}	+4 to +18	Vdc
Operating temperature range	T_A	−25 to +70	°C
Storage temperature range	T_{STG}	−65 to +150	°C

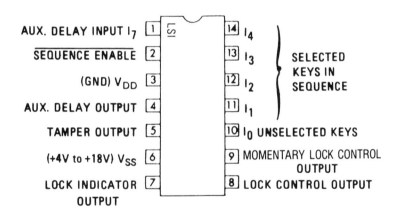

TOP VIEW
STANDARD 14 PIN DIP

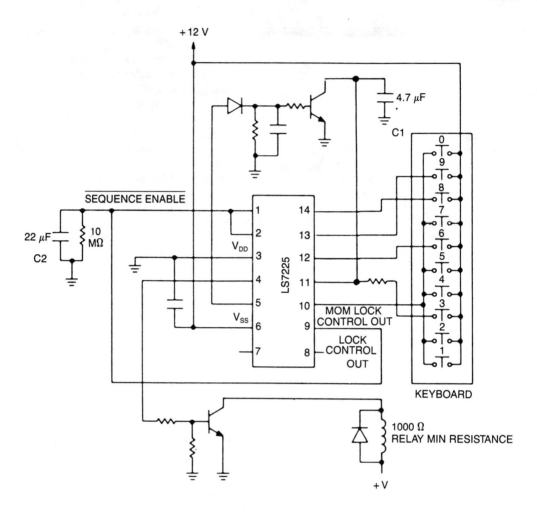

Typical application for independent control of combination (input) time and "UNLOCK" time.

C-1 determines input time.
C-2 determines "UNLOCK" time.

Note: With this configuration one tamper pulse is transmitted at the start of "UNLOCK" time.

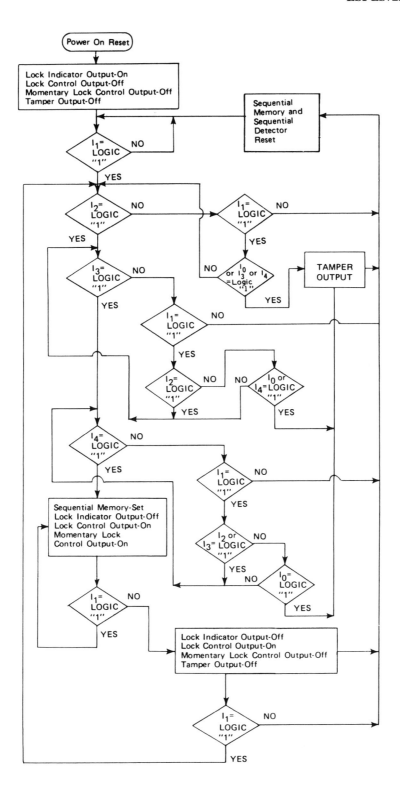

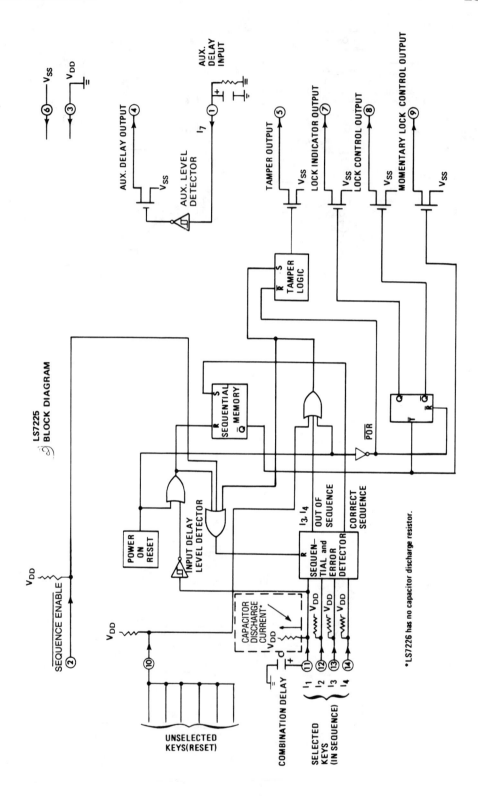

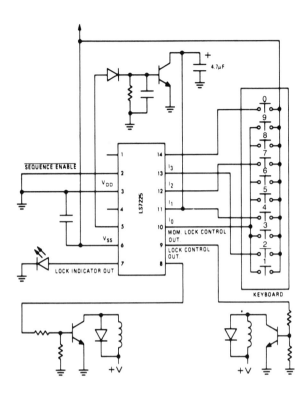

A typical circuit is shown in the schematic diagram. When input I1 (pin 11) goes high, the circuit is ready to accept the unlocking input sequence at I2, I3 and I4 (pins 12, 13, and 14 respectively). If the keys associated with these inputs are depressed exactly in sequence of I1, I2, I3 and I4, the lock control output (pin 8) will become on, the momentary lock control output (pin 9) will be on until input I1 (pin 11) becomes low. The ON state of the lock control will be indicated by the OFF condition of the lock indicator output (pin 7) which will render the LED off (an indication of unlock condition). If the keys are depressed in any sequence other than as described above, the internal "sequential detector" will be reset and the entire sequence must be repeated. The lock control output is turned off by repeating the input sequence. The momentary lock control output goes high each time the correct sequence is entered. The specific code shown is 4720.

TYPICAL APPLICATION OF LS7225 IN MACHINE OR AREA ACCESS

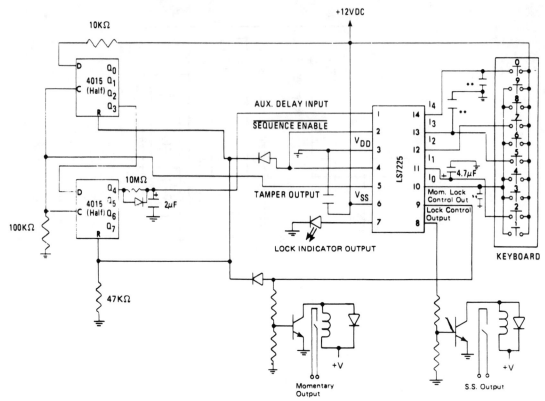

TYPICAL TAMPER LOCK DISABLE APPLICATION

NON COMBINATION LOCK / COMBINATION UNLOCK CIRCUITS
DUAL COIL LATCHING RELAY LOCK

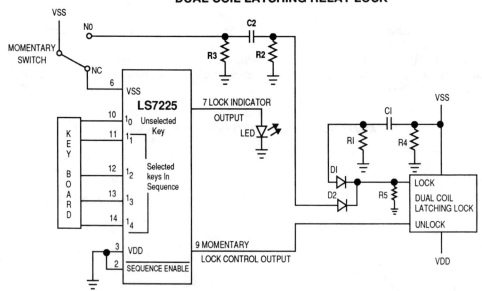

(circuit bottom of page 268)

This circuit depicts a method for allowing a single momentary pushbutton to lock the lock, and only the correct keyboard sequential sequence can unlock the lock. Upon application of power, the lock indicator light comes on and a locking pulse is generated by the R1/C1 circuit, which causes the lock to latch locked. Upon application of the correct key-depression sequence, a momentary lock-control output pulse appears on Pin 9 of the IC and unlocks the lock. The indicator light also goes out. By activating the momentary switch, a locking pulse generated by the R2/C2 network causes the lock to latch lock and also removes power from the LS7225. When the momentary switch is allowed to return to its normal position, the LS7225 will power-up again and the lock indicator will light, indicating that the lock is locked. R3 is needed to discharge C2 when the momentary lock is in the normal position. R4 is used to discharge C1 when power is removed. Diodes D1, D2, and resistor R5 are used to logically combine the power-up locking pulse and the pushbutton locking pulse together.

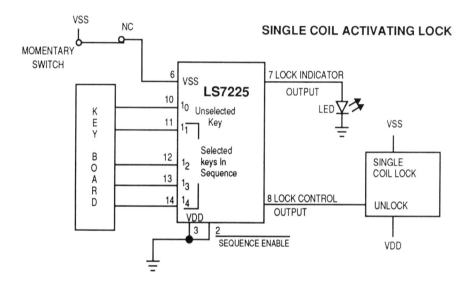

This circuit powers-up with the indicator lamp on, indicating that the lock is in the locked state. Upon application of the correct key-depression sequence, the lock unlocks and the indicator lamp goes off, indicating the lock is in the unlocked state. When the normally closed momentary switch is opened, the power is removed from the LS7225 and the lock will become locked. When the momentary returns to its normal position, the LS7225 will power up again and the indicator lamp will come on, indicating the lock is in its locked state.

LSI
LS7228/LS7229
Address Decoder/Two Pushbutton Digital Lock

FEATURES
- Stand-alone lock logic
- 9-bit code determined by 9 parallel inputs
- Two options of code input available:
  LS7228—dual-train pulsed input
  LS7229—two momentary switches
- Out-of-sequence disabling circuit
- Current source lock-control output
- Externally controlled delay to set maximum interpulse time
- Single power supply operation (2.5 V to 15.0 V)
- Low standby current (15 µA maximum)
- 16-pin dual-in-line plastic package
- Cascadable

MAXIMUM RATINGS

PARAMETER	SYMBOL	VALUE	UNITS
Storage temperature	T_{STG}	−65 to +150	°C
Operating temperature	T_A	−25 to +70	°C
Voltage (any pin to V_{SS})	V_{MAX}	−30 to +0.5	V

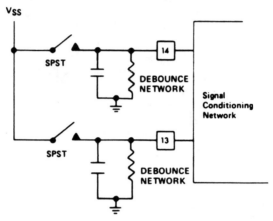

LS7228 may be used with manual single pole single throw switches if debounce filters are added to inputs 13 & 14. (As shown)

```
(leading bit) CODE 1  [1]        [16] VDD (GND)
              CODE 2  [2]        [15] COM
              CODE 3  [3]        [14] ZERO'S
              CODE 4  [4]        [13] ONE'S
              CODE 5  [5]        [12] EXTERNAL R/C
              CODE 6  [6]        [11] OUTPUT
              CODE 7  [7]        [10] VSS (+2.5V to +15V)
              CODE 8  [8]        [9]  CODE 9 (END BIT)
```

TOP VIEW
Standard 16 pin DIP

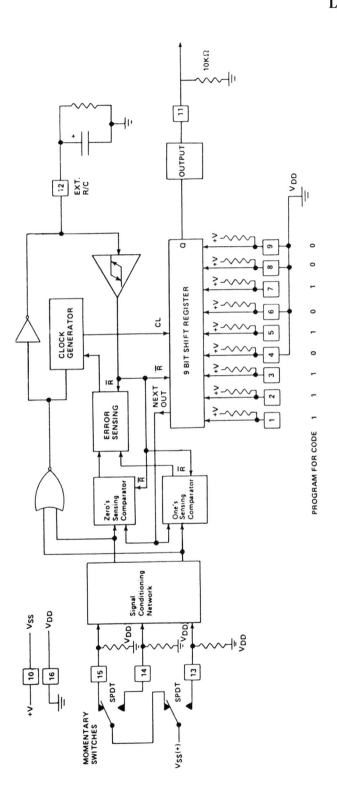

LS7229 BLOCK DIAGRAM

272 LSI LS7228/7229

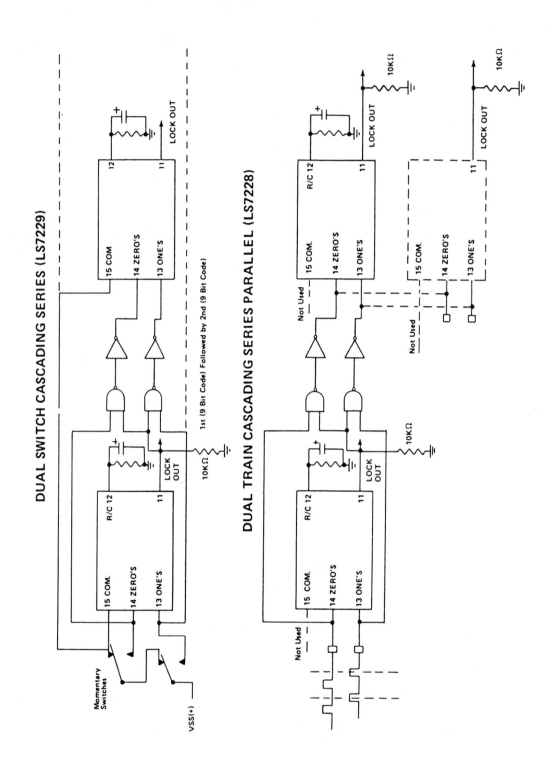

Raytheon
RC4805
Precision High-Speed Latching Comparator

FEATURES
- 22-nS propagation delay
- Low offset voltage: 100 µA
- Low offset current: 15 nA
- TTL-compatible latch
- TTL output

ABSOLUTE MAXIMUM RATINGS

Supply voltage	+5.5 V/−16.5 V
Differential input voltage	3 V
Internal power dissipation	500 mW
Input voltage	±4 V
Storage temperature range	−65 °C to +150 °C
Operating temperature range	
RM4805	−55 °C to +125 °C
RC4805	0 °C to +70 °C
Lead soldering temperature (60 sec)	+300 °C

ORDERING INFORMATION

Part Number	Package	Operating Temperature Range
RC4805EN	N	0°C to +70°C
RC4805N	N	0°C to +70°C
RM4805D	D	-55°C to +125°C
RM4805D/883B	D	-55°C to +125°C
RM4805AD	D	-55°C to +125°C
RM4805AD/883B	D	-55°C to +125°C
RM4805T	T	-55°C to +125°C
RM4805T/883B	T	-55°C to +125°C
RM4805AT	T	-55°C to +125°C
RM4805AT/883B	T	-55°C to +125°C

Notes:
/883B suffix denotes Mil-Std-883, Level B processing
N = 8-lead plastic DIP
D = 8 lead ceramic DIP
T = 8-lead metal can (TO-99)
Contact a Raytheon sales office or representative for ordering information on special package/temperature range combinations.

CONNECTION INFORMATION

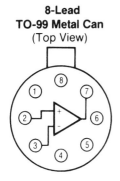

8-Lead TO-99 Metal Can (Top View)

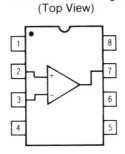

8-Lead Dual In-Line Package (Top View)

Pin	Function
1	Ground
2	+ Input
3	-Input
4	$-V_S$
5	NC
6	Latch Enable
7	V_{OUT}
8	$+V_S$

TYPICAL APPLICATIONS

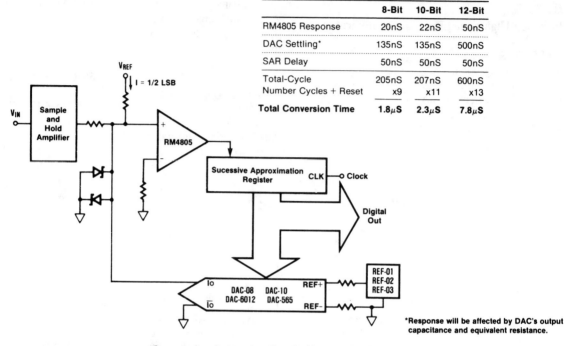

	Conversion Time		
	8-Bit	10-Bit	12-Bit
RM4805 Response	20nS	22nS	50nS
DAC Settling*	135nS	135nS	500nS
SAR Delay	50nS	50nS	50nS
Total-Cycle	205nS	207nS	600nS
Number Cycles + Reset	x9	x11	x13
Total Conversion Time	**1.8µS**	**2.3µS**	**7.8µS**

*Response will be affected by DAC's output capacitance and equivalent resistance.

Successive Approximation 8, 10, or 12-bit Resolution

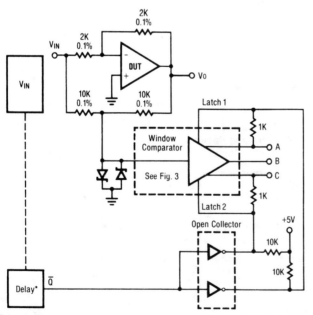

Op Amp Settling Time Tester

*Delay should equal the settling time specification minus 30nS minus appropriate guard band

LOG

Raytheon RC4805

The settling time tester uses the precision latching window comparator to automate op amp settling time testing. If the DUT is not settled by the end of the time delay, the A output is latched low.

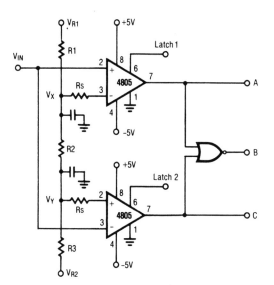

V_{IN} ($V_X > V_Y$)*	A	B	C
$V_{IN} > V_X$	1	0	0
$V_X > V_{IN} > V_Y$	0	1	0
$V_Y > V_{IN}$	0	0	1

*Both latches low

Precision Latching Window Comparator (Detail)

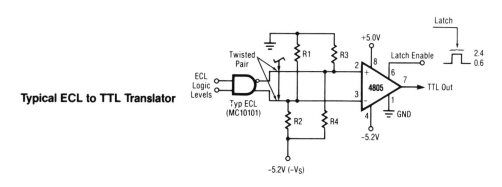

Typical ECL to TTL Translator

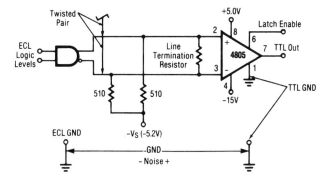

Notes:
1. Common mode range of 4805 is –8.0V to +2.0V.
2. The 4805 can stand –3.0V, +5.0V of GND noise from the ECL GND to the TTL GND.

ECL to TTL Translator With Extended Common Mode Range

276 Raytheon RC4805

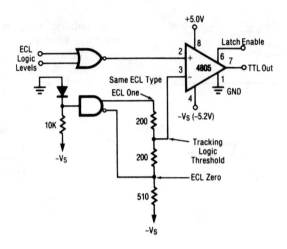

Single-Ended ECL to TTL Translator With Tracking ECL Reference

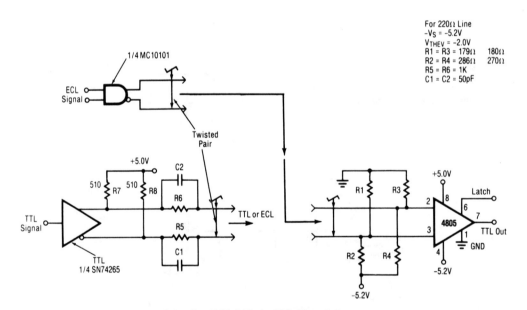

Adaptive ECL-TTL to TTL Translator

Raytheon LM139/139A, 339/339A
Single-Supply Quad Comparators

FEATURES

- Input common-mode voltage range includes ground
- Wide single-supply voltage range: 2 V to 36 V
- Output compatible with TTL, DTL, ECL, MOS, and CMOS logic systems
- Very low supply current drain (0.8 mA), independent of supply voltage

CONNECTION INFORMATION

14-Lead Dual In-Line Package (Top View)

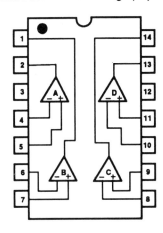

Pin	Function	Pin	Function
1	Output B	8	-Input C
2	Output A	9	+Input C
3	$+V_S$	10	-Input D
4	-Input A	11	+Input D
5	+Input A	12	Ground
6	-Input B	13	Output D
7	+Input B	14	Output C

ABSOLUTE MAXIMUM RATINGS

Supply voltage, $+V_S$	+36 V or ±18 V
Differential input voltage	36 V
Input voltage range	-0.3 to $+36$ V[2]
Output short circuit to ground[1]	Continuous
Input current ($V_{IN} < -0.3$ V)[2]	50 mA
Operating temperature range	
LM139	$-55\,°C$ to $+125\,°C$
LM339	$0\,°C$ to $+70\,°C$
Storage temperature range	$-65\,°C$ to $+150\,°C$
Lead soldering temperature (SO-14; 10 sec)	$+260\,°C$
Lead soldering temperature (DIP; 60 sec)	$+300\,°C$

ORDERING INFORMATION

Part Number	Package	Operating Temperature Range
LM339M	M	0°C to +70°C
LM339N	N	0°C to +70°C
LM339AM	M	0°C to +70°C
LM339AN	N	0°C to +70°C
LM139D	D	-55°C to +125°C
LM139D/883B	D	-55°C to +125°C
LM139AD	D	-55°C to +125°C
LM139AD/883B	D	-55°C to +125°C

Notes:
/883B suffix denotes Mil-Std-883, Level B processing
N = 14-lead plastic DIP
D = 14 lead ceramic DIP
M = 14-lead plastic SOIC
Contact a Raytheon sales office or representative for ordering information on special package/temperature range combinations.

TYPICAL APPLICATIONS—SINGLE SUPPLY ($+V_S = +15V$)

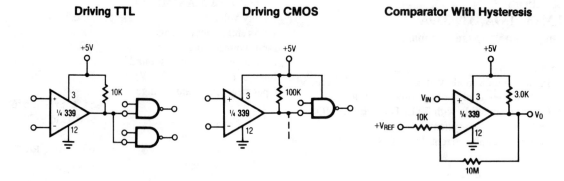

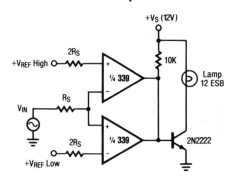

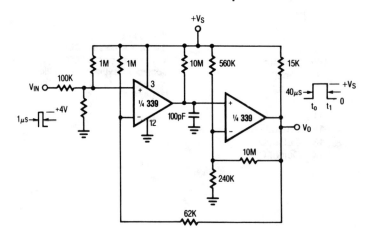

Zero Crossing Detector (Single Power Supply)

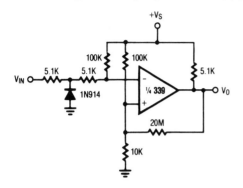

Low Frequency Op Amp

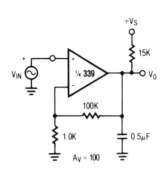

TTL to MOS Logic Converter

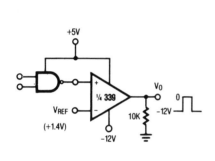

Pulse Generator

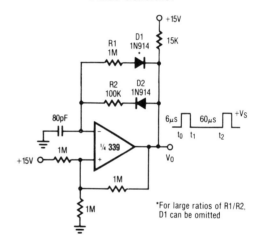

*For large ratios of R1/R2, D1 can be omitted

Zero Crossing Detector

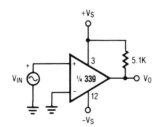

Comparator With a Negative Reference

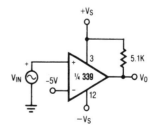

Raytheon LP165/365
Micropower Programmable Quad Comparator

FEATURES
- Single programming resistor tailors power, input currents, speed, and output current characteristics
- Uncommitted emitters allow logic interface flexibility
- Wide supply voltage range or dual supplies (4 V to 36 V, or ± 2 V to ± 18 V)
- Input common-mode range includes ground in single-supply applications
- Low power consumption (10 μW per comparator at $V_S = 5\ V$, $I_{SET} = 0.5\ \mu A$)

ABSOLUTE MAXIMUM RATINGS

Supply voltage	36 V or ± 18 V
Differential input voltage	36 V
Input voltage	-0.3 V to $+36$ V (single supply)*
Output short circuit duration to V_E	Indefinite**
Storage temperature range	$-65\,°C$ to $+150\,°C$
Operating temperature range	
LP165	$-55\,°C$ to $+125\,°C$
LP365/LP365A	$0\,°C$ to $+70\,°C$
Lead soldering temperature (60 sec)	$+300\,°C$

*The input voltage is not allowed to go 0.3 V above $+V_S$ or -0.3 V below $-V_S$ as this will turn on a parasitic transistor causing large currents to flow through the device.

**Short circuits from the output to $+V_S$ might cause excessive heating and eventual destruction. The current in the output leads and the V_E lead should not be allowed to exceed 30 mA. The output should not be shorted to $-V_S$ if $V_E \geq (-V_S) + 7$ V.

ORDERING INFORMATION

Part Number	Package	Operating Temperature Range
LP365N	N	0°C to +70°C
LP365AN	N	0°C to +70°C
LP165D	D	-55°C to +125°C
LP165D/883B	D	-55°C to +125°C

Notes:
/883B suffix denotes Mil-Std-883, Level B processing
N = 16-lead plastic DIP
D = 16-lead ceramic DIP
Contact a Raytheon sales office or representative for ordering information on special package/temperature range combinations.

CONNECTION INFORMATION
16-Lead DIP (Top View)

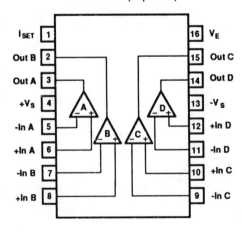

Pin	Function
1	Set Current
2	Output B
3	Output A
4	$+V_S$
5	-Input A
6	+Input A
7	-Input B
8	+Input B
9	-Input C
10	+Input D
11	-Input D
12	+Input D
13	$-V_S$
14	Output D
15	Output C
16	Emitter Common

LOG

TYPICAL APPLICATIONS

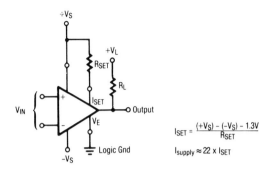

$$I_{SET} = \frac{(+V_S) - (-V_S) - 1.3V}{R_{SET}}$$

$I_{supply} \approx 22 \times I_{SET}$

Split Supply With Logic Output

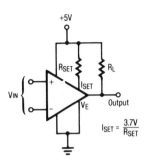

$$I_{SET} = \frac{3.7V}{R_{SET}}$$

TTL Supply — TTL Output

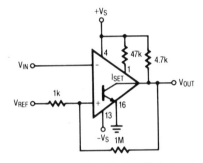

It is a good practice to add a few millivolts of positive feedback to prevent oscillation when the input voltage is near the threshold.

Ordinary Hysteresis

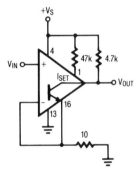

Positive feedback from the emitter can also prevent oscillations when V_{IN} is near the threshold. Can only be used with one section of four.

Hysteresis From Emitter

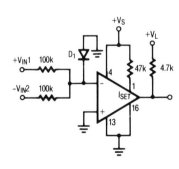

D_1 = Small signal Schottky or low V_D equivalent

Opposite Polarity Magnitude Comparator (Single Supply)

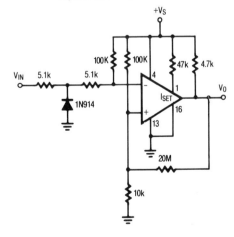

Zero Crossing Detector (Single Supply)

282   Raytheon LP165/365   LOG

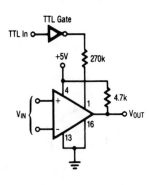

Chip Disable (TTL)

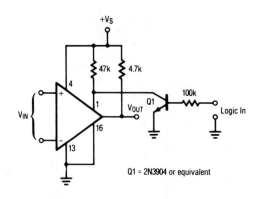

Chip Disable (Transistor)

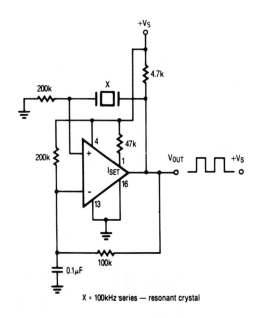

Crystal Controlled Oscillator (Single Supply)

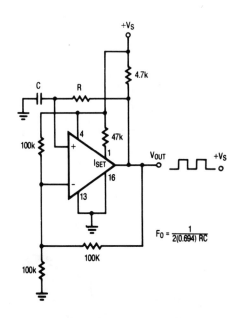

Squarewave Oscillator

$$F_O = \frac{1}{2(0.694)RC}$$

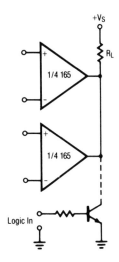

Wired-OR Outputs

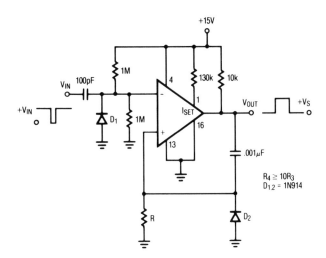

One Shot Multivibrator

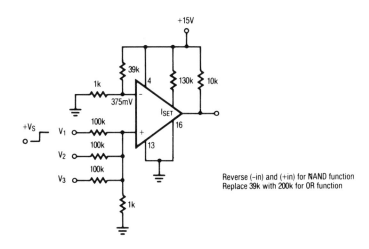

3 Input AND Gate

Raytheon LP165/365

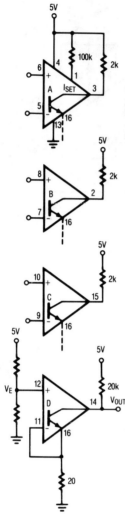

If you choose V_E = 25mV, 75mV, or 125mV, then V_{OUT} will fall if 1/3, 2/3 or all of the other three outputs are low.

Voting Comparator

Comparators B, C, and D do not respond until activated by the signal applied to comparator A.

Level Sensitive Strobe

CHAPTER 6

MICROCOMPUTER PERIPHERALS

Analog Devices
1B31
Wide-Bandwidth Strain-Gage Signal Conditioner

FEATURES
- Low cost
- Complete signal-conditioning solution
- Small package: 28-pin double DIP
- Internal half-bridge completion resistors
- Remote sensing
- High accuracy
  Low drift: ±0.25 μV/°C
  Low noise: 0.3 μV p-p
- Low nonlinearity: ±0.005% max.
- High CMR: 140 dB min (60 Hz, G=1000 V/V)
- Programmable bridge excitation: +4 V to +15 V
- Adjustable low-pass filter: f_c = 10 Hz to 20 kHz

APPLICATIONS
- Measurement of: strain, torque, force, pressure
- Instrumentation: indicators, recorders, controllers
- Data acquisition systems
- Microcomputer analog I/O

DESIGN FEATURES AND USER BENEFITS

Ease of use Direct transducer interface with minimum external parts required, convenient offset and span adjustment capability.

Half-Bridge Completion Matched resistor pair tracking to ±5 ppm/°C max. for half-bridge strain-gage applications.

Remote Sensing Voltage drops across the excitation lead-wires are compensated by the regulated supply, making 6-wire load-cell interfacing straightforward.

Programmable transducer Excitation Excitation source preset for +10 Vdc operation without external components. User-programmable from a +4 V to +15 Vdc to optimize transducer performance.

Adjustable Low-Pass Filter The two-pole active filter (f_c = 1 kHz) reduces noise bandwidth and aliasing errors with provisions for external adjustment of cutoff frequency (10 Hz to 20 kHz).

Analog Devices 1B31

SPECIFICATIONS (typical @ +25°C and $V_S = \pm 15V$ unless otherwise noted)

Model	1B31AN	1B31SD†
GAIN[1]		
Gain Range	2 to 5000V/V	*
Gain Equation	$R_G = \dfrac{80k\Omega}{G-2}$	*
Gain Equation Accuracy, $G \leq 1000V/V$	±3%	*
Gain Temperature Coefficient[2]	±15ppm/°C (±25ppm/°C max)	*
Nonlinearity	±0.005% max	*
OFFSET VOLTAGES[1]		
Total Offset Voltage, Referred to Input		
Initial, @ +25°C (Adjustable to Zero)		
G = 2V/V	±2mV (±10mV max)	*
G = 1000V/V	±50µV (±200µV max)	*
Warm-Up Drift, 5 min., G = 1000V/V	Within ±1µV of final value	*
vs. Temperature		
G = 2V/V	±25µV/°C (±50µV/°C max)	*
G = 1000V/V	±0.25µV/°C (±2µV/°C max)	*
At Other Gains	$\left(\pm 2 \pm \dfrac{100}{G}\right)\mu V/°C$	*
vs. Supply		
G = 2V/V	±50µV/V	*
G = 1000V/V	±0.5µV/V	*
Output Offset Adjust Range	±10V min	*
INPUT BIAS CURRENT		
Initial @25°C	±10nA (±50nA max)	*
vs. Temperature	±25pA/°C	*
INPUT DIFFERENCE CURRENT		
Initial @ +25°C	±5nA (±20nA max)	*
vs. Temperature	±10pA/°C	*
INPUT IMPEDANCE		
Differential	1GΩ∥4pF	*
Common Mode	1GΩ∥4pF	*
INPUT VOLTAGE RANGE		
Linear Differential Input (V_D)	±5V	*
Maximum CMV Input	$\pm\left(12 - \dfrac{G \times V_D}{4}\right)V$ max	*
CMR, 1kΩ Source Imbalance		
G = 2V/V, dc to 60Hz	86dB	*
G = 100V/V to 5000V/V		
1kHz Bandwidth[3]		
@ dc to 60Hz	110dB min	*
10Hz Bandwidth[4]		
@ dc	110dB min	*
@ 60Hz	140dB min	*
INPUT NOISE		
Voltage, G = 1000V/V		
0.1Hz to 10Hz	0.3µV p-p	*
10Hz to 100Hz	1µV p-p	*
Current, G = 1000V/V		
0.1Hz to 10Hz	60pA p-p	*
10Hz to 100Hz	100pA p-p	*
RATED OUTPUT[1]		
Voltage, 2kΩ Load, min	±10V	*
Current	±5mA	*
Impedance, dc to 2Hz, G = 2V/V to 1000V/V	0.5Ω	*
Load Capacitance	1000pF	*
Output Short-Circuit Duration	Indefinite	*
DYNAMIC RESPONSE[1]		
Small Signal Bandwidth −3dB, G = 2V/V to 1000V/V	1kHz	*
Slew Rate	0.05V/µs	*
Full Power	350Hz	*
Settling Time, G = 2V/V to 1000V/V, ±10V Output, Step to ±0.1%	2ms	*
LOW PASS FILTER		
Number of Poles	2	*
Gain (Pass Band)	−2V/V	*
Cutoff Frequency (−3dB Point)	1kHz	*
Roll-Off	40dB/decade	*

OUTLINE DIMENSIONS

Dimensions shown in inches and (mm).

Plastic Package (N)

Ceramic Package (D)

PIN DESIGNATIONS

PIN	FUNCTION	PIN	FUNCTION
1	+INPUT	15	−V_S
2	−INPUT	16	COMMON
3	GAIN	17	+V_S
4	GAIN	18	+V_S REGULATOR
8	V_{OUT} (UNFILTERED)	19	REF OUT
9	INPUT OFFSET ADJ.	20	REF IN
10	INPUT OFFSET ADJ.	21	EXCITATION ADJ.
11	OUTPUT OFFSET ADJ.	25	HALF-BRIDGE COMP.
12	BANDWIDTH ADJ. 1	26	SENSE LOW
13	BANDWIDTH ADJ. 2	27	SENSE HIGH
14	V_{OUT} (FILTERED)	28	V_{EXC} OUT

NOTES:
1. LEAD NO. 1 IDENTIFIED BY DOT OR NOTCH.

MIC

Analog Devices 1B31

Model	1B31AN	1B31SD†
BRIDGE EXCITATION		
Regulator Input Voltage Range	+9.5V to +28V	*
Output Voltage Range	+4V to +15V	*
Regulator Input/Output Voltage Differential	+3V to +24V	*
Output Current[5]	100mA max	*
Regulation, Output Voltage vs. Supply	±0.05%/V	*
Load Regulation, I_L = 1mA to 50mA	±0.1%	*
Output Voltage vs. Temperature	±0.004%/°C	*
Output Noise, 10Hz to 1kHz[6]	200µV p-p	*
Reference Voltage (Internal)	+6.8V ±5%	*
Internal Half-Bridge Completion		
Nominal Resistor Value	20kΩ ±1%	*
Temperature Tracking	±5ppm/°C max	*
POWER SUPPLY		
Voltage, Rated Performance	±15V dc	*
Voltage, Operating	±12V to ±18V dc	*
Current, Quiescent[7]	+10mA	*
ENVIRONMENTAL		
Temperature Range		
Rated Performance	−40°C to +85°C	−55°C to +125°C
Operating	−40°C to +85°C	−55°C to +125°C
Storage	−40°C to +100°C	−65°C to +150°C
Relative Humidity	0 to 95% @ +60°C	*
CASE SIZE	0.83" × 1.64" × 0.25" (21.1 × 41.7 × 6.350mm) max	0.81" × 1.57" × 0.23" (20.6 × 40.0 × 5.72mm)

NOTES
*Specifications same as 1B31AN.
†SD grade available in Spring 1988.
[1] Specifications referred to the filtered output at Pin 14.
[2] Exclusive of external gain settling resistor.
[3] Unadjusted filter setting.
[4] Filter cutoff frequency set with external capacitors.
[5] Derate from +50°C as shown in Figure 14.
[6] 4.7µF capacitor from $V_{REF\,IN}$ (Pin 20) to COMM.
[7] Excluding bridge excitation's current, and with no loading on the output.

Specifications subject to change without notice.

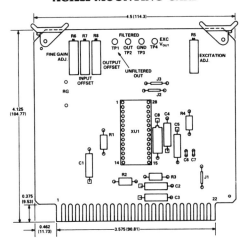

AC1222 MOUNTING CARD

AC1222 CONNECTOR DESIGNATION

PIN	FUNCTION	PIN	FUNCTION
1	+INPUT	T	V_{EXC} OUT
2	−INPUT	U	SENSE HIGH
3	N/C	V	SENSE LOW
4	GAIN (3)	W	HALF-BRIDGE COMP.
5	GAIN (4)	X	REF OUT
6	V_{OUT} (UNFILTERED)	Y	REF IN
9	INPUT OFFSET ADJ. (9)	Z	EXC. ADJ.
10	INPUT OFFSET ADJ. (10)		
11	OUTPUT OFFSET ADJ.		
12	BANDWIDTH ADJ. 1		
13	BANDWIDTH ADJ. 2		
14	V_{OUT} (FILTERED)		
19	−V_S		
20	COMMON		
21	+V_S		
22	+V_S REG		

The AC1222 mounting card is available for the 1B31. The AC1222 is an edge connector card with a 28-pin socket for plugging in the 1B31. In addition, it has provisions for installing the gain resistor and adjusting the bridge excitation voltage and cutoff frequency. Adjustment potentiometers for offset, fine gain and excitation are also provided. The AC1222 comes with a Cinch 251-22-30-160 (or equivalent) edge connector.

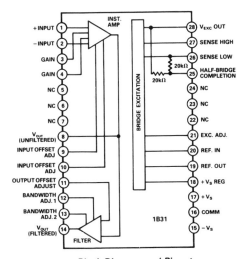

Block Diagram and Pinout

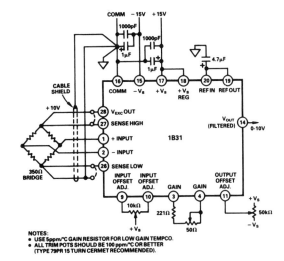

NOTES:
- USE 5ppm/°C GAIN RESISTOR FOR LOW GAIN TEMPCO.
- ALL TRIM POTS SHOULD BE 100 ppm/°C OR BETTER (TYPE 79PR 15 TURN CERMET RECOMMENDED).

Typical Application

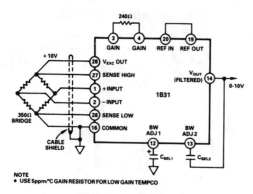

Narrow Bandwidth Application

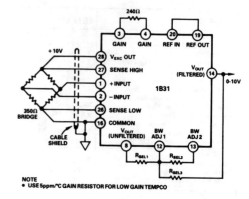

Wide Bandwidth Application

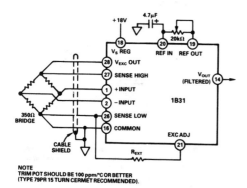

Constant Voltage Excitation: +10V to +15V Range

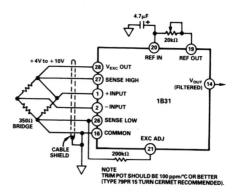

Constant Voltage Excitation: +4V to +10V Range

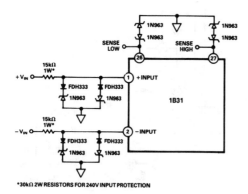

115V Input Protection for 1B31

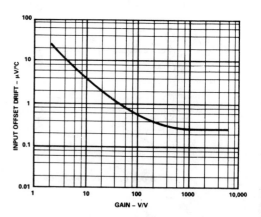

Total Input Offset Drift vs. Gain

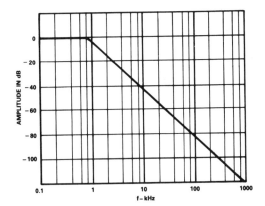

Filter Amplitude Response vs. Frequency

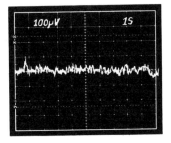

a. Bandwidth = 0.1Hz to 10Hz

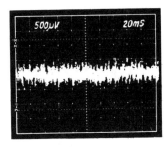

b. Bandwidth = 0.1Hz to 1kHz

Voltage Noise, RTO @ G = 1000V/V

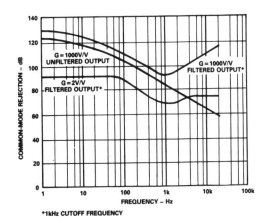

Common-Mode Rejection vs. Frequency and Gain

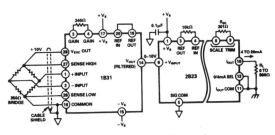

Isolated 4-20mA Transmitter

Analog Devices
1B32
Bridge-Transducer Signal Conditioner

FEATURES
- Low cost
- Complete signal-conditioning solution
- Small package: 28-pin double DIP
- Internal thin-film gain network
- High accuracy
  Low input offset tempco: $\pm 0.07\ \mu V/°C$
  Low gain tempco: ± 2 ppm/°C
- Low nonlinearity: $\pm 0.005\%$ max.
- High CMR: 14 dB min. (60 Hz, G=1000 V/V)
- Programmable bridge excitation: +4 V to +15 V
- Remote sensing
- Low-pass filter ($f_c = 4$ Hz)

APPLICATIONS
- Weigh scales
- Instrumentation: indicators, recorders, controllers
- Data-acquisition systems
- Microcomputer analog I/O

DESIGN FEATURES AND USER BENEFITS
Pin-Strappable Gain The internal resistor network can be pin-strapped for gains of 500 V/V and 333.3 V/V for 2 mV/V and 3 mV/V load cells. The tracking network guarantees a gain tempco of ± 6 ppm/°C max.

Custom Trimmable Network For volume applications, the 1B32 can be supplied with a custom laser trimmed gain network. Contact factory for further information.

Wide-Range Zero Suppression The output can be offset by ± 10 V for nulling out a dead load or to do a tare adjustment.

Remote Sensing Voltage drops across the excitation lead-wires are compensated by the regulated supply, making 6-wire load-cell interfacing straightforward.

Programmable Transducer Excitation The excitation source is preset for +10 Vdc operation without external components. It is user-programmable for a +4 V to +15 Vdc range (@ 100 mA) to optimize transducer performance.

Low-Pass Filter The three-pole active filter ($f_c = 4$ Hz) reduces 60-Hz line noise and improves system signal-to-noise ratio.

MIC

SPECIFICATIONS (typical @ +25°C and $V_S = \pm 15V$ unless otherwise noted)

Model	1B32AN
GAIN	
Gain Range	100V/V to 5000V/V
Internal Gain Setting	333.3V/V and 500V/V
Gain Equation	$1 + \dfrac{R_F}{R_I}$
Gain Equation Accuracy[1]	±0.1%
Gain Temperature Coefficient[2]	±2ppm/°C (±6ppm/°C max)
Gain Nonlinearity	±0.005% max
OFFSET VOLTAGES	
Total Offset Voltage, RTI	
Initial, @ +25°C, G = 1000V/V	±40μV
Warm-Up Drift, G = 1000V/V, 10 min	Within ±1μV
vs. Temperature (−25°C to +85°C)	
G = 1000V/V	±0.07μV/°C (±0.2μV/°C max)
At Other Gains	$\pm \left(0.06 + \dfrac{15}{G}\right)\mu V/°C$
Output Offset Adjust Range	±10V
INPUT BIAS CURRENT	
Initial @ 25°C	±3nA
vs. Temperature (−25°C to +85°C)	±50pA/°C
INPUT DIFFERENCE CURRENT	
Initial @ +25°C	±3nA
vs. Temperature (−25°C to +85°C)	±10pA/°C
INPUT RESISTANCE	
Differential	100MΩ
Common Mode	100MΩ
INPUT VOLTAGE RANGE	
Linear Differential Input	±0.1V
Maximum Differential Input	+5V
CMV Input Range	0 to +7.5V
CMR, 1kΩ Source Imbalance[3]	
G = 100V/V to 5000V/V @ dc	86dB
G = 100V/V, @ 60Hz	120dB
G = 1000V/V, @ 60Hz	140dB min
INPUT NOISE	
Voltage, G = 1000V/V	
0.1Hz to 10Hz	1μV p-p
Current, G = 1000V/V	
0.1Hz to 10Hz	3pA p-p
RATED OUTPUT	
Voltage, 2kΩ Load, min	±10V
Current	±5mA
Impedance, dc to 2Hz, G = 100V/V	0.6Ω
Load Capacitance	500pF
Output Short Circuit Duration (to Ground)	Indefinite
DYNAMIC RESPONSE	
Small Signal Bandwidth	
−3dB Gain Accuracy, G = 100V/V	4Hz
G = 1000V/V	3.5Hz
Slew Rate	20V/sec
Full Power	0.5Hz
Settling Time, G = 100V/V, ±10V Output	2sec
Step to ±0.1%	
LOW PASS FILTER	
Number of Poles	3
Cutoff Frequency (−3dB Point)	4Hz
Roll-Off	60dB/decade

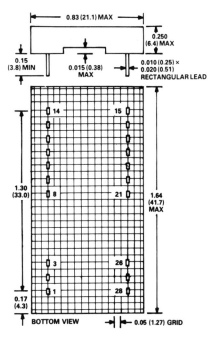

OUTLINE DIMENSIONS
Dimensions shown in inches and (mm).

PIN DESIGNATIONS

PIN	FUNCTION	PIN	FUNCTION
1	+INPUT	15	−V_S
2	−INPUT	16	COMM
3	INPUT OFFSET ADJ	17	+V_S
4	NC	18	+V_S REG
5	NC	19	REF OUT
6	NC	20	REF IN
7	NC	21	EXC ADJ
8	SIGNAL COMM	22	NC
9	EXT GAIN SET	23	NC
10	333.3 GAIN	24	NC
11	500 GAIN	25	NC
12	GAIN SENSE	26	SENSE LOW
13	GAIN COMM	27	SENSE HIGH
14	V_{OUT}	28	V_{EXC} OUT

Model	1B32AN
BRIDGE EXCITATION	
Regulator Input Voltage Range	+9.5V to +28V
Output Voltage Range	+4V to +15V
Regulator Input/Output Voltage Differential	+3V to +24V
Output Current[4]	100mA max
Regulation, Output Voltage vs. Supply	±0.05%/V
Load Regulation, I_L = 1mA to 50mA	±0.1%
Output Voltage vs. Temperature (−25°C to +85°C)	±40ppm/°C
Output Noise, 0.1Hz to 10Hz[5]	300µV p-p
Reference Voltage (Internal)	+6.8V ±5%
Sense & Excitation Lead Resistance	10Ω max
POWER SUPPLY	
Voltage, Rated Performance	±15V dc
Voltage, Operating	±12V to ±18V dc
Current, Quiescent[6]	+4mA, −1mA
ENVIRONMENTAL	
Temperature Range	
Rated Performance	−25°C to +85°C
Operating	−40°C to +85°C
Storage	−40°C to +100°C
Relative Humidity	0 to 95%, Noncondensing, @ +60°C
CASE SIZE	0.83" × 1.64" × 0.25" (21.1 × 41.7 × 6.35mm) max

NOTES
[1] Using internal network for gain.
[2] For pin-strapped gain. The tempco of the individual thin-film resistors is ±50ppm/°C max.
[3] 3V p-p 60Hz common-mode signal used in test setup.
[4] Derate 2mA/°C from +50°C.
[5] 4.7µF capacitor from REF IN (Pin 20) to COMM.
[6] Excluding bridge excitation current and with no loading on the output.
Specifications subject to change without notice.

AC1224 MOUNTING CARD

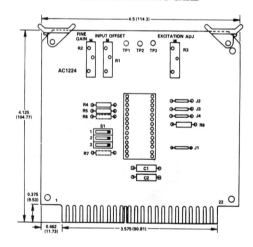

AC1224 GAIN SETTINGS VIA SWITCH S1

GAIN	S1-1	S1-2	S1-3
333	CLOSED	OPEN	CLOSED
500	OPEN	CLOSED	CLOSED
EXTERNAL	OPEN	OPEN	OPEN

AC1224 CONNECTOR DESIGNATIONS

PIN	FUNCTION	PIN	FUNCTION
T	V_{EXC} OUT	1	+INPUT
U	SENSE HIGH	2	−INPUT
V	SENSE LOW	12	V_{OUT}
X	REF OUT	19	$-V_S$
Y	REF IN	20	COMM
Z	EXC ADJ	21	$+V_S$
		22	$+V_S$ REG

The AC1224 mounting card is available for the 1B32. The AC1224 is an edge connector card with a socket for plugging in the 1B32. In addition it has provisions for switch selecting internal gains as well as installing gain resistors. Adjustment pots for offset, fine gain and excitation are also provided. The AC1224 comes with a Cinch 251-22-30-160 (or equivalent) edge connector.

MIC

Analog Devices 1B32

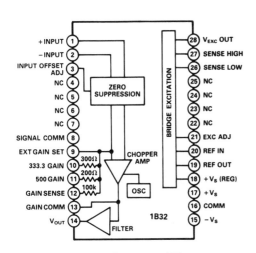

1B32 Block Diagram and Pinout

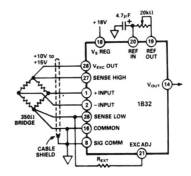

Constant Voltage Excitation: +10V to +15V Range

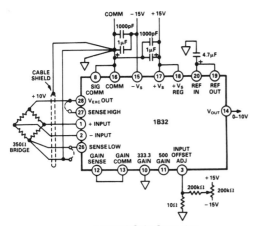

Internal Gain Strapping

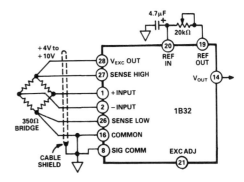

Constant Voltage Excitation: +4V to +10V Range

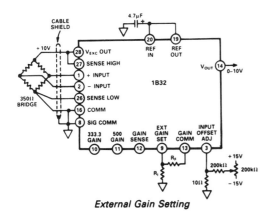

External Gain Setting

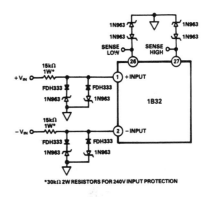

115V Input Protection

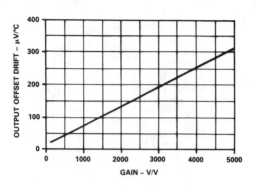

Total Output Offset Drift vs. Gain

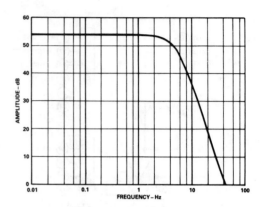

Filter Amplitude Response vs. Frequency, G = 500

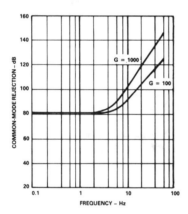

Common-Mode Rejection vs. Frequency

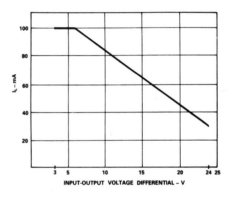

Excitation Source Load Current vs. Input-Output Voltage Differential, ≤25°C

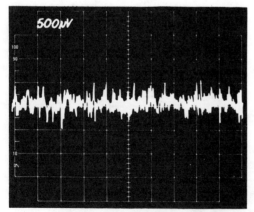

Voltage Noise, 0.1Hz to 10Hz, G = 1000

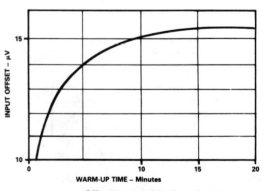

Offset Voltage RTI, Turn-On Drift

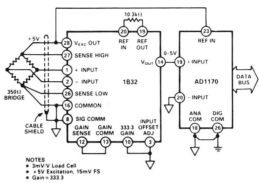

Auto-Calibrating Data Acquisition Using 1B32 and AD1170

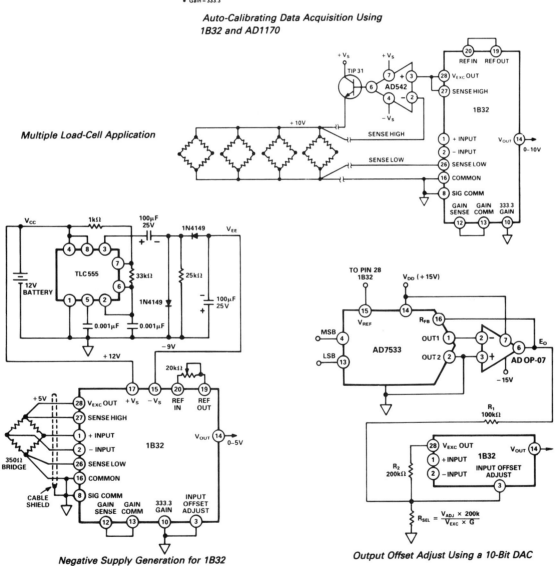

Multiple Load-Cell Application

Negative Supply Generation for 1B32

Output Offset Adjust Using a 10-Bit DAC

Raytheon DAC-4881
High-Performance Microprocessor-Compatible Complete 12-Bit D/A Converter

FEATURES
- High speed op amp for voltage output
- Precision trimmed thin film resistors
- Voltage reference: buried zener, 10 ppm/°C typical
- Input latches for microprocessor compatibility
- Internal ac compensation
- Nonlinearity: less than 1/4 LSB
- Differential nonlinearity: less than 1/2 LSB
- Monotonicity guaranteed over temperature range
- Settling time: 250 nS (current output)
- Settling time: 2 μS (voltage output)
- High-compliance complementary-current outputs
- Input codes: binary, complementary binary, offset binary, complementary offset binary
- Voltage output ranges—0 to +10 V, 0 to +5 V, ±2.5 V, ±5 V, ±10 V
- Direct interface to major logic families
- Direct interface to 8- and 16-bit buses
- Operates with ±12 V to ±15 V supplies
- Low power dissipation: 350 mW
- Monolithic
- Metal/ceramic package
- 883B processing available

ABSOLUTE MAXIMUM RATINGS

Supply voltage	±16.5 V
Logic input voltages	-5 V to $-V_s + 33$ V
I_o and $\overline{I}_o$ voltages	-5 V to $+12$ V
Reference input voltage	$-V_s$ to $+V_s$
Reference input current	2 mA
Storage temperature range	-65°C to $+150$°C
Lead soldering temperature (60 S)	$+300$°C

CONNECTION INFORMATION

28-Lead Ceramic Side-Brazed Dual In-Line Package
(Top View)

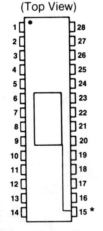

Note: Package lid ac grounded to $-V_S$

Pin	Function
1	$\overline{CS}$
2	$\overline{ADH}$
3	Bit 1 (MSB)
4–13	Input Bits
14	Bit 12 (LSB)
15	$-V_S$
16	V_{OUT}
17	Ref In
18	Bip Off
19	10V Span
20	20V Span
21	Sum Node
22	$\overline{I}_O$
23	I_O
24	Gnd
25	$+V_S$
26	Gain Adj
27	Ref Out
28	$\overline{ADL}$

ORDERING INFORMATION

Part Number	Package	Operating Temperature Range
DAC-4881FS	S	0°C to +70°C
DAC-4881DS	S	0°C to +70°C
DAC-4881BS	S	-55°C to +125°C
DAC-4881BS/883B	S	-55°C to +125°C

Notes:
/883B suffix denotes Mil-Std-883, Level B processing
S = 28-lead ceramic sidebrazed DIP
Contact a Raytheon sales office or representative for ordering information on special package/temperature range combinations.

FUNCTIONAL BLOCK DIAGRAM

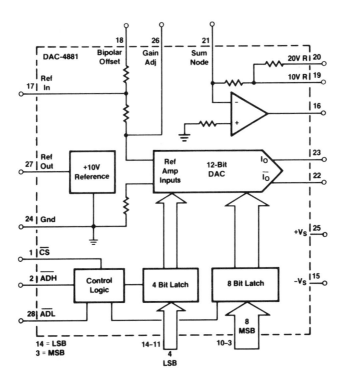

Calibration Procedure:
1. Set inputs to all zeros
2. Adjust offset until V_{OUT} equals 0V
3. Set inputs to all ones
4. Adjust gain until V_{OUT} equals correct full scale value

*Optional — reduces reference noise
**Optional — improves settling time (see table for values)

Format	Output Scale	B1	B2	B3	B4	B5	B6	B7	B8	B9	B10	B11	B12	I_O(mA)	$\overline{I_O}$(mA)	V_{OUT}
Straight Binary. Unipolar with True Input Code. True Zero Output	Positive Full Scale	1	1	1	1	1	1	1	1	1	1	1	1	3.999	0.000	9.9976
	Positive Full Scale — LSB	1	1	1	1	1	1	1	1	1	1	1	0	3.998	0.001	9.9951
	LSB	0	0	0	0	0	0	0	0	0	0	0	1	0.0001	3.998	0.0024
	Zero Scale	0	0	0	0	0	0	0	0	0	0	0	0	0.000	3.999	0.0000
Complementary Binary. Unipolar with Complementary Input Code. True Zero Output	Positive Full Scale	0	0	0	0	0	0	0	0	0	0	0	0	0.000	3.999	9.9976
	Positive Full Scale — LSB	0	0	0	0	0	0	0	0	0	0	0	1	0.001	3.998	9.9951
	LSB	1	1	1	1	1	1	1	1	1	1	1	0	3.998	0.001	0.0024
	Zero Scale	1	1	1	1	1	1	1	1	1	1	1	1	3.999	0.000	0.0000

Stand-Alone, 0 to −10V, 12-Bit Straight Binary With Gain and Offset Adjust Connections

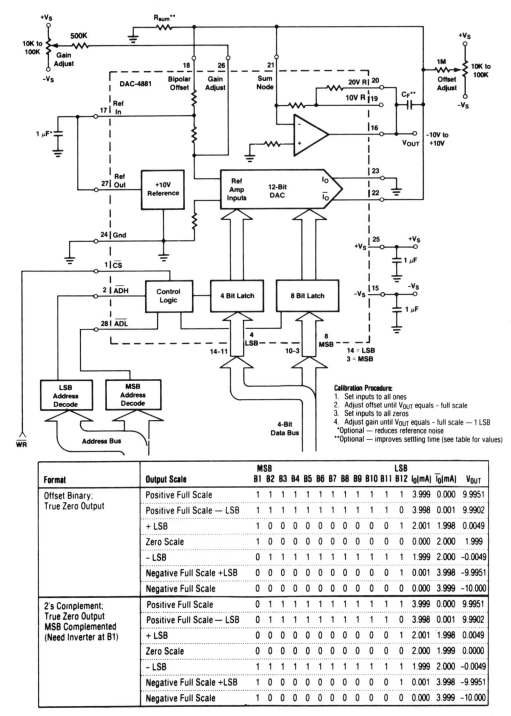

Microprocessor Interface, 8-Bit Data Bus, =10V to −10V Output With Complementary Binary Input (All Zeros Equal − Full Scale)

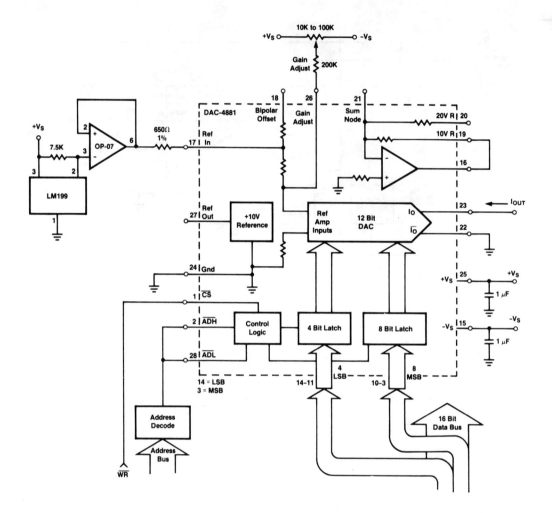

Calibration Procedure:
Zero scale error is entirely leakage current — no adjustment necessary
1. Set inputs to all ones
2. Adjust gain until I_0 equals correct full scale value

Microprocessor Interface, 16-Bit Data Bus, 0 to −4 mA Output With Straight Binary Input and External Reference

Raytheon DAC-4888
8-Bit D/A Converter With Microprocessor Interface Latches

FEATURES
- High speed op amp for voltage output
- Tracking thin-film resistors
- Voltage reference: bandgap, 25 ppm/°C
- Input latches for microprocessor compatibility
- Internal ac compensation
- Nonlinearity: ±1/4 LSB max. over temperature range
- Monotonic: differential nonlinearity ±1/3 LSB max. over temperature range
- Settling time: 150 nS (current output)
- Settling time: 1.4 µS (voltage output)
- High-compliance complementary-current outputs
- Input codes: binary, complementary binary, offset binary, complementary offset binary
- Voltage output ranges: 0 to +10 V, 0 to +5 V, ±2.5 V, ±5 V, ±10 V
- Direct interface to major logic families
- Direct interface to 4- and 8-bit buses
- Operates with ±12-V to ±15-V supplies
- Low power dissipation: 330 mW
- Monolithic
- Ceramic package
- 883B processing available

ABSOLUTE MAXIMUM RATINGS

Supply voltage	±18 V
Logic input voltages	−5 V to $-V_s$ +36 V
I_o and $\overline{I}_o$ voltages	−5 V to +12 V
Reference input voltage	$-V_s$ to $+V_s$
Reference input current	2 mA
Storage temperature range	−65°C to +150°C
Lead soldering temperature (60 s)	+300°C

ORDERING INFORMATION

Part Number	Package	Operating Temperature Range
DAC-4888FD	D	0°C to +70°C
DAC-4888DD	D	0°C to +70°C
DAC-4888BD	D	-55°C to +125°C
DAC-4888BD/883B	D	-55°C to +125°C

Notes:
/883B suffix denotes Mil-Std-883, Level B processing
D = 24-lead ceramic DIP
Contact a Raytheon sales office or representative for ordering information on special package/temperature range combinations.

CONNECTION INFORMATION
24-Pin Ceramic Dual In-Line Package
(Top View)

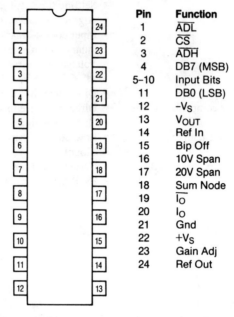

Pin	Function
1	$\overline{ADL}$
2	$\overline{CS}$
3	$\overline{ADH}$
4	DB7 (MSB)
5–10	Input Bits
11	DB0 (LSB)
12	$-V_S$
13	V_{OUT}
14	Ref In
15	Bip Off
16	10V Span
17	20V Span
18	Sum Node
19	$\overline{I_O}$
20	I_O
21	Gnd
22	$+V_S$
23	Gain Adj
24	Ref Out

FUNCTIONAL BLOCK DIAGRAM

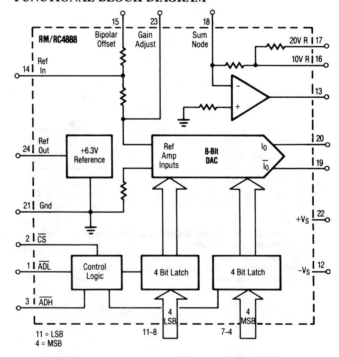

☐ MIC Raytheon DAC-4888

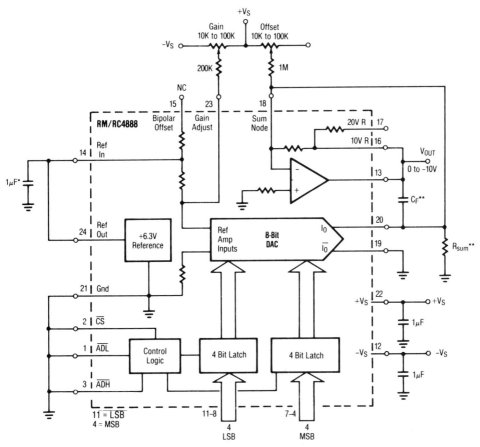

Calibration Procedure:
1. Set inputs to all zeros
2. Adjust offset until V_{OUT} equals 0V
3. Set inputs to all ones
4. Adjust gain until V_{OUT} equals correct full scale value

*Optional — reduces reference noise
**Optional — improves settling time (see table for values)

Format	Output Scale	MSB DB7	DB6	DB5	DB4	DB3	DB2	DB1	LSB DB0	I_O (mA)	$\overline{I_O}$ (mA)	V_{OUT}
Straight Binary: Unipolar With True Input Code. True Zero Output	Positive Full Scale	1	1	1	1	1	1	1	1	3.999	0.000	9.9609
	Positive Full Scale − LSB	1	1	1	1	1	1	1	0	3.984	0.001	9.9219
	LSB	0	0	0	0	0	0	0	1	0.0001	3.984	0.0391
	Zero Scale	0	0	0	0	0	0	0	0	0.000	3.999	0.0000
Complementary Binary: Unipolar With Complementary Input Code. True Zero Output	Positive Full Scale	0	0	0	0	0	0	0	0	0.000	3.999	9.9609
	Positive Full Scale − LSB	0	0	0	0	0	0	0	1	0.001	3.984	9.9219
	LSB	1	1	1	1	1	1	1	0	3.984	0.001	0.0391
	Zero Scale	1	1	1	1	1	1	1	1	3.999	0.000	0.0000

Stand-Alone, 0 to +10V, 8-Bit Straight Binary With Gain and Offset Adjust Connections

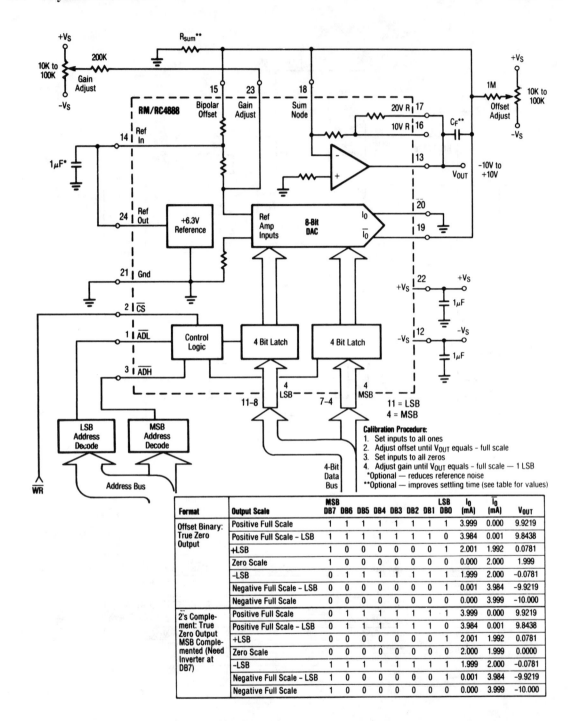

Microprocessor Interface, 4-Bit Data Bus, +10V to −10V Output With Complementary Binary Input (All Zeros Equal — Full Scale)

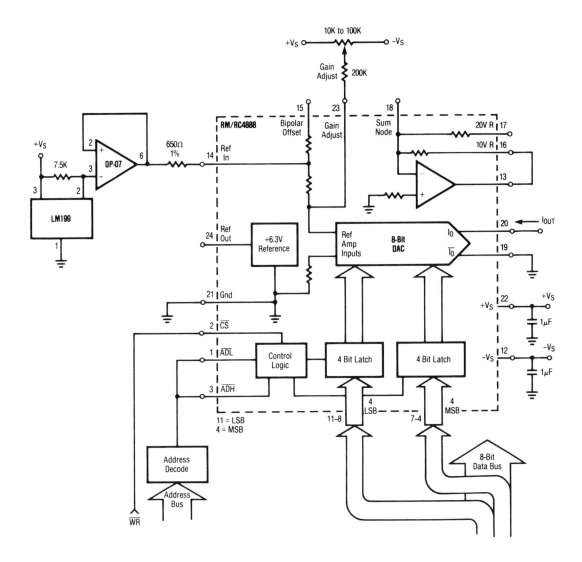

Calibration Procedure:
Zero scale error is entirely leakage current — no adjustment necessary
1. Set inputs to all ones
2. Adjust gain until I_0 equals correct full scale value

Microprocessor Interface, 8-Bit Data Bus, 0 to −4mA Output With Straight Binary Input and External Reference

Silicon Systems
SSI 73K222/K222L V.22, V.21, Bell 212A
Single-Chip Modem

FEATURES

- One-chip CCITT V.22, V.21, Bell 212A and 103 standard compatible modem data pump
- Full-duplex operation at 0-300 bit/s (FSK) or 600 and 1200 bit/s (DPSK)
- Pin and software compatible with other SSI K-Series 1-chip modems
- Interfaces directly with standard microprocessors (8048, 80C51 typical)
- Serial (22-pin DIP) or parallel (28-pin DIP) microprocessor bus for control
- Serial port for data transfer
- Both synchronous and asynchronous modes of operation including V.22 extended overspeed
- Call progress, carrier, precise answer tone (2100 or 2225 Hz), and long loop detectors
- DTMF, and 550- or 1800-Hz guard tone generators
- Test modes available: ALB, DL, RDL, Mark, Space, Alternating bit patterns
- Precise automatic gain control allows 45-dB dynamic range
- Space-efficient 22- or 28-pin DIP packages
- CMOS technology for low power consumption using 30 mW @ 5 V or 180 mW @ 12 V
- Single +5 V (73K222L) or +12 V (73K222) versions

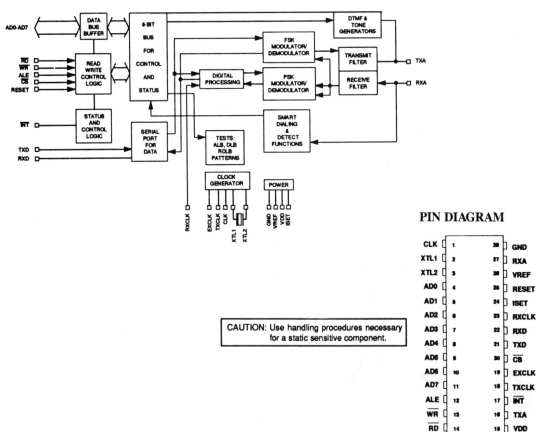

CAUTION: Use handling procedures necessary for a static sensitive component.

PIN DESCRIPTION

POWER

NAME	28-PIN	22-PIN	TYPE	DESCRIPTION
GND	28	1	I	System Ground.
VDD	15	11	I	Power supply input, 12V +10%, -20% (73K222) or 5V ±10% (73K222L). Bypass with .1 and 22 µF capacitors to GND.
VREF	26	21	O	An internally generated reference voltage. Bypass with .1 µF capacitor to ground.
ISET	24	19	I	Chip current reference. Sets bias current for op-amps. The chip current is set by connecting this pin to VDD through a 2 MΩ resistor. ISET should be bypassed to GND with a .1 µF capacitor.

PARALLEL MICROPROCESSOR INTERFACE

NAME	28-PIN	22-PIN	TYPE	DESCRIPTION
ALE	12	-	I	Address latch enable. The falling edge of ALE latches the address on AD0-AD2 and the chip select on $\overline{CS}$.
AD0-AD7	4-11	-	I/O	Address/data bus. These bidirectional tri-state multiplexed lines carry information to and from the internal registers.
$\overline{CS}$	20	-	I	Chip select. A low on this pin during the falling edge of ALE allows a read cycle or a write cycle to occur. AD0-AD7 will not be driven and no registers will be written if $\overline{CS}$ (latched) is not active. The state of $\overline{CS}$ is latched on the falling edge of ALE.
CLK	1	2	O	Output clock. This pin is selectable under processor control to be either the crystal frequency (for use as a processor clock) or 16 x the data rate for use as a baud rate clock in DPSK modes only. The pin defaults to the crystal frequency on reset.
$\overline{INT}$	17	13	O	Interrupt. This open drain output signal is used to inform the processor that a detect flag has occurred. The processor must then read the detect register to determine which detect triggered the interrupt. $\overline{INT}$ will stay low until the processor reads the detect register or does a full reset.
$\overline{RD}$	14	-	I	Read. A low requests a read of the SSI 73K222 internal registers. Data cannot be output unless both $\overline{RD}$ and the latched $\overline{CS}$ are active or low.
RESET	25	20	I	Reset. An active high signal on this pin will put the chip into an inactive state. All control register bits (CR0, CR1, Tone) will be reset. The output of the CLK pin will be set to the crystal frequency. An internal pull down resistor permits power on reset using a capacitor to VDD.
$\overline{WR}$	13	-	I	Write. A low on this informs the SSI 73K222 that data is available on AD0-AD7 for writing into an internal register. Data is latched on the rising edge of $\overline{WR}$. No data is written unless both $\overline{WR}$ and the latched $\overline{CS}$ are low.

DTE USER

NAME	28-PIN	22-PIN	TYPE	DESCRIPTION
EXCLK	19	15	I	External Clock. This signal is used in synchronous transmission when the external timing option has been selected. In the external timing mode the rising edge of EXCLK is used to strobe synchronous DPSK transmit data applied to on the TXD pin. Also used for serial control interface.
RXCLK	23	18	O	Receive Clock. The falling edge of this clock output is coincident with the transitions in the serial received data output. The rising edge of RXCLK can be used to latch the valid output data. RXCLK will be valid as long as a carrier is present.
RXD	22	17	O	Received Data Output. Serial receive data is available on this pin. The data is always valid on the rising edge of RXCLK when in synchronous mode. RXD will output constant marks if no carrier is detected.
TXCLK	18	14	O	Transmit Clock. This signal is used in synchronous transmission to latch serial input data on the TXD pin. Data must be provided so that valid data is available on the rising edge of the TXCLK. The transmit clock is derived from different sources depending upon the synchronization mode selection. In Internal Mode the clock is generated internally. In External Mode TXCLK is phase locked to the EXCLK pin. In Slave Mode TXCLK is phase locked to the RXCLK pin. TXCLK is always active.
TXD	21	16	I	Transmit Data Input. Serial data for transmission is applied on this pin. In synchronous modes, the data must be valid on the rising edge of the TXCLK clock. In asynchronous modes (1200/600 bit/s or 300 baud) no clocking is necessary. DPSK data must be 1200/600 bit/s +1%, -2.5% or +2.3%, -2.5 % in extended overspeed mode.

ANALOG INTERFACE AND OSCILLATOR

NAME	28-PIN	22-PIN	TYPE	DESCRIPTION
RXA	27	22	I	Received modulated analog signal input from the telephone line interface.
TXA	16	12	O	Transmit analog output to the telephone line interface.
XTL1 XTL2	2 3	3 4	I I	These pins are for the internal crystal oscillator requiring a 11.0592 MHz parallel mode crystal. Load capacitors should be connected from XTL1 and XTL2 to Ground. XTL2 can also be driven from an external clock.

SERIAL MICROPROCESSOR INTERFACE

A0-A2	-	5-7	I	Register Address Selection. These lines carry register addresses and should be valid during any read or write operation.
DATA	-	8	I/O	Serial Control Data. Data for a read/write operation is clocked in or out on the falling edge of the EXCLK pin. The direction of data flow is controlled by the $\overline{RD}$ pin. $\overline{RD}$ low outputs data. $\overline{RD}$ high inputs data.
$\overline{RD}$	-	10	I	Read. A low on this input informs the SSI 73K222 that data or status information is being read by the processor. The falling edge of the $\overline{RD}$ signal will initiate a read from the addressed register. The $\overline{RD}$ signal must continue for eight falling edges of EXCLK in order to read all eight bits of the referenced register. Read data is provided LSB first. Data will not be output unless the $\overline{RD}$ signal is active.
$\overline{WR}$	-	9	I	Write. A low on this input informs the SSI 73K222 that data or status information has been shifted in through the DATA pin and is available for writing to an internal register. The normal procedure for a write is to shift in data LSB first on the DATA pin for eight consecutive falling edges of EXCLK and then to pulse $\overline{WR}$ low. Data is written on the rising edge of $\overline{WR}$.
Note:	\multicolumn{4}{l	}{ In the serial, 22-pin version, the pins AD0-AD7, ALE and $\overline{CS}$ are removed and replaced with the pins; A0, A1, A2, DATA, and an unconnected pin. Also, the $\overline{RD}$ and $\overline{WR}$ controls are used differently. The serial control mode is provided in the 28-pin version by tying ALE high and $\overline{CS}$ low. In this configuration AD7 becomes DATA and AD0, AD1 and AD2 become A0, A1 and A2, respectively. }

ELECTRICAL SPECIFICATIONS

ABSOLUTE MAXIMUM RATINGS

PARAMETER	RATING	UNIT
VDD Supply Voltage	14	V
Storage Temperature	-65 to 150	°C
Soldering Temperature (10 sec.)	260	°C
Applied Voltage	-0.3 to VDD+0.3	V
Note: All inputs and outputs are protected from static charge using built-in, industry standard protection devices and all outputs are short-circuit protected.		

RECOMMENDED OPERATING CONDITIONS

PARAMETER	CONDITIONS	MIN	NOM	MAX	UNITS
VDD Supply voltage		4.5	5	5.5	V
TA, Operating Free-Air Temperature		-40		+85	°C
Clock Variation	(11.0592 MHz) Crystal or external clock	-0.01		+0.01	%
External Components (Refer to Application section for placement.)					
VREF Bypass Capacitor	(External to GND)	0.1			µF
Bias setting resistor	(Placed between VDD and ISET pins)	1.8	2	2.2	MΩ
ISET Bypass Capacitor	(ISET pin to GND)	0.1			µF
VDD Bypass Capacitor 1	(External to GND)	0.1			µF
VDD Bypass Capacitor 2	(External to GND)	22			µF
XTL1 Load Capacitor	Depends on crystal characteristics; from pin to GND			40	pF
XTL2 Load Capacitor				20	

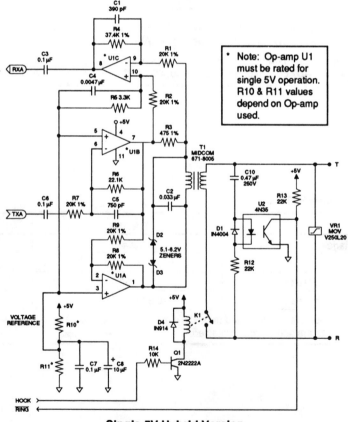

Single 5V Hybrid Version

MIC

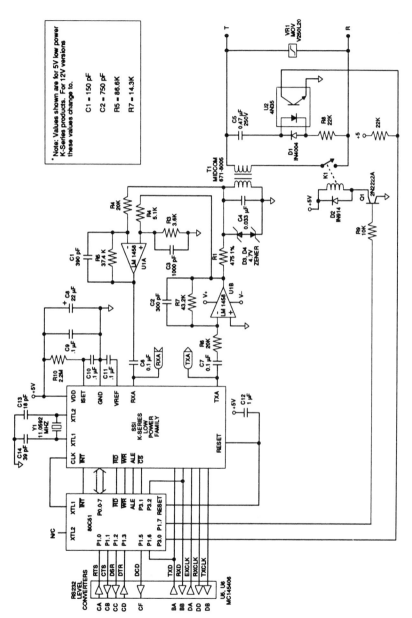

Basic Box Modem with Dual-Supply Hybrid

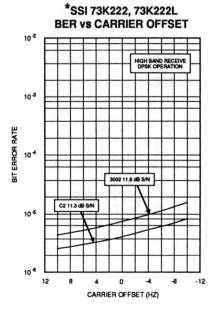

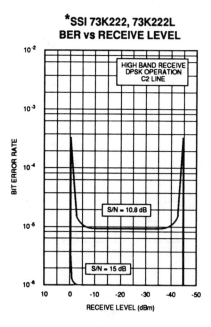
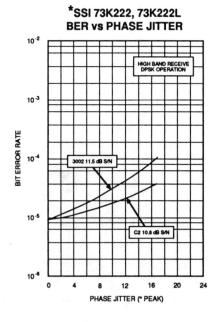

* = "EQ On" Indicates bit CR1 D4 is set for additional phase equalization.

PACKAGE PIN DESIGNATIONS

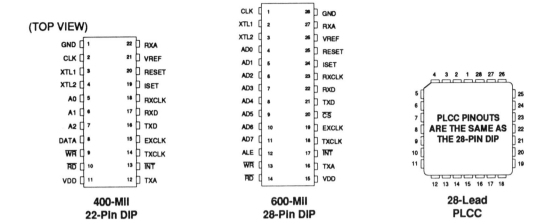

400-Mil
22-Pin DIP

600-Mil
28-Pin DIP

28-Lead
PLCC

ORDERING INFORMATION

PART DESCRIPTION	ORDER NO.	PKG. MARK

SSI 73K222 with Parallel Bus Interface
28-Pin 12 Volt Supply
    Plastic Dual-In-Line
    Plastic Leaded Chip Carrier
	SSI 73K222 – IP
	SSI 73K222 – IH
		73K222 – IP
		73K222 – IH

28-Pin 5 Volt Supply
    Plastic Dual-In-Line
    Plastic Leaded Chip Carrier
	SSI 73K222L – IP
	SSI 73K222L – IH
		73K222L – IP
		73K222L – IH

SSI 73K222 with Serial Interface 22-Pin 12 Volt Supply Plastic Dual-In-Line	SSI 73K222S – IP	73K222S – IP

22-Pin 5 Volt Supply
    Plastic Dual-In-Line
    Ceramic Dual-In-Line
	SSI 73K222SL – IP
	SSI 73K222SL – IC
		73K222SL – IP
		73K222SL – IC

No responsibility is assumed by Silicon Systems for use of this product nor for any infringements of patents and trademarks or other rights of third parties resulting from its use. No license is granted under any patents, patent rights or trademarks of Silicon Systems. Silicon Systems reserves the right to make changes in specifications at any time without notice. Accordingly, the reader is cautioned to verify that the data sheet is current before placing orders.

Silicon Systems
SSI 73K322L CCITT V.23, V.22, V.21
Single-Chip Modem

FEATURES
- One-chip CCITT V.23, V.22, and V.21 standard compatible modem data pump
- Full-duplex operation at 0-300 bit/s (FSK) or 600 and 1200 bit/s (DPSK) or 0-1200 bit/s (FSK) forward channel, with or without 0-75 bit/s back channel
- Interfaces directly with standard microprocessors (8048, 80C51 typical)
- Serial (22-pin DIP) or parallel microprocessor bus (28-pin DIP) for control
- Serial port for data transfer
- Both synchronous and asynchronous modes of operation
- Call progress, carrier, precise answer-tone (2100 Hz), calling-tone (1300 Hz), and FSK mark detectors
- DTMF and 550- or 1800-Hz guard-tone generators
- Test modes available: ALB, DL, RDL, Mark, Space, Alternating bit patterns
- Precise automatic gain control allows 45-dB dynamic range
- Space-efficient 22- or 28-pin DIP packages
- CMOS technology for low power consumption using 30 mW @ 5 V from a single power supply

BLOCK DIAGRAM

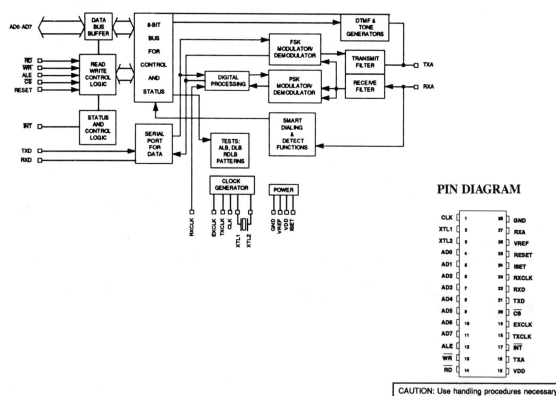

PIN DIAGRAM

CAUTION: Use handling procedures necessary for a static sensitive component.

PIN DESCRIPTION

POWER

NAME	28-PIN	22-PIN	TYPE	DESCRIPTION
GND	28	1	I	System Ground.
VDD	15	11	I	Power supply input, 5V ±10%. Bypass with .1 and 22 µF capacitors to GND.
VREF	26	21	O	An internally generated reference voltage. Bypass with .1 µF capacitor to GND.
ISET	24	19	I	Chip current reference. Sets bias current for op-amps. The chip current is set by connecting this pin to VDD through a 2 MΩ resistor. ISET should be bypassed to GND with a .1 µF capacitor.

PARALLEL MICROPROCESSOR INTERFACE

NAME	28-PIN	22-PIN	TYPE	DESCRIPTION
ALE	12	-	I	Address latch enable. The falling edge of ALE latches the address on AD0-AD2 and the chip select on $\overline{CS}$.
AD0-AD7	4-11	-	I/O	Address/data bus. These bidirectional tri-state multiplexed lines carry information to and from the internal registers.
$\overline{CS}$	20	-	I	Chip select. A low on this pin during the falling edge of ALE allows a read cycle or a write cycle to occur. AD0-AD7 will not be driven and no registers will be written if $\overline{CS}$ (latched) is not active. The state of $\overline{CS}$ is latched on the falling edge of ALE.
CLK	1	2	O	Output clock. This pin is selectable under processor control to be either the crystal frequency (for use as a processor clock) or 16 x the data rate for use as a baud rate clock in DPSK modes only. The pin defaults to the crystal frequency on reset.
$\overline{INT}$	17	13	O	Interrupt. This open drain output signal is used to inform the processor that a detect flag has occurred. The processor must then read the detect register to determine which detect triggered the interrupt. $\overline{INT}$ will stay low until the processor reads the detect register or does a full reset.
$\overline{RD}$	14	-	I	Read. A low requests a read of the SSI 73K322L internal registers. Data cannot be output unless both $\overline{RD}$ and the latched $\overline{CS}$ are active or low.
RESET	25	20	I	Reset. An active high signal on this pin will put the chip into an inactive state. All control register bits (CR0, CR1, Tone) will be reset. The output of the CLK pin will be set to the crystal frequency. An internal pull down resistor permits power on reset using a capacitor to VDD.
$\overline{WR}$	13	-	I	Write. A low on this informs the SSI 73K322L that data is available on AD0-AD7 for writing into an internal register. Data is latched on the rising edge of $\overline{WR}$. No data is written unless both $\overline{WR}$ and the latched $\overline{CS}$ are low.

SERIAL MICROPROCESSOR INTERFACE

A0-A2	-	5-7	I	Register Address Selection. These lines carry register addresses and should be valid during any read or write operation.
DATA	-	8	I/O	Serial Control Data. Data for a read/write operation is clocked in or out on the falling edge of the EXCLK pin. The direction of data flow is controlled by the $\overline{RD}$ pin. $\overline{RD}$ low outputs data. $\overline{RD}$ high inputs data.
$\overline{RD}$	-	10	I	Read. A low on this input informs the SSI 73K322L that data or status information is being read by the processor. The falling edge of the $\overline{RD}$ signal will initiate a read from the addressed register. The $\overline{RD}$ signal must continue for eight falling edges of EXCLK in order to read all eight bits of the referenced register. Read data is provided LSB first. Data will not be output unless the $\overline{RD}$ signal is active.
$\overline{WR}$	-	9	I	Write. A low on this input informs the SSI 73K322L that data or status information has been shifted in through the DATA pin and is available for writing to an internal register. The normal procedure for a write is to shift in data LSB first on the DATA pin for eight consecutive falling edges of EXCLK and then to pulse $\overline{WR}$ low. Data is written on the rising edge of $\overline{WR}$.
Note:	\multicolumn{4}{l	}{In the serial, 22-pin version, the pins AD0-AD7, ALE and $\overline{CS}$ are removed and replaced with the pins; A0, A1, A2, DATA, and an unconnected pin. Also, the $\overline{RD}$ and $\overline{WR}$ controls are used differently.}
	\multicolumn{4}{l	}{The serial control mode is provided in the 28-pin version by tying ALE high and $\overline{CS}$ low. In this configuration AD7 becomes DATA and AD0, AD1 and AD2 become A0, A1 and A2, respectively.}

DTE USER INTERFACE

NAME	28-PIN	22-PIN	TYPE	DESCRIPTION
EXCLK	19	15	I	External Clock. This signal is used in synchronous DPSK transmission when the external timing option has been selected. In the external timing mode the rising edge of EXCLK is used to strobe synchronous DPSK transmit data available on the TXD pin. Also used for serial control interface.

RS-232 INTERFACE (Continued)

NAME	28-PIN	22-PIN	TYPE	DESCRIPTION
RXCLK	23	18	O	Receive Clock. The falling edge of this clock output is coincident with the transitions in the serial received DPSK data output. The rising edge of RXCLK can be used to latch the valid output data. RXCLK will be valid as long as a carrier is present. In V.23 or V.21 mode a clock which is 16 x 1200 or 16 x 300 Hz baud data rate is output, respectively.
RXD	22	17	O	Received Data Output. Serial receive data is available on this pin. The data is always valid on the rising edge of RXCLK when in synchronous mode. RXD will output constant marks if no carrier is detected.
TXCLK	18	14	O	Transmit Clock. This signal is used in synchronous DPSK transmission to latch serial input data on the TXD pin. Data must be provided so that valid data is available on the rising edge of the TXCLK. The transmit clock is derived from different sources depending upon the synchronization mode selection. In Internal Mode the clock is 1200 Hz generated internally. In External Mode TXCLK is phase locked to the EXCLK pin. In Slave Mode TXCLK is phase locked to the RXCLK pin. TXCLK is always active. In V.23 or V.21 mode the output is a 16 x 1200 or 16 x 300 Hz baud clock, respectively.
TXD	21	16	I	Transmit Data Input. Serial data for transmission is applied on this pin. In synchronous modes, the data must be valid on the rising edge of the TXCLK clock. In asynchronous modes (1200 or 300 baud) no clocking is necessary. DPSK must be 1200/600 bit/s +1%, -2.5% or +2.3%, -2.5 % in extended overspeed mode.

ANALOG INTERFACE AND OSCILLATOR

RXA	27	22	I	Received modulated analog signal input from the telephone line interface.
TXA	16	12	O	Transmit analog output to the telephone line interface.
XTL1 XTL2	2 3	3 4	I I	These pins are for the internal crystal oscillator requiring a 11.0592 MHz parallel mode crystal and two load capacitors to Ground. XTL2 can also be driven from an external clock.

ELECTRICAL SPECIFICATIONS

ABSOLUTE MAXIMUM RATINGS

PARAMETER	RATING	UNIT
VDD Supply Voltage	14	V
Storage Temperature	-65 to 150	°C
Soldering Temperature (10 sec.)	260	°C
Applied Voltage	-0.3 to VDD+0.3	V

Note: All inputs and outputs are protected from static charge using built-in, industry standard protection devices and all outputs are short-circuit protected.

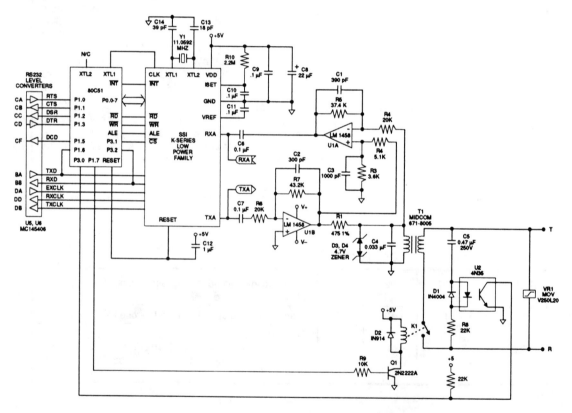

Basic Box Modem with Dual-Supply Hybrid

RECOMMENDED OPERATING CONDITIONS

PARAMETER	CONDITIONS	MIN	NOM	MAX	UNITS
VDD Supply voltage		4.5	5	5.5	V
TA, Operating Free-Air Temp.		-40		+85	°C
Clock Variation	(11.0592 MHz) Crystal or external clock	-0.01		+0.01	%
External Components (Refer to Application section for placement.)					
VREF Bypass Capacitor	(External to GND)	0.1			µF
Bias setting resistor	(Placed between VDD and ISET pins)	1.8	2	2.2	MΩ
ISET Bypass Capacitor	(ISET pin to GND)	0.1			µF
VDD Bypass Capacitor 1	(External to GND)	0.1			µF
VDD Bypass Capacitor 2	(External to GND)	22			µF
XTL1 Load Capacitor	Depends on crystal characteristics;			40	pF
XTL2 Load Capacitor	from pin to GND			20	

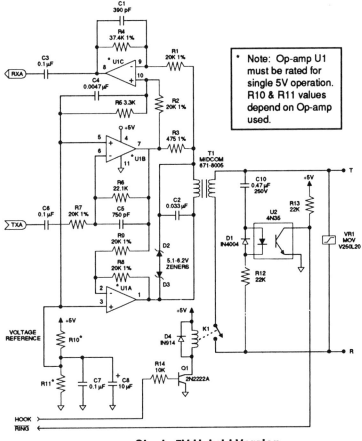

Single 5V Hybrid Version

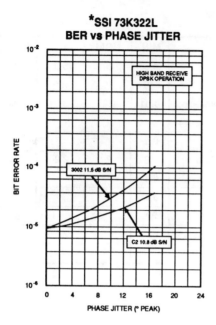

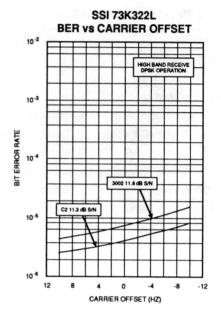

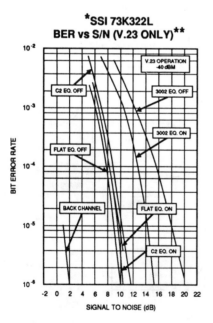

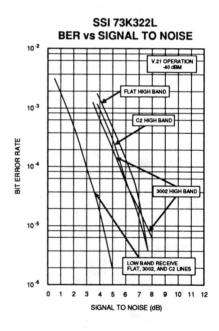

* = "EQ On" Indicates bit CR1 D4 is set for additional phase equalization.

** = 73K302L performance is similar to that of the 73K322L. V.23 operation corresponds to Bell 202.

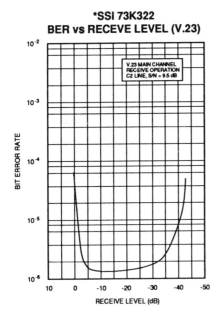

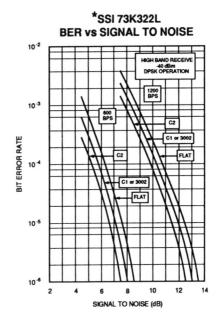

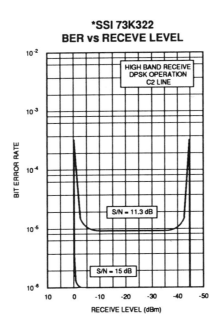

PACKAGE PIN DESIGNATIONS

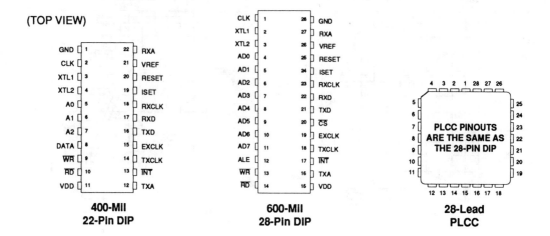

ORDERING INFORMATION

PART DESCRIPTION	ORDER NO.	PKG. MARK
SSI 73K322L with Parallel Bus Interface 28-Pin 5 Volt Supply		
Plastic Dual-In-Line	SSI 73K322L - IP	73K322L - IP
Plastic Leaded Chip Carrier	SSI 73K322L - IH	73K322L - IH
SSI 73K322L with Serial Interface 22-Pin 5 Volt Supply		
Plastic Dual-In-Line	SSI 73K322SL - IP	73K322SL - IP
	SSI 73K322SL - IC	73K322SL - IC

No responsibility is assumed by Silicon Systems for use of this product nor for any infringements of patents and trademarks or other rights of third parties resulting from its use. No license is granted under any patents, patent rights or trademarks of Silicon Systems. Silicon Systems reserves the right to make changes in specifications at any time without notice. Accordingly, the reader is cautioned to verify that the data sheet is current before placing orders.

Silicon Systems
SSI 73K224L
V.22 bis/V.22/V.21, Bell 212A/103
Single-Chip Modem

FEATURES

- One-chip multi-mode V.22 bis/V.22/V.21 and Bell 212A/103-compatible modem data pump
- FSK (300 bit/s), DPSK (600, 1200 bit/s), or QAM (2400 bit/s) encoding
- Pin- and software-compatible with other SSI K-Series 1-chip modems

□ MIC   Silicon Systems SSI 73K224L   323

- Interfaces directly with standard microprocessors (8048, 80C51 typical)
- Parallel microprocessor bus (28-pin DIP, 32- and 44-pin PLCC) for control
- Selectable asynch/synch with internal buffer/debuffer and scrambler/descrambler functions
- All synchronous and asynchronous operating modes (internal, external, slave)
- Adaptive equalization for optimum performance over all lines
- Programmable transmit attenuation (16 dB, 1 dB steps), selectable receive boost (+12 dB)
- Call progress, carrier, answer tone, unscrambled mark, S1, and signal-quality monitors
- DTMF, answer, and guard-tone generators
- Test modes available: ALB, DL, RDL, Mark, Space, Alternating bit, S1 pattern
- CMOS technology for low power consumption (125 mW @ 5 V) with power down mode (30 mW @ 5 V)
- TTL and CMOS-compatible inputs and outputs

BLOCK DIAGRAM

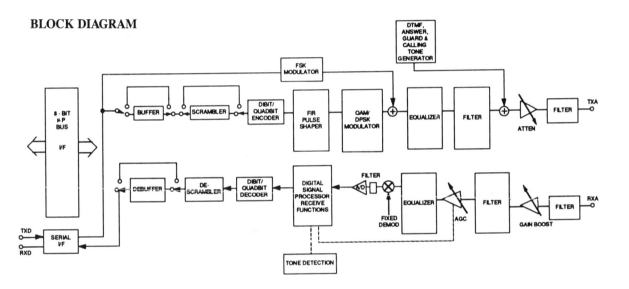

PIN DESCRIPTION

POWER

NAME	28-PIN	32-PIN	44-PIN	TYPE	DESCRIPTION
GND	28	32	44	I	System Ground.
VDD	15	17	23	I	Power supply input, 5V -5% +10%. Bypass with .1 µF and 22 µF capacitors to GND.
VREF	26	30	42	O	An internally generated reference voltage. Bypass with .1 µF capacitor to GND.
ISET	24	27	36	I	Chip current reference. Sets bias current for op-amps. The chip current is set by connecting this pin to VDD through a 2 MΩ resistor. Iset should be bypassed to GND with a .1 µF capacitor.

PARALLEL MICROPROCESSOR INTERFACE

NAME	28-PIN	32-PIN	44-PIN	TYPE	DESCRIPTION
ALE	12	14	20	I	Address latch enable. The falling edge of ALE latches the address on AD0-AD2 and the chip select on $\overline{CS}$.
AD0-AD7	4-11	4, 6-12	4, 9-15	I/O	Address/data bus. These bidirectional tri-state multi-plexed lines carry information to and from the internal registers.
$\overline{CS}$	20	23	32	I	Chip select. A low on this pin allows a read cycle or a write cycle to occur. AD0-AD7 will not be driven and no registers will be written if $\overline{CS}$ (latched) is not active. $\overline{CS}$ is latched on the falling edge of ALE.
CLK	1	1	1	O	Output clock. This pin is selectable under processor control to be either the crystal frequency (for use as a processor clock) or 16 x the data rate for use as a baud rate clock in QAM/DPSK modes only. The pin defaults to the crystal frequency on reset.
$\overline{INT}$	17	19	25	O	Interrupt. This open drain weak pullup, output signal is used to inform the processor that a detect flag has occurred. The processor must then read the detect register to determine which detect triggered the interrupt. $\overline{INT}$ will stay active until the processor reads the detect register or does a full reset.
$\overline{RD}$	14	16	22	I	Read. A low requests a read of the SSI 73K224L internal registers. Data cannot be output unless both $\overline{RD}$ and the latched $\overline{CS}$ are active or low.
RESET	25	28	37	I	Reset. An active high signal on this pin will put the chip into an inactive state. All control register bits (CR0, CR1, CR2, CR3, Tone) will be reset. The output of the CLK pin will be set to the crystal frequency. An internal pull down resistor permits power on reset using a capacitor to VDD.
$\overline{WR}$	13	15	21	I	Write. A low on this informs the SSI 73K224L that data is available on AD0-AD7 for writing into an internal register. Data is latched on the rising edge of $\overline{WR}$. No data is written unless both $\overline{WR}$ and the latched $\overline{CS}$ are active (low).

Note: The serial control mode is provided in the 28-pin version by tying ALE high and $\overline{CS}$ low. In this configuration AD7 becomes DATA and AD0, AD1 and AD2 become A0, A1 and A2, respectively.

ANALOG INTERFACE AND OSCILLATOR

NAME	28-PIN	32-PIN	44-PIN	TYPE	DESCRIPTION
RXA	27	32	43	I	Received modulated analog signal input from the phone line.
TXA	16	18	24	O	Transmit analog output to the phone line.
XTL1 XTL2	2 3	2 3	2 3	I I	These pins are for the internal crystal oscillator requiring a 11.0592 MHz parallel mode crystal. Two capacitors from these pins to ground are also required for proper crystal operation. Consult crystal manufacturer for proper values. XTL2 can also be driven from an external clock.

DTE USER INTERFACE

NAME	28-PIN	32-PIN	44-PIN	TYPE	DESCRIPTION
EXCLK	19	22	31	I	External Clock. This signal is used in synchronous transmission when the external timing option has been selected. In the external timing mode the rising edge of EXCLK is used to strobe synchronous DPSK transmit data available on the TXD pin. Also used for serial control interface.
RXCLK	23	26	35	O	Receive Clock. The falling edge of this clock output is coincident with the transitions in the serial received data output. The rising edge of RXCLK can be used to latch the valid output data. RXCLK will be active as long as a carrier is present.
RXD	22	25	34	O	Received Digital Data Output. Serial receive data is available on this pin. The data is always valid on the rising edge of RXCLK when in synchronous mode. RXD will output constant marks if no carrier is detected.
TXCLK	18	20	26	O	Transmit Clock. This signal is used in synchronous transmission to latch serial input data on the TXD pin. Data must be provided so that valid data is available on the rising edge of the TXCLK. The transmit clock is derived from different sources depending upon the synchronization mode selection. In Internal Mode the clock is generated internally. In External Mode TXCLK is phase locked to the EXCLK pin. In Slave Mode TXCLK is phase locked to the RXCLK pin. TXCLK is always active.
TXD	21	24	33	I	Transmit Digital Data Input. Serial data for transmission is input on this pin. In synchronous modes, the data must be valid on the rising edge of the TXCLK clock. In asynchronous modes (2400/1200/600 bit/s or 300 baud) no clocking is necessary. DPSK data must be +1%, -2.5% or +2.3%, -2.5 % in extended overspeed mode.

ELECTRICAL SPECIFICATIONS

ABSOLUTE MAXIMUM RATINGS

PARAMETER	RATING	UNIT
VDD Supply Voltage	14	V
Storage Temperature	-65 to 150	°C
Soldering Temperature (10 sec.)	260	°C
Applied Voltage	-0.3 to VDD+0.3	V

Note: All inputs and outputs are protected from static charge using built-in, industry standard protection devices and all outputs are short-circuit protected.

RECOMMENDED OPERATING CONDITIONS

PARAMETER	CONDITIONS	MIN	NOM	MAX	UNITS
VDD Supply voltage		4.75	5	5.5	V
External Components (Refer to Application section for placement.)					
VREF Bypass capacitor	(VREF to GND)	0.1			µF
Bias setting resistor	(Placed between VDD and ISET pins)	1.8	2	2.2	MΩ
ISET Bypass capacitor	(ISET pin to GND)	0.1			µF
VDD Bypass capacitor 1	(VDD to GND)	0.1			µF
VDD Bypass capacitor 2	(VDD to GND)	22			µF
XTL1 Load Capacitance	Depends on crystal requirements			40	pF
XTL2 Load Capacitance	Depends on crystal requirements			20	pF
Clock Variation	(11.0592 MHz) Crystal or external clock	-0.01		+0.01	%
TA, Operating Free-Air Temperature		-40		55	°C

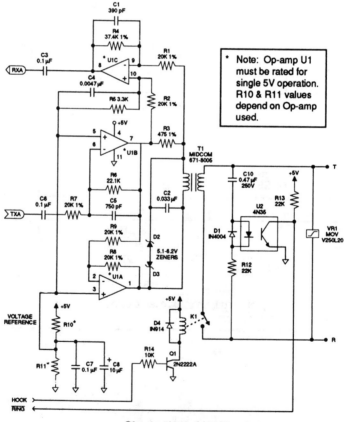

Single 5V Hybrid Version

MIC

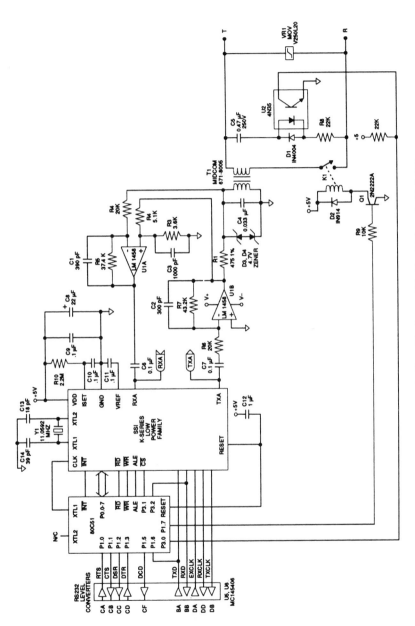

Basic Box Modem with Dual-Supply Hybrid

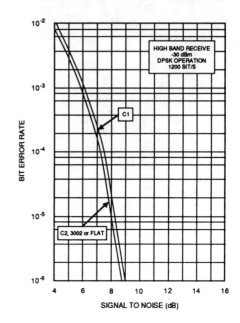
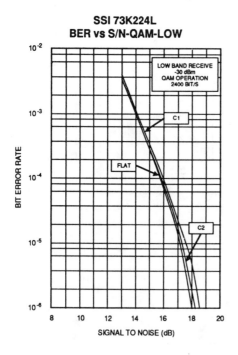
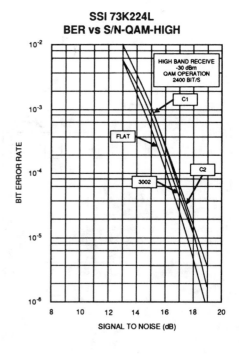

MIC

Silicon Systems SSI 73K224L

PACKAGE PIN DESIGNATIONS

(TOP VIEW)

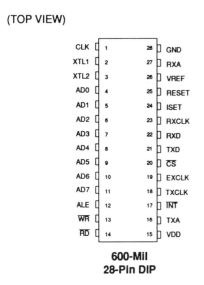

600-Mil
28-Pin DIP

32, 44-Lead PLCC

CAUTION: Use handling procedures necessary for a static sensitive component.

ORDERING INFORMATION

PART DESCRIPTION	ORDER NO.	PKG. MARK
SSI 73K224L with Parallel Bus Interface		
28-Pin Plastic Dual-In-Line	SSI 73K224L – CP	73K224L – CP
32-Pin Plastic Leaded Chip Carrier	SSI 73K224L – 32CH	73K224L – 32CH
44-Pin Plastic Leaded Chip Carrier	SSI 73K224L – CH	73K224L – CH

No responsibility is assumed by Silicon Systems for use of this product nor for any infringements of patents and trademarks or other rights of third parties resulting from its use. No license is granted under any patents, patent rights or trademarks of Silicon Systems. Silicon Systems reserves the right to make changes in specifications at any time without notice. Accordingly, the reader is cautioned to verify that the data sheet is current before placing orders.

Silicon Systems
SSI 73K324L
CCITT V.22bis, V.22, V.21, V.23
Single-Chip Modem

FEATURES

- One-chip multi-mode CCITT V.22bis, V.22, V.21, V.23 compatible modem data pump
- FSK (75, 300/1200 bit/s), DPSK (600, 1200 bit/s), or QAM (2400 bit/s) encoding
- Pin and software compatible with other SSI K-Series family one-chip modems
- Interfaces directly with standard microprocessors (8048, 80C51 typical)
- Serial or parallel microprocessor bus for control
- Selectable asynch/synch with internal buffer/debuffer and scrambler/descrambler functions
- All synchronous (internal, external, slave) and asynchronous operating modes
- Adaptive equalization for optimum performance over all lines
- Programmable transmit attenuation (15-dB and 1-dB steps), and selectable receive boost (+12 dB)
- Call progress, carrier, answer tone, unscrambled mark, S1, SCT (900 Hz) calling tone (1300 Hz), and signal-quality monitors
- DTMF, answer, calling, SCT, and guard-tone generators
- Test modes available: ALB, DL, RDL, Mark, Space, and Alternating bit patterns
- CMOS technology for low power consumption
- 4-wire full-duplex operation in all modes

BLOCK DIAGRAM

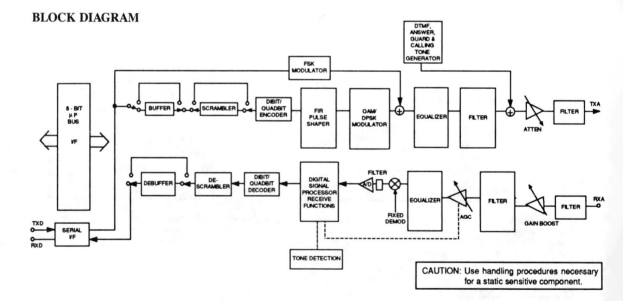

CAUTION: Use handling procedures necessary for a static sensitive component.

☐ MIC

PIN DESCRIPTION

POWER

NAME	28-PIN	32-PIN	TYPE	DESCRIPTION
GND	28	32	I	System Ground.
VDD	15	17	I	Power supply input, 5V ±10%. Bypass with .1 and 22 µF capacitors to GND.
VREF	26	30	O	An internally generated reference voltage. Bypass with .1 µF capacitor to GND.
ISET	24	27	I	Chip current reference. Sets bias current for op-amps. The chip current is set by connecting this pin to VDD through a 2 MΩ resistor. Iset should be bypassed to GND with a .1 µF capacitor.

PARALLEL MICROPROCESSOR INTERFACE

NAME	28-PIN	32-PIN	TYPE	DESCRIPTION
ALE	12	14	I	Address latch enable. The falling edge of ALE latches the address on AD0-AD2 and the chip select on $\overline{CS}$.
AD0-AD7	4-11	4, 6-12	I/O	Address/data bus. These bidirectional tri-state multi-plexed lines carry information to and from the internal registers.
$\overline{CS}$	20	23	I	Chip select. A low on this pin allows a read cycle or a write cycle to occur. AD0-AD7 will not be driven and no registers will be written if $\overline{CS}$ (latched) is not active. $\overline{CS}$ is latched on the falling edge of ALE.
CLK	1	1	O	Output clock. This pin is selectable under processor control to be either the crystal frequency (for use as a processor clock) or 16 x the data rate for use as a baud rate clock in QAM/DPSK modes only. The pin defaults to the crystal frequency on reset.
$\overline{INT}$	17	19	O	Interrupt. This open drain weak pullup, output signal is used to inform the processor that a detect flag has occurred. The processor must then read the detect register to determine which detect triggered the interrupt. $\overline{INT}$ will stay active until the processor reads the detect register or does a full reset.
$\overline{RD}$	14	16	I	Read. A low requests a read of the SSI 73K324L internal registers. Data cannot be output unless both $\overline{RD}$ and the latched $\overline{CS}$ are active or low.
RESET	25	28	I	Reset. An active signal high on this pin will put the chip into an inactive state. All control register bits (CR0, CR1, CR2, CR3, Tone) will be reset. The output of the CLK pin will be set to the crystal frequency. An internal pull down resistor permits power on reset using a capacitor to VDD.
$\overline{WR}$	13	15	I	Write. A low on this informs the SSI 73K324L that data is available on AD0-AD7 for writing into an internal register. Data is latched on the rising edge of $\overline{WR}$. No data is written unless both $\overline{WR}$ and the latched $\overline{CS}$ are low.
Note:				The serial control mode is provided in the 28-pin version by tying ALE high and $\overline{CS}$ low. In this configuration AD7 becomes DATA and AD0, AD1 and AD2 become A0, A1 and A2, respectively.

RS-232 INTERFACE

NAME	28-PIN	32-PIN	TYPE	DESCRIPTION
EXCLK	19	22	I	External Clock. This signal is used in synchronous transmission when the external timing option has been selected. In the external timing mode the rising edge of EXCLK is used to strobe synchronous DPSK transmit data available on the TXD pin. Also used for serial control interface.
RXCLK	23	26	O	Receive Clock Tristate. The falling edge of this clock output is coincident with the transitions in the serial received DPSK data output. The rising edge of RXCLK can be used to latch the valid output data. RXCLK will be valid as long as a carrier is present. In V.23 or V.21 mode a clock which is 16 x 1200/75 or 16 x 300 Hz data rate is output, respectively.
RXD	22	25	O	Received Data Output. Serial receive data is available on this pin. The data is always valid on the rising edge of RXCLK when in synchronous mode. RXD will output constant marks if no carrier is detected.
TXCLK	18	20	O	Transmit Clock Tristate. This signal is used in synchronous DPSK transmission to latch serial input data on the TXD pin. Data must be provided so that valid data is available on the rising edge of the TXCLK. The transmit clock is derived from different sources depending upon the synchronization mode selection. In Internal Mode the clock is 1200 Hz generated internally. In External Mode TXCLK is phase locked to the EXCLK pin. In Slave Mode TXCLK is phase locked to the RXCLK pin. TXCLK is always active. In V.23 or V.21 mode the output is a 16 x 1200/75 or 16 x 300 Hz clock, respectively.
TXD	21	24	I	Transmit Data Input. Serial data for transmission is input on this pin. In synchronous modes, the data must be valid on the rising edge of the TXCLK clock. In asynchronous modes (2400/1200/600 bit/s or 300 baud) no clocking is necessary. DPSK/QAM data must be +1%, -2.5% or +2.3%, -2.5 % in extended overspeed mode.

ANALOG INTERFACE

NAME	28-PIN	32-PIN	TYPE	DESCRIPTION
RXA	27	32	I	Received modulated analog signal input from the phone line.
TXA	16	18	O	Transmit analog output to the phone line.
XTL1 XTL2	2 3	2 3	I I	These pins are for the internal crystal oscillator requiring a 11.0592 MHz parallel mode crystal. Two capacitors from these pins to ground are also required for proper crystal operation. Consult crystal manufacturer for proper values. XTL2 can also be driven from an external clock.

ELECTRICAL SPECIFICATIONS

ABSOLUTE MAXIMUM RATINGS

PARAMETER	RATING	UNIT
VDD Supply Voltage	14	V
Storage Temperature	-65 to 150	°C
Soldering Temperature (10 sec.)	260	°C
Applied Voltage	-0.3 to VDD+0.3	V

Note: All inputs and outputs are protected from static charge using built-in, industry standard protection devices and all outputs are short-circuit protected.

RECOMMENDED OPERATING CONDITIONS

PARAMETER	CONDITIONS	MIN	NOM	MAX	UNITS
VDD Supply voltage		4.5	5	5.5	V
External Components (Refer to Application section for placement.)					
VREF Bypass capacitor	(VREF to GND)	0.1			µF
Bias setting resistor	(Placed between VDD and ISET pins)	1.8	2	2.2	MΩ
ISET Bypass capacitor	(ISET pin to GND)	0.1			µF
VDD Bypass capacitor 1	(VDD to GND)	0.1			µF
VDD Bypass capacitor 2	(VDD to GND)	22			µF
XTL1 Load Capacitance	Depends on crystal requirements			40	pF
XTL2 Load Capacitance	Depends on crystal requirements			20	pF
Clock Variation	(11.0592 MHz) Crystal or external clock	-0.01		+0.01	%
TA, Operating Free-Air Temperature		-40		55	°C

Silicon Systems SSI 73K324L

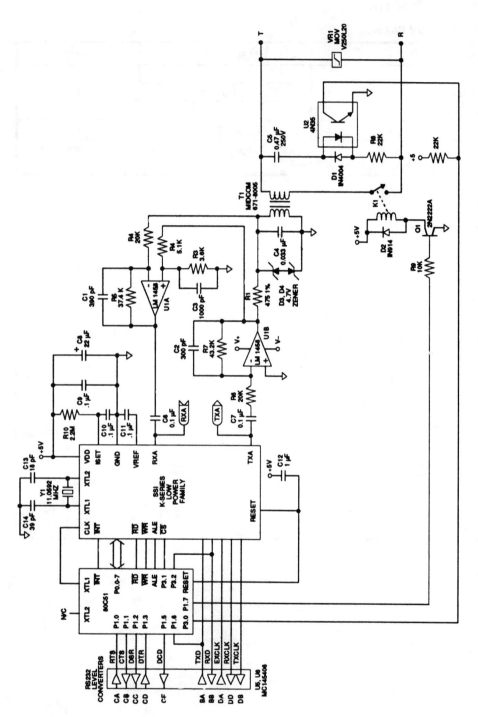

Basic Box Modem with Dual-Supply Hybrid

☐ **MIC**

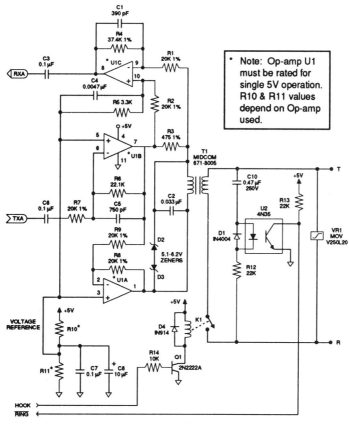

* Note: Op-amp U1 must be rated for single 5V operation. R10 & R11 values depend on Op-amp used.

Single 5V Hybrid Version

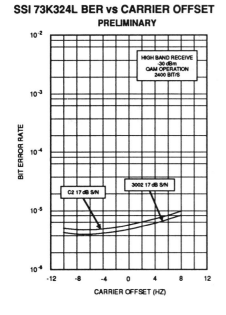

SSI 73K324L BER vs CARRIER OFFSET
PRELIMINARY

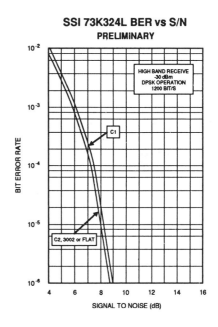

SSI 73K324L BER vs S/N
PRELIMINARY

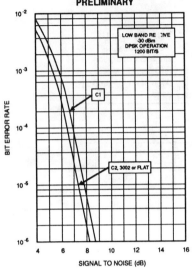

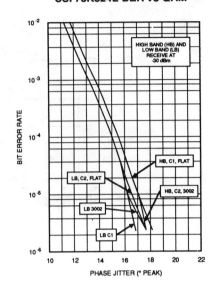

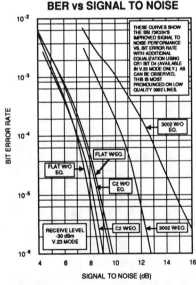

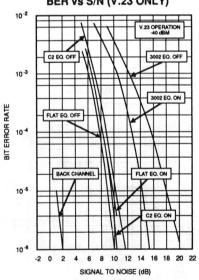

* = "EQ On" Indicates bit CR1 D4 is set for additional phase equalization.

** = 73K302L performance is similar to that of the 73K322L. V.23 operation corresponds to Bell 202.

PACKAGE PIN DESIGNATIONS
(TOP VIEW)

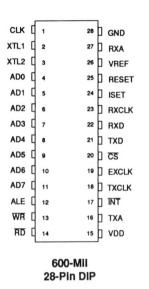

600-MII
28-Pin DIP

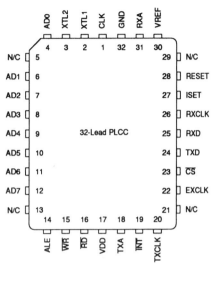

32-Lead PLCC

Advance Information: Indicates a product still in the design cycle, and any specifications are based on design goals only. Do not use for final design.

No responsibility is assumed by Silicon Systems for use of this product nor for any infringements of patents and trademarks or other rights of third parties resulting from its use. No license is granted under any patents, patent rights or trademarks of Silicon Systems. Silicon Systems reserves the right to make changes in specifications at any time without notice. Accordingly, the reader is cautioned to verify that the data sheet is current before placing orders.

Silicon Systems
Setting DTMF Levels for 1200 Bit/s K-Series Modems

Some applications of the K-series modems without output-level adjustment might require setting the DTMF-transmit level to something other than the normally transmitted level. This level is nominally about 5 dB higher than during data transmission. If the data is transmitted at -10 dBm, the DTMF levels will be at about -5 dBm, which is adequate in most applications.

The simplest way to change the relative levels of DTMF tones and data is to change the transmit gain during dialing. This can be accomplished as shown in the following figures. In this example, it is assumed that the DTMF tones are to be transmitted at a higher level than normal. Closing relay K2 will increase the gain of the transmit op amp and allow a higher DTMF tone level during dialing. If it is desired to decrease the DTMF level, the relay can be open for dialing and closed for data. The value of the shunt resistor, Rdtmf, will be relatively large compared to the resistor R1; therefore the precision of Rdtmf is not as critical as R1. This means an analog switch or similar device could be used instead of relay, with the on resistance of the switch not seriously affecting the tolerance of the gain setting.

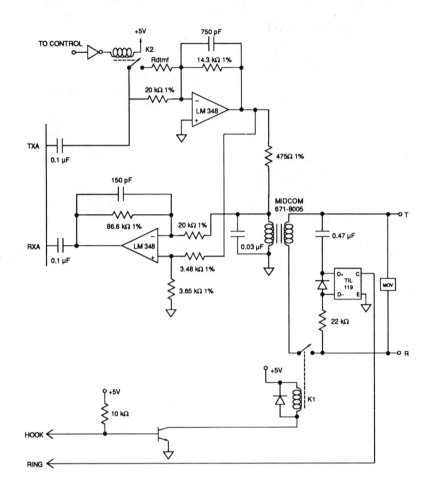

Silicon Systems
SSI 34P570
2-Channel Floppy-Disk Read/Write Device

FEATURES
- Single-chip read/write amplifier and read data-processing function
- Compatible with 8″, 5 1/4″, and 3 1/2″ drives
- Internal write and erase current sources, externally set
- Control signals are TTL compatible
- Schmitt trigger inputs for higher noise immunity on bused control signals
- TTL selectable write current boost
- Operates on +12-V and +5-V power supplies
- High-gain, low-noise, low-peak shift (0.3% typical) read-processing circuits

ORDERING INFORMATION

PART DESCRIPTION	ORDER NO.	PKG. MARK
SSI 34P570 28-Pin DIP	SSI 34P570-CP	34P570-CP
SSI 34P570 28-Pin PLCC	SSI 34P570-CH	34P570-CH

No responsibility is assumed by Silicon Systems for use of this product nor for any infringements of patents and trademarks or other rights of third parties resulting from its use. No license is granted under any patents, patent rights or trademarks of Silicon Systems. Silicon Systems reserves the right to make changes in specifications at any time without notice. Accordingly, the reader is cautioned to verify that the data sheet is current before placing orders.

PACKAGE PIN DESIGNATIONS
(TOP VIEW)

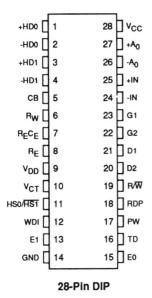

28-Pin DIP

THERMAL CHARACTERISTICS: $\emptyset$ ja

28-Pin	DIP	55°C/W
28-Pin	PLCC	65°C/W

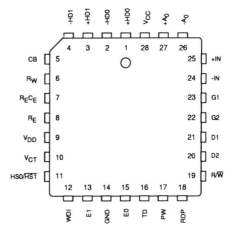

28-Pin PLCC

BLOCK DIAGRAM

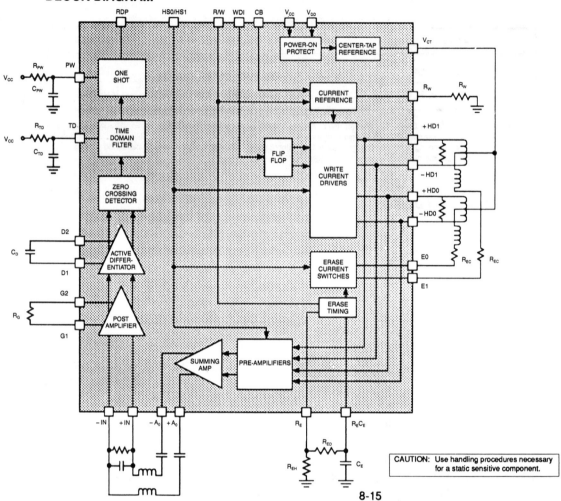

8-15

ABSOLUTE MAXIMUM RATINGS (Operating above absolute maximum ratings may damage the device.)

PARAMETER	RATING	UNIT
5 V Supply Voltage, V_{CC}	7	V
12 V Supply Voltage, V_{DD}	14	V
Storage Temperature	65 to +130	°C
Junction Operating Temperature	130	°C
Logic Input Voltage	-0.5 V to 7.0 V	dc
Lead Temperature (Soldering, 10 sec.)	260	°C
Power Dissipation	800	mW

POWER SUPPLY CURRENTS

PARAMETER	CONDITIONS	MIN	NOM	MAX	UNIT
I_{CC} - 5 V Supply Current	Read Mode			35	mA
	Write Mode			38	mA
I_{DD} - 12 V Supply Current	Read Mode			26	mA
	Write Mode (excluding Write & Erase currents)			24	mA

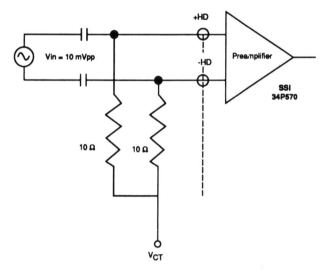

Preamplifier Characteristics

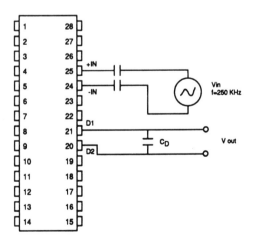

Postamplifier Differential Output Voltage Swing and Voltage Gain

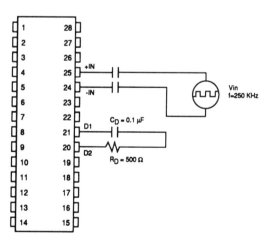

Postamplifier Threshold Differential Input Voltage

Silicon Systems
SSI 34R575
2- or 4-Channel Floppy Disk Read/Write Device

FEATURES
- Operates on +5-V and +12-V power supplies
- Two- or four-channel capability
- TTL-compatible control inputs
- Read/Write functions on one-chip
- Internal center-tap voltage source
- Supports all disk sizes
- Applicable to tape systems

ORDERING INFORMATION

PART DESCRIPTION	ORDER NO.	PKG. MARK
SSI 34R575 24-Pin DIP	SSI 34R575-4CP	34R575-4CP
SSI 34R575 18-Pin DIP	SSI 34R575-2CP	34R575-2CP

No responsibility is assumed by Silicon Systems for use of this product nor for any infringements of patents and trademarks or other rights of third parties resulting from its use. No license is granted under any patents, patent rights or trademarks of Silicon Systems. Silicon Systems reserves the right to make changes in specifications at any time without notice. Accordingly, the reader is cautioned to verify that the data sheet is current before placing orders.

PACKAGE PIN DESIGNATIONS

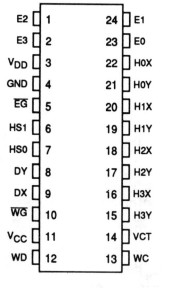

24-Pin DIP

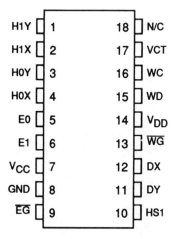

18-Pin DIP

MIC

Silicon Systems SSI 34R575

BLOCK DIAGRAM

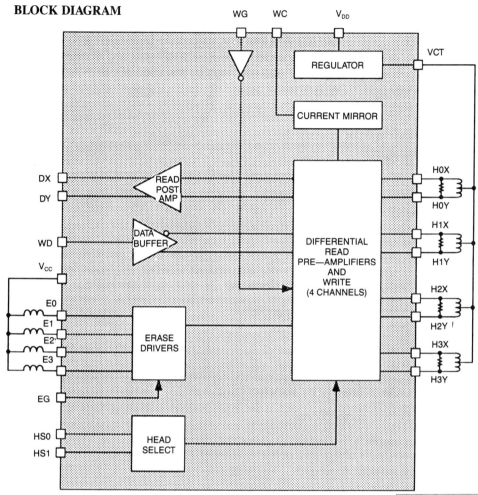

CAUTION: Use handling procedures necessary for a static sensitive component.

RECOMMENDED OPERATING CONDITIONS (0 °C < Ta < 50 °C, 4.7 V < V_{CC} < 5.3 V, 11 V < V_{DD} < 13 V)

PARAMETER	CONDITIONS	MIN	NOM	MAX	UNIT
Vcc Supply Current					
Read mode	Vcc MAX			15	mA
Write mode	Vcc MAX			35	mA
VDD Supply Current					
Read mode	VDD MAX			25	mA
Write mode	VDD MAX			15	mA
Write Current			5.5		mA

ABSOLUTE MAXIMUM RATINGS

(Operating above absolute maximum ratings may damage the device.)

PARAMETER		RATING	UNIT
DC Supply Voltage:	Vcc	6.0	V
	Vdd	14.0	V
Write Current		10	mA
Head Port Voltage		18.0	V
Digital Input Voltages:	DX, DY, HS0, HS1, WD	-0.3 to + 10	V
	$\overline{EG}$, $\overline{WG}$	-0.3 to V_{cc} + 0.3	V
DX, DY Output Current		-5	mA
VCT Output Current		-10	mA
Storage Temperature Range		-65 to + 150	°C
Junction Temperature		125	°C
Lead Temperature (Soldering, 10 sec.)		260	°C

Silicon Systems
SSI 34B580
Port-Expander Floppy-Disk Drive

FEATURES
- Reduces package count in flexible disk-drive systems
- Replaces bus interface and combinational logic devices between the SSI 34P570, on board microprocessor and mechanical interfaces
- Surface-mount available for further real-estate reduction
- Provides drive capability for mechanical and system interfaces

ORDERING INFORMATION

PART DESCRIPTION	ORDER NO.	PKG. MARK
SSI 34B580 28-Pin DIP	SSI 34B580-CP	34B580-CP
SSI 34B580 28-Pin PLCC	SSI 34B580-CH	34B580-CH

No responsibility is assumed by Silicon Systems for use of this product nor for any infringements of patents and trademarks or other rights of third parties resulting from its use. No license is granted under any patents, patent rights or trademarks of Silicon Systems. Silicon Systems reserves the right to make changes in specifications at any time without notice. Accordingly, the reader is cautioned to verify that the data sheet is current before placing orders.

PACKAGE PIN DESIGNATIONS
(TOP VIEW)

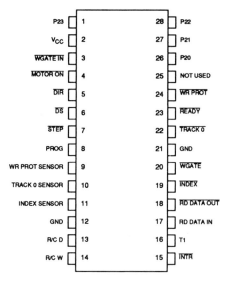

28-Pin DIP

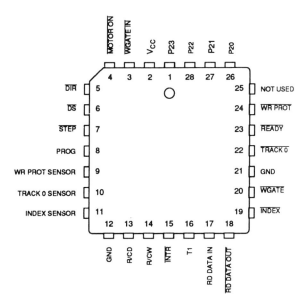

28-Pin PLCC

Silicon Systems SSI 34B580

BLOCK DIAGRAM

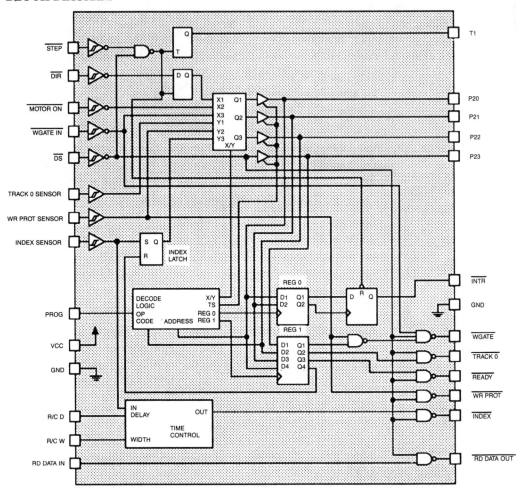

CAUTION: Use handling procedures necessary for a static sensitive component.

ABSOLUTE MAXIMUM RATINGS (All voltages referred to GND)
Operation above absolute maximum ratings may permanently damage the device.

PARAMETER	RATING	UNIT
DC Supply	+7	VDC
Voltage Range (any pin to GND)	-0.4 to +7	VDC
Power Dissipation	700	mW
Storage Temperature	-40 to +125	°C
Lead Temperature (10 sec soldering)	260	°C

INPUT TO PORT2		READ FROM PORT 2				4-BIT Input Port
OP Code P22	Addr. P20	P23	P22	P21	P20	
0	0	$\overline{DS}$	Index Sensor Latch	WR Sensor	Track 0 Sensor	B
0	1	$\overline{DS}$	$\overline{WGATEIN}$	$\overline{MOTORON}$	$\overline{DIR}$	A

Read Mode

INPUT TO PORT2		DATA PROCESSED FROM PORT 2				Index Latch Reset
OP Code P22	Addr. P20	$\overline{WGATE}$	$\overline{TRACK0}$	$\overline{READY}$	$\overline{INTR}$	
1	0	Z	$\overline{(P22 \cdot DS)}$	$\overline{(P21 \cdot DS)}$		P20
1	1				See Text	
Where Z = (P23 · WR PROT SENSOR) + $\overline{(DS \cdot WGATEIN)}$						

Write Mode

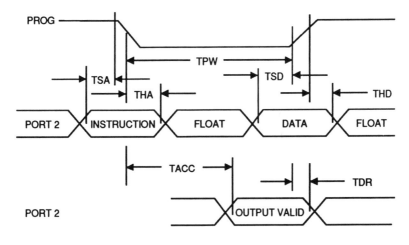

Timing Diagram

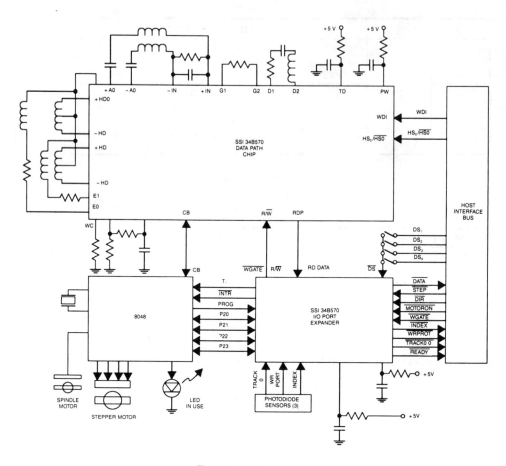

Typical Application

Silicon Systems
SSI 34D441
Data Synchronizer and Write-Precompensator Device

FEATURES
- Ideal for operation with NEC μPD7265A/μPD7265
- Fast acquisition analog PLL for precise read-data synchronization
- No adjustments or trims needed to external components
- Programmable-data rate, up to 1 Mbit/s
- Internal crystal-controlled oscillator
- Selectable write precompensation intervals
- Programmable-write clock
- DRQ (Data DMA Request) delay function
- Low-power CMOS, +5-V operation
- 28-pin PDIP and 28-pin PLCC

ORDERING INFORMATION

PART DESCRIPTION	ORDER NO.	PKG. MARK
SSI 34D441 28-pin PDIP	SSI 34D441-CP	34D441-CP
SSI 34D441 28-pin PLCC	SSI 34D441-CH	34D441-CH

No responsibility is assumed by Silicon Systems for use of this product nor for any infringements of patents and trademarks or other rights of third parties resulting from its use. No license is granted under any patents, patent rights or trademarks of Silicon Systems. Silicon Systems reserves the right to make changes in specifications at any time without notice. Accordingly, the reader is cautioned to verify that the data sheet is current before placing orders.

BLOCK DIAGRAM

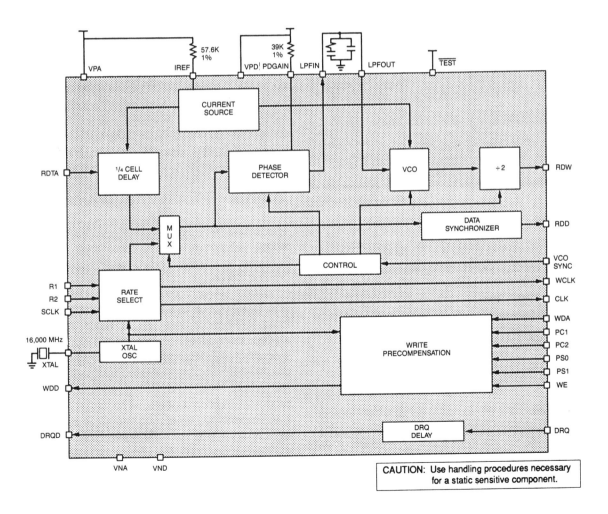

CAUTION: Use handling procedures necessary for a static sensitive component.

PACKAGE PIN DESIGNATIONS

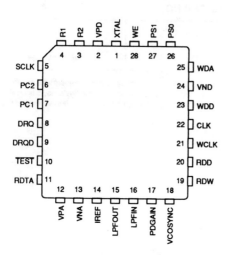

28-lead PLCC
PLCC pinouts are the same as the 28-pin DIP

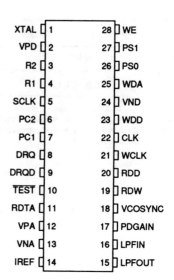

600-mil
28-pin DIP

THERMAL CHARACTERISTICS: θja

28-lead PLCC	55°C/W
28-pin PDIP	65°C/W

ABSOLUTE MAXIMUM RATINGS

PARAMETER	RATING	UNIT
Storage Temperature	-40 to +120	°C
Ambient Operating Temperature, TA	0 to +70	°C
Supply Voltages, VPD, VPA	-0.5 to +7.0	VDC
Voltage Applied to Logic inputs	-0.5 to +7.0	VDC
Voltage Supplied to Logic Outputs	-0.5 to +5.5	VDC
Maximum Power Dissipation	750	mW

RECOMMENDED OPERATING CONDITIONS

PARAMETER	CONDITIONS	MIN	NOM	MAX	UNIT
Ambient Temperature, TA		0		70	°C
Power Supply Voltage, VPD, VPA		4.75	5	5.25	VDC
High Level Input Voltage, VIH	Power supply = 4.75V	2.0			V
Input Current High, IIH	Power supply = 4.75V, VIH = 2.4V			20	µA
Low Level Input Voltage, VIL	Power supply = 4.75V			0.8	V
Input Current Low, IIH	Power supply = 5.25V, VIL = 0.4V			-20	µA
High Level Output Voltage, VOH	Power supply = 4.75V, IOH = 4 mA	2.4			V
Low Level Output Voltage All others, VOL	Power supply = 4.75V, IOL = 8 mA			0.4	V
Short Circuit Output Current WDD only IOS (to positive supply)	Power supply = 5.25V	20		150	mA

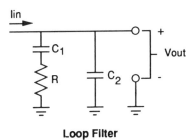

Loop Filter

DATA RATE	LOCK TIME	LOOP FILTER
125 kHz	192 µs	R = 10 kΩ, C_1 = 6800 pF, C_2 = 360 pF
250 kHz	96 µs	R = 10 kΩ, C_1 = 3300 pF, C_2 = 180 pF
500 kHz	46 µs	R = 11 kΩ, C_1 = 1500 pF, C_2 = 82 pF
1 MHz	24 µs	R = 13 kΩ, C_1 = 680 pF, C_2 = 39 pF

Silicon Systems SSI 34D441

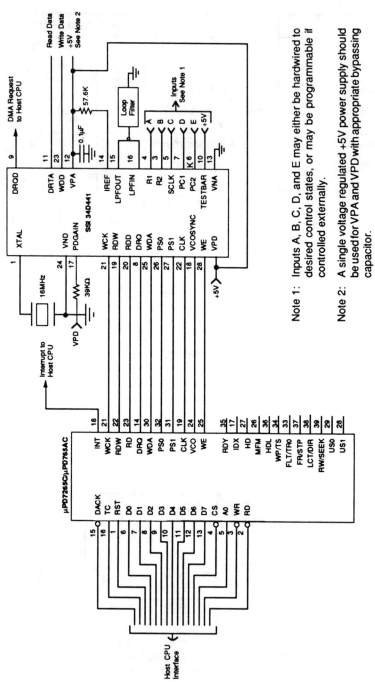

Application Diagram

Note 1: Inputs A, B, C, D, and E may either be hardwired to desired control states, or may be programmable if controlled externally.

Note 2: A single voltage regulated +5V power supply should be used for VPA and VPD with appropriate bypassing capacitor.

CHAPTER 7

POWER SUPPLIES TEST EQUIPMENT, AND INSTRUMENTS

Allegro
2429
Fluid Detector

FEATURES
- High output current
- ac or dc output
- Single-wire probe
- Low external parts count
- Internal voltage regulator
- Reverse voltage protection

ABSOLUTE MAXIMUM RATINGS

Supply Voltage, V_{CC}
   (continuous) .... -50 V to +16 V
   (1 hr. at +25°C) .... 24 V
   (10 µs) .... 50 V
Output Voltage, V_{OUT}
   (ULN2429A) .... 30 V
   (ULN2429A-1) .... 50 V
Output Current, I_{OUT}
   (continuous) .... 700 mA
   (1 hr. at +25°C) .... 1.0 A
Package Power Dissipation, P_D .... 1.33 W*
Operating Temperature Range,
   T_A .... -40°C to +85°C
Storage Temperature Range,
   T_S .... -65°C to +150°C

* Derate at the rate of 16.67 mW/°C above T_A = +70°C.

Always order by complete part number:

Part Number	Maximum V_{OUT}
ULN2429A	30 V
ULN2429A-1	50 V

ELECTRICAL CHARACTERISTICS at $T_A = +25°C$, $V_{CC} = V_{OUT} = 12$ V (unless otherwise specified).

Characteristic	Symbol	Test Pin	Test Conditions	Limits			Units
				Min.	Typ.	Max.	
Supply Voltage Range	V_{CC}	13	Operating	10	—	16	V
Supply Current	I_{CC}	13	$V_{CC} = 16$ V	—	—	10	mA
Oscillator Output Voltage	V_{OSC}	6	$R_L = 18$ kΩ	—	3.0	—	V_{PP}
Output ON Voltage	V_{OUT}	1, 14	$R_L \geq 30$ kΩ, $I_{OUT} = 500$ mA	—	0.9	1.5	V
Output OFF Current	I_{OUT}	1, 14	$R_L \leq 10$ kΩ, $V_{OUT} = V_{OUT}$(max)	—	—	100	µA
Oscillator Frequency	f_{OSC}	6	$R_L = 18$ kΩ	—	2.4	—	kHz

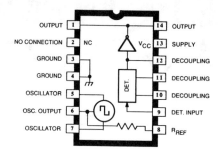

CIRCUIT SCHEMATIC

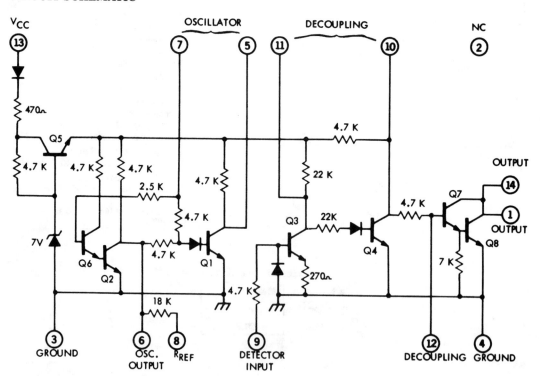

TEST CIRCUIT

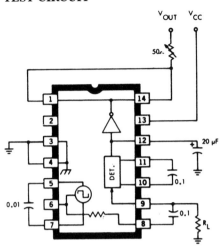

TYPICAL APPLICATIONS

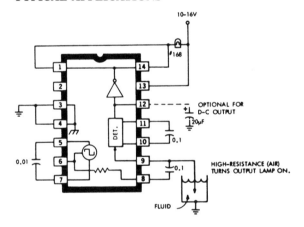

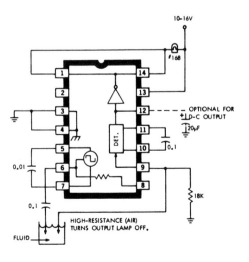

Allegro 2453, 2454, 2455
Automotive Lamp Monitors

FEATURES
- No standby power
- Integral to wiring assembly
- Fail-safe
- Reverse-voltage protected
- Internal transient protection
- DIP or SOIC plastic packages

ABSOLUTE MAXIMUM RATINGS
at +25°C Free-Air Temperature

Supply Voltage, V_{CC} .... 30 V
Peak Supply Voltage, V_{CC}(100 ms) ... 80 V
Peak Reverse Voltage, V_R .... 30 V
Output Current, I_{OUT} .... 35 mA
Package Power Dissipation, P_D . **See Graph**
Operating Temperature Range,
  T_A .... -40°C to +85°C
Storage Temperature Range,
  T_S .... -65°C to +150°C

Always order by complete part number:

Part Number	Function	Style
ULQ2453M	Dual Comparator	8-Pin Mini-DIP
ULN2454L	Dual Comparator with OR Output	8-Lead SOIC
ULN2454M	Dual Comparator with OR Output	8-Pin Mini-DIP
ULN2455A	Quad Comparator	14-Pin DIP
ULN2455L	Quad Comparator	14-Lead SOIC

PIN OUT & FUNCTIONAL BLOCK DIAGRAMS

ULQ2453M

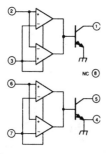

ULN2454L and ULN2454M

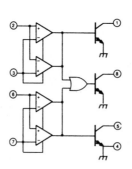

ULN2455A and ULN2455L

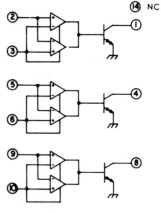

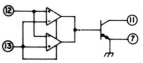

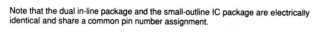

Note that the dual in-line package and the small-outline IC package are electrically identical and share a common pin number assignment.

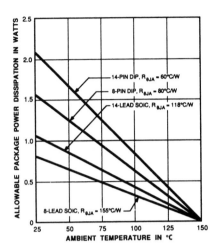

SIMPLIFIED SCHEMATIC
(SINGLE DIFFERENTIAL SENSE AMPLIFIER)

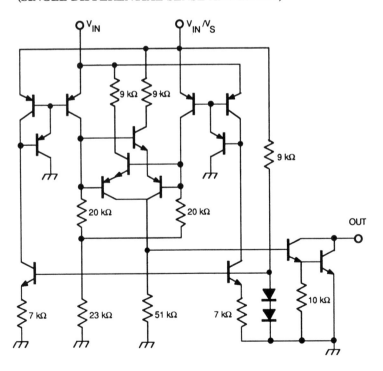

ELECTRICAL CHARACTERISTICS at $T_A = +25°C$, $V_{CC} = V_{IN} = 10$ to 16 V (unless otherwise noted).

Characteristic	Symbol	Test Conditions	Min.	Typ.	Max.	Units
Output Leakage Current	I_{CEX}	$V_{OUT} = 80$ V, $\Delta V_{IN} < 7$ mV	—	—	100	µA
Output Saturation Voltage	$V_{CE(SAT)}$	$I_{OUT} = 5$ mA, $\Delta V_{IN} > 20$ mV	—	0.8	1.0	V
		$I_{OUT} = 30$ mA, $\Delta V_{IN} > 20$ mV	—	1.4	2.0	V
Differential Switch Voltage	ΔV_{IN}	$V_{IN} - V_{IN}/V_S$	7.0	13	20	mV
Input Current	I_{IN}	$\Delta V_{IN} = V_{IN} - V_{IN}/V_S = +30$ mV	150	300	800	µA
	I_{IN}/I_S	$\Delta V_{IN} = V_{IN} - V_{IN}/V_S = -30$ mV	0.5	1.7	3.5	mA

BASIC BRIDGE MONITORING SYSTEM

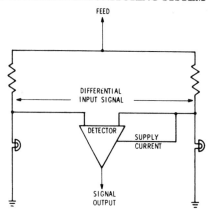

TYPICAL SWITCH CHARACTERISTICS

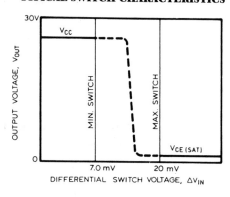

POWER SUPPLY SUPERVISORY CIRCUIT

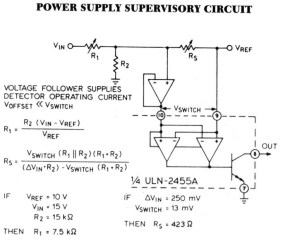

QUAD LAMP MONITOR

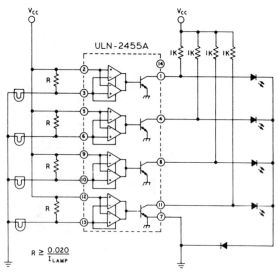

PLANAR GAS-DISCHARGE DISPLAY DRIVER

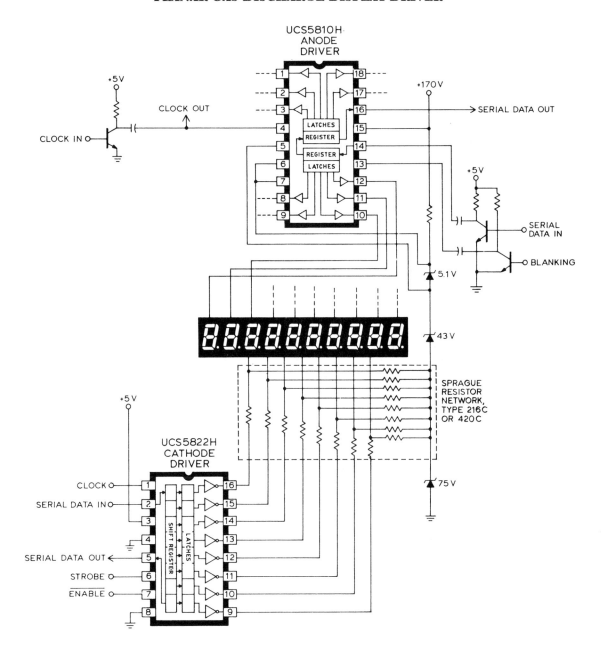

ALPHANUMERIC VACUUM-FLUORESCENT DISPLAY DRIVERS

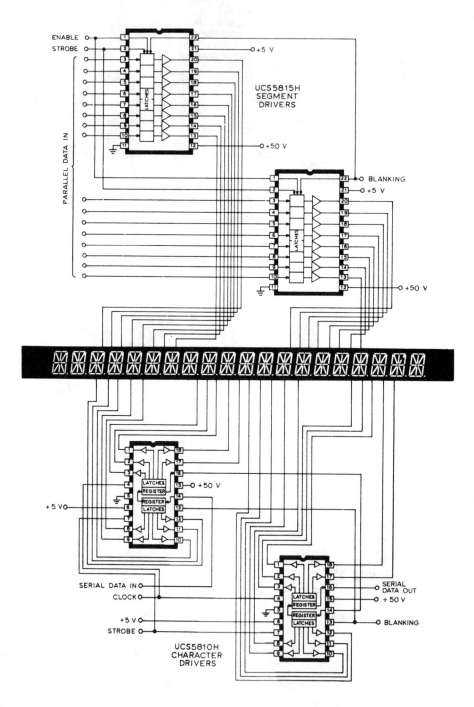

Allegro 3501
Linear-Output Hall-Effect Sensors

Utilizing the Hall effect for sensing a magnetic field, UGN3501U and UGN3501UA integrated circuits provide a linear single-ended output that is a function of magnetic-field intensity.

These devices can sense relatively small changes in a magnetic field—changes that are too small to operate a Hall effect switch. They can be capacitively coupled to an amplifier, to boost the output to a higher level.

The UGN3501U/UA include a Hall cell, linear amplifier, emitter-follower output, and a voltage regulator. Integrating the Hall cell and the amplifier into one monolithic device minimizes problems related to the handling of millivolt analog signals.

Both devices are rated for continuous operation over the temperature range of 0°C to +70°C and over a supply-voltage range of 8 V to 12 V.

FEATURES
- Excellent sensitivity
- Flat response to 25 kHz (typ.)
- Internal voltage regulation
- Excellent temperature stability

ABSOLUTE MAXIMUM RATINGS

Supply Voltage, V_{CC} .... **16 V**
Output Current, I_{OUT} .... **4 mA**
Magnetic Flux Density, B .... **Unlimited**
Operating Temperature Range,
   T_A .... **0°C to +70°C**
Storage Temperature Range,
   T_S .... **-65°C to +150°C**

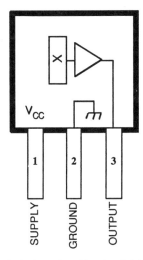

Pinning is shown viewed from branded side.

FUNCTIONAL BLOCK DIAGRAM

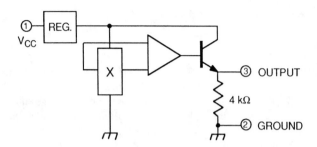

ELECTRICAL CHARACTERISTICS at $T_A = +25°C$, $V_{CC} = 12$ V

Characteristic	Symbol	Test Conditions	Min.	Typ.	Max.	Units
Operating Voltage	V_{CC}		8.0	—	12	V
Supply Current	I_{CC}	$V_{CC} = 12$ V	—	10	20	mA
Quiescent Output Voltage	V_{OUT}	B = 0 G, Note 1	2.5	3.6	5.0	V
Sensitivity	ΔV_{OUT}	B = 1000 G, Notes 1, 2	0.35	0.7	—	mV/G
Frequency Response	BW	$f_H - f_L$ at -3 dB	—	25	—	kHz
Broadband Output Noise	e_n	f = 10 Hz to 10 kHz	—	0.1	—	mV
Output Resistance	R_0		—	100	—	Ω

NOTE 1. All output voltage measurements are made with a voltmeter having an input impedance of 10 kΩ or greater.
NOTE 2. Magnetic flux density is measured at the most sensitive area of the device, which is 0.017" ± 0.001" (0.43 ± 0.03 mm) below the branded side of the "U" package.

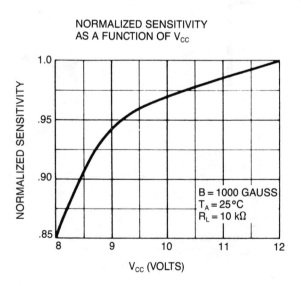

NORMALIZED SENSITIVITY AS A FUNCTION OF V_{CC}

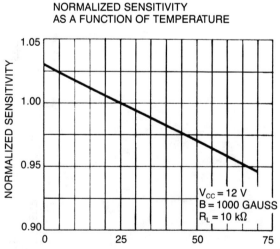

NORMALIZED SENSITIVITY AS A FUNCTION OF TEMPERATURE

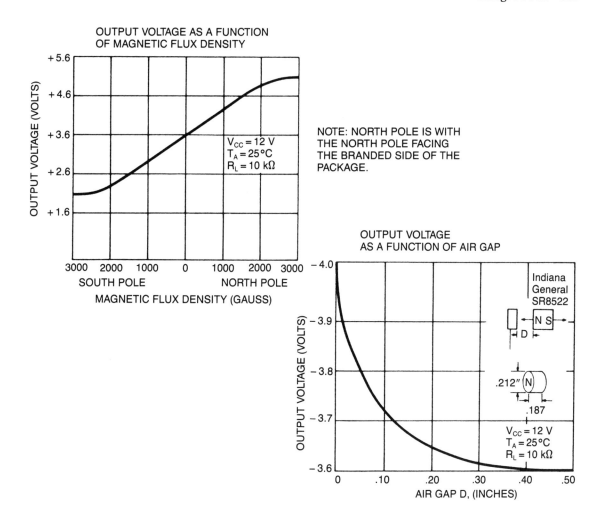
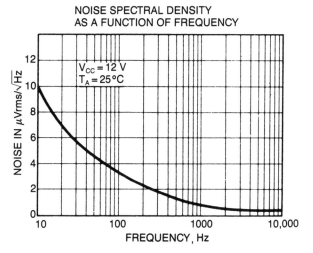

TYPICAL APPLICATIONS

SENSITIVE PROXIMITY DETECTOR

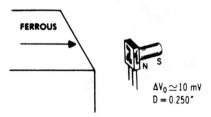

FERROUS METAL SENSOR

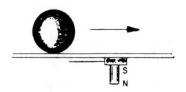

LOBE OR COG SENSOR

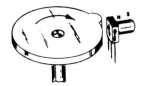

NOTCH OR HOLE SENSOR

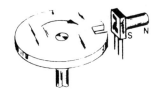

Allegro 3503
Ratiometric, Linear Hall-Effect Sensors

Type UGN3503U/UA and UGS3503U/UA Hall-effect sensors accurately track extremely small changes in magnetic flux density—changes that are generally too small to operate Hall-effect switches.

As motion detectors, gear-tooth sensors, and proximity detectors, they are magnetically driven mirrors of mechanical events. As sensitive monitors of electromagnets, they can effectively measure a system's performance with negligible system loading, while providing isolation from contaminated and electrically noisy environments.

Each Hall-effect integrated circuit includes a Hall-sensing element, linear amplifier, and emitter-follower output stage. Problems associated with handling tiny analog signals are minimized by having the Hall cell and amplifier on a single chip.

The UGN3503U and UGN3503UA are rated for continuous operation over the temperature range of $-20\,°C$ to $+85\,°C$. The UGS3503U and UGS3503UA operate over an extended temperature range of $-40\,°C$ to $+125\,°C$.

FEATURES
- Extremely sensitive
- Flat response to 23 kHz
- Low-noise output
- 4.5-V to 6-V operation
- Magnetically optimized package

ABSOLUTE MAXIMUM RATINGS

Supply Voltage, V_{CC} .... 8 V
Magnetic Flux Density, B ...... Unlimited
Operating Temperature Range, T_A
    UGN3503U/UA .... -20°C to +85°C
    UGS3503U/UA ...... -40°C to +125°C
Storage Temperature Range,
    T_S .... -65°C to +150°C

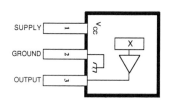

Pinning is shown viewed from branded side.

FUNCTIONAL BLOCK DIAGRAM

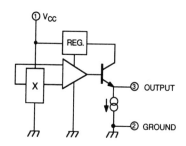

ELECTRICAL CHARACTERISTICS at $T_A = 25°C$, $V_{CC} = 5$ V

Characteristic	Symbol	Test Conditions	Min.	Typ.	Max.	Units
Operating Voltage	V_{CC}		4.5	—	6.0	V
Supply Current	I_{CC}		—	9.0	14	mA
Quiescent Output Voltage	V_{OUT}	B = 0 G	2.25	2.50	2.75	V
Sensitivity	ΔV_{OUT}	B = 0 G to ±900 G	0.75	1.30	1.72	mV/G
Bandwidth (-3 dB)	BW		—	23	—	kHz
Broadband Output Noise	V_{out}	BW = 10 Hz to 10 kHz	—	90	—	μV
Output Resistance	R_{OUT}		—	50	—	Ω

All output-voltage measurements are made with a voltmeter having an input impedance of at least 10 kΩ.
Magnetic flux density is measured at most sensitive area of device located 0.016" ±0.002" (0.41 mm ±0.05 mm) below the branded face of the 'U' package.

OUTPUT VOLTAGE AS A FUNCTION OF TEMPERATURE

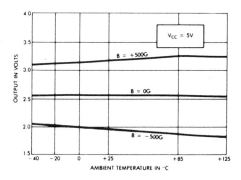

OUTPUT NOISE AS A FUNCTION OF FREQUENCY

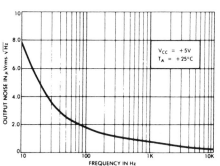

SUPPLY CURRENT AS A FUNCTION OF SUPPLY VOLTAGE

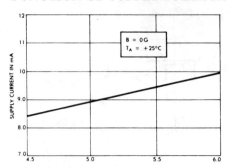

DEVICE SENSITIVITY AS A FUNCTION OF SUPPLY VOLTAGE

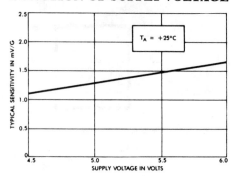

OUTPUT NULL VOLTAGE AS A FUNCTION OF SUPPLY VOLTAGE

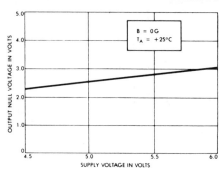

LINEARITY AND SYMMETRY AS A FUNCTION OF SUPPLY VOLTAGE

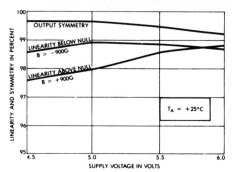

NOTCH SENSOR

GEAR TOOTH SENSOR

CURRENT MONITOR

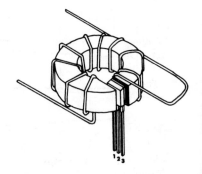

SENSOR LOCATIONS

SUFFIX "U"

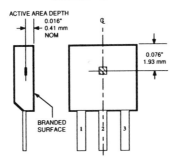

SUFFIX "UA"

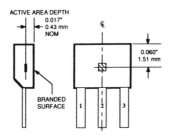

Allegro 3311 and 3312
Precision Light Sensors with Calibrated Current Amplifiers

Direct replacements for photocells and phototransistors, the ULN3311T and ULN3312T precision light sensors are two-terminal monolithic integrated circuits that linearly convert light level into electrical current. The light-controlled current sources are linear over a wide range of supply voltages and light levels and require no external calibration.

Each precision light sensor (PLS) consists of a photodiode and a calibrated current amplifier. The design of the amplifier allows derivation of its supply current from the same terminal as the photodiode cathode and amplifier output. Because this supply current is a linear function of the photodiode current, it acts as part of the signal current. On-chip resistor-trimming techniques are used during manufacture to adjust each PLS to specified sensitivity. A 100-μW/cm^2 GaAIAs LED emission provides the light source for this calibration.

The ULN3311T and ULN3312T are supplied in an inexpensive clear plastic package. Both devices are rated for operation over the temperature range of $-40\,°$C to $+70\,°$C.

FEATURES
- Two-terminal operation
- Linear over a wide range
- Precalibrated
- Wide operating-voltage range
- High output

ABSOLUTE MAXIMUM RATINGS

Supply Voltage, V_{CC} .... **24 V**
Operating Temperature Range,
   T_A .... **$-40\,°$C to $+70\,°$C**
Storage Temperature Range,
   T_S .... **$-55\,°$C to $+110\,°$C**

SCHEMATIC

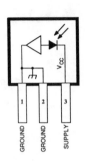

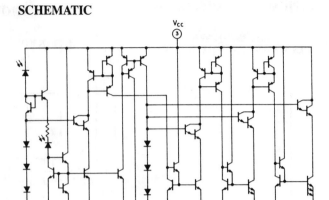

UNPACKAGED CHIP
0.053" × 0.077" (1.35 mm × 1.96 mm)

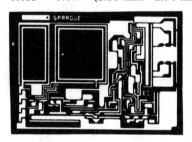

ELECTRICAL CHARACTERISTICS at $T_A = +25°C$, $V_{CC} = 12$ V

Characteristics	Device Type	Limits			Units
		Min.	Typ.	Max.	
Initial Accuracy at 100 µW/cm²	Both	—	—	±7.2	%
Sensitivity	ULN3311T	280	—	350	nA/µW/cm²
	ULN3312T	350	—	420	nA/µW/cm²
Operating Voltage Range	Both	2.7	12	24	V
Output Linearity, 10 to 10k µW/cm²	Both	—	—	±7.2	%
Dark Current	Both	—	—	100	nA
Power Supply Rejection, $(\Delta I_o/I_o)\Delta V$	Both	40	50	—	dB
Temperature Coefficient of Sensitivity	Both	—	3500	—	ppM/°C

NOTE: Light source is an infrared LED with a peak output wavelength of 880 nm.

TYPICAL CHARACTERISTICS

RELATIVE SPECTRAL RESPONSE AS A FUNCTION OF WAVELENGTH OF LIGHT

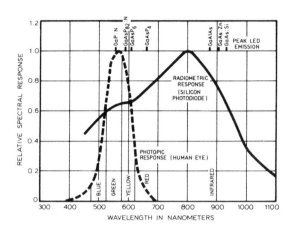

OUTPUT CURRENT AS A FUNCTION OF ILLUMINANCE

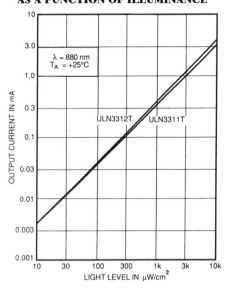

PROPABATION DELAY AS A FUNCTION OF ILLUMINANCE

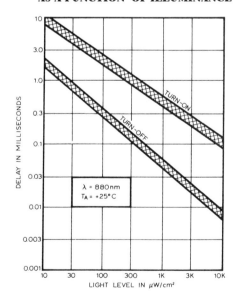

OUTPUT CURRENT AS A FUNCTION OF SUPPLY VOLTAGE

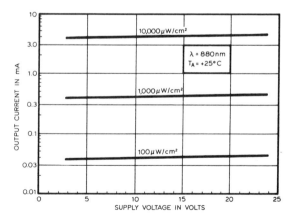

LIGHT-LEVEL DETECTOR USING PLS

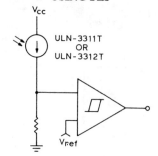

DIFFERENTIAL EDGE DETECTOR

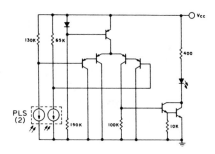

SENSOR-CENTER LOCATION

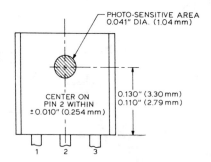

Analog Devices AD834
500-MHz Four-Quadrant Multiplier

FEATURES
- dc to >500-MHz operation
- Differential ±1-V full-scale inputs
- Differential ±4-mA full-scale output current
- Low distortion (≤0.05% for 0-dBm input)
- Supply voltages from ±4 V to ±9 V
- Low power (280 mW typical at $V_S = \pm 5$ V)

APPLICATIONS
- High-speed real-time computation
- Wideband modulation and gain control
- Signal correlation and RF power measurement
- Voltage-controlled filters and oscillators
- Linear keyers for high-resolution television
- Wideband true RMS

PRODUCT HIGHLIGHTS

1. The AD834 combines high static accuracy (low input and output offsets and accurate scale factor) with very high bandwidth. As a four-quadrant multiplier or squarer, the response extends from dc to an upper frequency limited mainly by packaging and external board layout considerations. A large signal bandwidth of over 500 MHz is attainable under optimum conditions.

2. The AD834 can be used in many high-speed nonlinear operations, such as square rooting, analog division, vector addition and rms-to-dc conversion. In these modes, the bandwidth is limited by the external active components.

3. Special design techniques result in low distortion levels (better than −60 dB on either input) at high frequencies and low signal feedthrough (typically −65 dB up to 20 MHz).

4. The AD834 exhibits low differential phase error over the input range—typically 0.08° at 5 MHz and 0.8° at 50 MHz. The large signal transient response is free from overshoot, and has an intrinsic rise time of 500 ps, typically settling to within 1% in under 5 ns.

5. The nonloading, high impedance, differential inputs simplify the application of the AD834.

ABSOLUTE MAXIMUM RATINGS[1]

Supply Voltage ($+V_S$ to $-V_S$)................18V
Internal Power Dissipation...................500mW
Input Voltages (X1, X2, Y1, Y2)................$+V_S$
Operating Temperature Range
    AD834J..........................0 to +70°C
    AD834A......................−40°C to +85°C
    AD834S/883B................−55°C to +125°C
Storage Temperature Range Q..........−65°C to +150°C
Storage Temperature Range R, N........−65°C to +125°C
Lead Temperature, Soldering 60sec.............+300°C

NOTE
[1]Stresses above those listed under "Absolute Maximum Ratings" may cause permanent damage to the device. This is a stress rating only and functional operation of the device at these or any other conditions above those indicated in the operational section of this specification is not implied. Exposure to absolute maximum rating conditions for extended periods may affect device reliability.

Analog Devices AD834

SPECIFICATIONS ($T_A = +25°C$ and $\pm V_S = \pm 5$ V, unless otherwise noted; dBm assumes 50 Ω load.)

Model	Conditions	AD834J Min	AD834J Typ	AD834J Max	AD834A, S Min	AD834A, S Typ	AD834A, S Max	Units
MULTIPLIER PERFORMANCE								
Transfer Function			$W = \dfrac{XY}{(1V)^2} \times 4mA$			$W = \dfrac{XY}{(1V)^2} \times 4mA$		
Total Error[1] (Figure 6)	$-1V \leq X, Y < +1V$		±0.5	±2		±0.5	±2	% FS
vs. Temperature	T_{min} to T_{max}					±1.5	±3	% FS
vs. Supplies[2]	±4V to ±6V		0.1	0.3		0.1	0.3	% FS/V
Linearity[3]			±0.5	±1		±0.5	±1	% FS
Bandwidth[4]	See Figure 5	500			500			MHz
Feedthrough, X	X=±1V, Y=Nulled		0.2	0.3		0.2	0.3	% FS
Feedthrough, Y	X=Nulled, Y=±1V		0.1	0.2		0.1	0.2	% FS
AC Feedthrough, X[5]	X=0dBm, Y=Nulled							
	f=10MHz		−65			−65		dB
	f=100MHz		−50			−50		dB
AC Feedthrough, Y[5]	X=Nulled, Y=0dBm							
	f=10MHz		−70			−70		dB
	f=100MHz		−50			−50		dB
INPUTS (X1, X2, Y1, Y2)								
Full Scale Range	Differential		±1			±1		V
Clipping Level	Differential	±1.1	±1.3		±1.1	±1.3		V
Input Resistance	Differential		25			25		kΩ
Offset Voltage			0.5	3		0.5	3	mV
vs. Temperature	T_{min} to T_{max}		10			10		μV/°C
				4			4	mV
vs. Supplies[2]	±4V to ±6V		100	300		100	300	μV/V
Bias Current			45			45		μA
Common Mode Rejection	f≤100kHz; 1V p-p		70			70		dB
Nonlinearity, X	Y=1V; X=±1V		0.2	0.5		0.2	0.5	% FS
Nonlinearity, Y	X=1V; Y=±1V		0.1	0.3		0.1	0.3	% FS
Distortion, X	X=0dBm, Y=1V							
	f=10MHz		−60			−60		dB
	f=100MHz		−44			−44		dB
Distortion, Y	X=1V, Y=0dBM							
	f=10MHz		−65			−65		dB
	f=100MHz		−50			−50		dB
OUTPUTS (W1, W2)								
Zero Signal Current	Each Output		8.5			8.5		mA
Differential Offset	X=0, Y=0		±20	±60		±20	±60	μA
vs. Temperature	T_{min} to T_{max}		40			40		nA/°C
							±60	μA
Scaling Current	Differential	3.96	4	4.04	3.96	4	4.04	mA
Output Compliance		4.75		9	4.75		9	V
Noise Spectral Density	f=10Hz to 1MHz Outputs into 50Ω Load		16			16		nV/$\sqrt{Hz}$
POWER SUPPLIES								
Operating Range		±4		±9	±4		±9	V
Quiescent Current[6]	T_{min} to T_{max}							
$+V_S$			11	14		11	14	mA
$-V_S$			28	35		28	35	mA
TEMPERATURE RANGE								
Operating, Rated Performance								
Commercial (0 to +70°C)			AD834J					
Military (−55°C to +125°C)						AD834S		
Industrial (−40°C to +85°C)						AD834A		
PACKAGE OPTIONS								
8-Pin SOIC (R)			AD834JR					
8-Pin Cerdip (Q)						AD834AQ		
8-Pin Plastic DIP (N)			AD834JN			AD834SQ/883B		

NOTE
[1] Error is defined as the maximum deviation from the ideal output, and expressed as a percentage of the full scale output.
[2] Both supplies taken simultaneously; sinusoidal input at f≤10kHz.
[3] Linearity is defined as residual error after compensating for input offset voltage, output offset current and scaling current errors.
[4] Bandwidth is guaranteed when configured in squarer mode. See Figure 5.
[5] Sine input; relative to full scale output; zero input port nulled; represents feedthrough of the fundamental.
[6] Negative supply current is equal to the sum of positive supply current, the signal currents into each output, W1 and W2, and the input bias currents.

Specifications in **boldface** are tested on all production units at final electrical test. Results from those tests are used to calculate outgoing quality levels.
Specifications subject to change without notice.

CONNECTION DIAGRAM

Small Outline (R) Package
Plastic DIP (N) Package
Cerdip (Q) Package

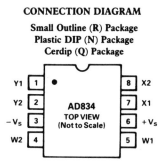

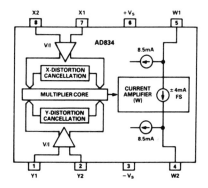

AD834 Functional Block Diagram

TYPICAL CHARACTERISTICS

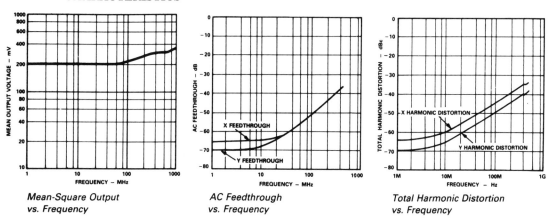

Mean-Square Output vs. Frequency

AC Feedthrough vs. Frequency

Total Harmonic Distortion vs. Frequency

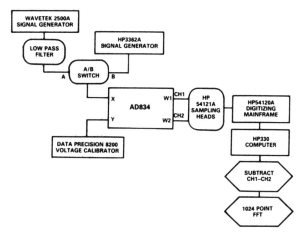

Test Configuration for Measuring ac Feedthrough and Total Harmonic Distortion

Analog Devices AD834

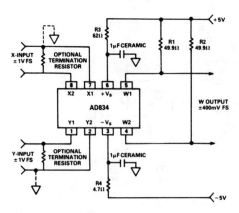

Basic Connections for Wideband Operation

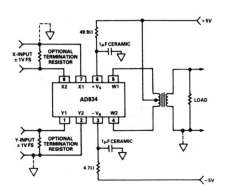

Transformer–Coupled Output

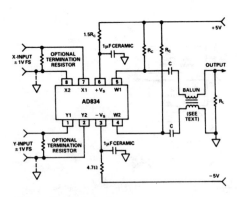

Using a Balun at the Output

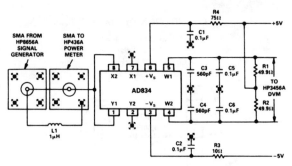

Bandwidth Test Circuit

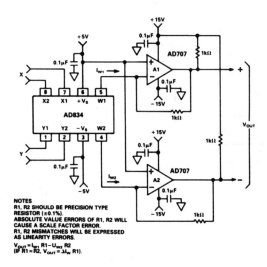

Low Frequency Test Circuit

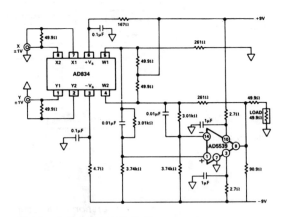

Wideband dc-Coupled Multiplier

Analog Devices AD834

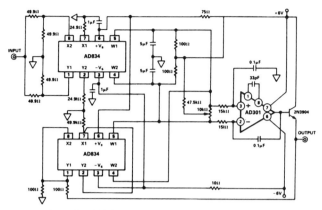

Connections for Wideband rms Measurement

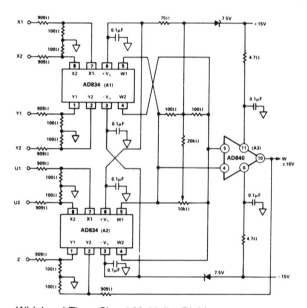

Wideband Three Signal Multiplier/Divider

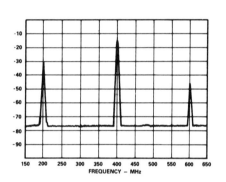

Output Spectrum for Configuration of Figure 14

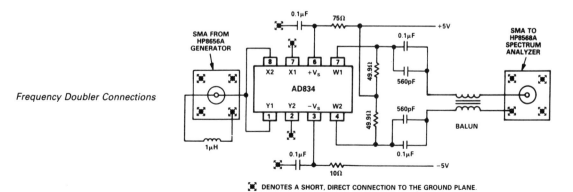

Frequency Doubler Connections

Analog Devices
AD532
Internally Trimmed Integrated Circuit Multiplier

FEATURES
- Pretrimmed to ±1.0% (AD532K)
- No external components required
- Guaranteed ±1.0% max. 4-quadrant error (AD532K)
- Diff. inputs for $(X_1-X_2)(Y_1-Y_2)/10$ V transfer function
- Monolithic construction, low cost

APPLICATIONS
- Multiplication, division, squaring, square rooting
- Algebraic computation
- Power measurements
- Instrumentation applications
- Available in chip form

GUARANTEED PERFORMANCE OVER TEMPERATURE
The AD532J and AD532K are specified for maximum multiplying errors of ±2% and ±1% of full scale, respectively at +25°C, and are rated for operation from 0 to +70°C. The AD532S has a maximum multiplying error of ±1% of full scale at +25°C; it is also 100% tested to guarantee a maximum error of ±4% at the extended operating temperature limits of −55°C and +125°C. All devices are available in either the hermetically sealed TO-100 metal can, TO-116 ceramic DIP, or LCC packages. J, K and S grade chips are also available.

ADVANTAGES OF ON-THE-CHIP TRIMMING OF THE MONOLITHIC AD532
1. True ratiometric trim for improved power supply rejection.
2. Reduced power requirements because no networks across supplies are required.
3. More reliable because standard monolithic assembly techniques can be used, rather than more complex hybrid approaches.
4. High-impedance X and Y inputs with negligible circuit loading.
5. Differential X and Y inputs for noise rejection and additional computational flexibility.

SPECIFICATIONS (@ +25°C, $V_S = \pm15$ V, $R \geq 2$ kΩ V_{OS} grounded)

Model	AD532J Min	AD532J Typ	AD532J Max	AD532K Min	AD532K Typ	AD532K Max	AD532S Min	AD532S Typ	AD532S Max	Units
MULTIPLIER PERFORMANCE										
Transfer Function		$\frac{(X_1-X_2)(Y_1-Y_2)}{10V}$			$\frac{(X_1-X_2)(Y_1-Y_2)}{10V}$			$\frac{(X_1-X_2)(Y_1-Y_2)}{10V}$		
Total Error ($-10V \leq X, Y \leq +10V$)		±1.5	±2.0		±0.7	±1.0		±0.5	±1.0	%
T_A = min to max		±2.5			±1.5				±4.0	%
Total Error vs Temperature		±0.04			±0.03			±0.01	±0.04	%/°C
Supply Rejection (±15V ±10%)		±0.05			±0.05			±0.05		%/%
Nonlinearity, X (X = 20V pk-pk, Y = 10V)		±0.8			±0.5			±0.5		%
Nonlinearity, Y (Y = 20V pk-pk, X = 10V)		±0.3			±0.2			±0.2		%
Feedthrough, X (Y Nulled, X = 20V pk-pk 50Hz)		50	200		30	100		30	100	mV
Feedthrough, Y (X Nulled, Y = 20V pk-pk 50Hz)		30	150		25	80		25	80	mV
Feedthrough vs. Temp.		2.0			1.0			1.0		mV p-p/°C
Feedthrough vs. Power Supply		±0.25			±0.25			±0.25		mV/%
DYNAMICS										
Small Signal BW (V_{OUT} = 0.1 rms)		1			1			1		MHz
1% Amplitude Error		75			75			75		kHz
Slew Rate (V_{OUT} 20 pk-pk)		45			45			45		V/μs
Settling Time (to 2%, ΔV_{OUT} = 20V)		1			1			1		μs
NOISE										
Wideband Noise f = 5Hz to 10kHz		0.6			0.6			0.6		mV (rms)
f = 5Hz to 5MHz		3.0			3.0			3.0		mV (rms)

Analog Devices AD532

Model	AD532J Min	AD532J Typ	AD532J Max	AD532K Min	AD532K Typ	AD532K Max	AD532S Min	AD532S Typ	AD532S Max	Units
MULTIPLIER PERFORMANCE										
OUTPUT										
Output Voltage Swing	±10	±13		±10	±13		±10	±13		V
Output Impedance (f≤1kHz)		1			1			1		Ω
Output Offset Voltage		±40				±30			±30	mV
Output Offset Voltage vs. Temp.		0.7			0.7			2.0		mV/°C
Output Offset Voltage vs. Supply		±2.5			±2.5			±2.5		mV/%
INPUT AMPLIFIERS (X, Y and Z)										
Signal Voltage Range (Diff. or CM Operating Diff)		±10			±10			±10		V
CMRR	40			50			50			dB
Input Bias Current										
X, Y Inputs		3			1.5	4		1.5	4	μA
X, Y Inputs T_{min} to T_{max}		10			8			8		μA
Z Input		±10			±5	±15		±5	±15	μA
Z Input T_{min} to T_{max}		±30			±25			±25		μA
Offset Current		±0.3			±0.1			±0.1		μA
Differential Resistance		10			10			10		MΩ
DIVIDER PERFORMANCE										
Transfer Function ($X_1 > X_2$)		$10V Z/(X_1-X_2)$			$10V Z/(X_1-X_2)$			$10V Z/(X_1-X_2)$		
Total Error										
($V_X = -10V, -10V \le V_Z \le +10V$)		±2			±1			±1		%
($V_X = -1V, -10V \le V_Z \le +10V$)		±4			±3			±3		%
SQUARE PERFORMANCE										
Transfer Function		$\frac{(X_1-X_2)^2}{10V}$			$\frac{(X_1-X_2)^2}{10V}$			$\frac{(X_1-X_2)^2}{10V}$		
Total Error		±0.8			±0.4			±0.4		%
SQUARE-ROOTER PERFORMANCE										
Transfer Function		$-\sqrt{10VZ}$			$-\sqrt{10VZ}$			$-\sqrt{10VZ}$		
Total Error ($0V \le V_Z \le 10V$)		±1.5			±1.0			±1.0		%
POWER SUPPLY SPECIFICATIONS										
Supply Voltage										
Rated Performance		±15			±15			±15		V
Operating	±10		±18	±10		±18	±10		±22	V
Supply Current										
Quiescent		4	6		4	6		4	6	mA
PACKAGE OPTIONS										
TO-116 (D-14)	AD532JD			AD532KD			AD532SD			
TO-100 (H-10a)	AD532JH			AD532KH			AD532SH			
LCC (E-20A)								AD532SE		

NOTE
Specifications subject to change without notice.

Specifications shown in boldface are tested on all production units at final electrical test. Results from those tests are used to calculate outgoing quality levels. All min and max specifications are guaranteed, although only those shown in boldface are tested on all production units.

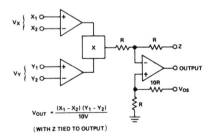

$$V_{OUT} = \frac{(X_1 - X_2)(Y_1 - Y_2)}{10V}$$

(WITH Z TIED TO OUTPUT)

Functional Block Diagram

AD532 PIN CONFIGURATIONS

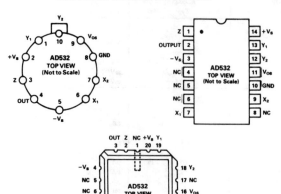

NC ARE NO CONNECT PINS

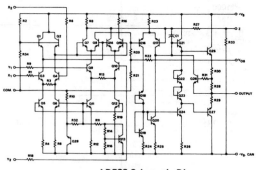

AD532 Schematic Diagram

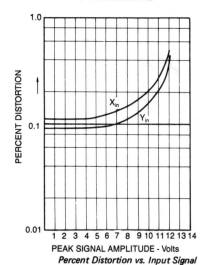

Percent Distortion vs. Input Signal

Percent Distortion vs. Frequency

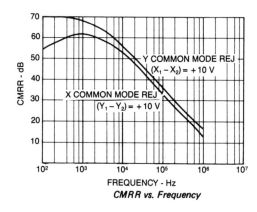

CMRR vs. Frequency

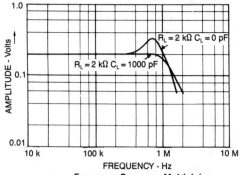

Frequency Response, Multiplying

Analog Devices AD532

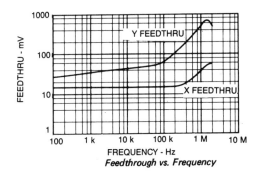
Feedthrough vs. Frequency

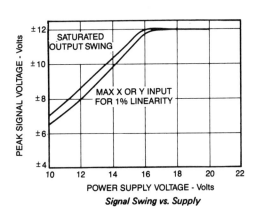
Signal Swing vs. Supply

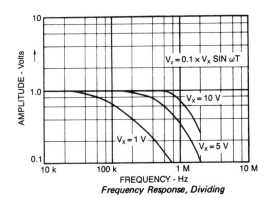
Frequency Response, Dividing

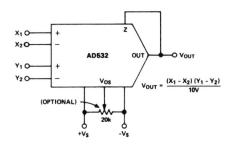
Multiplier Connection

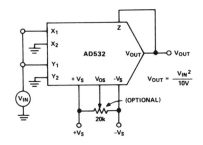
Squarer Connection

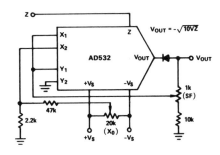
Square Rooter Connection

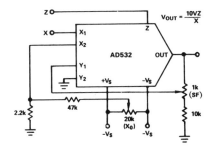
Divider Connection

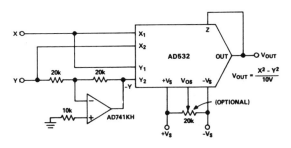
Differential of Squares Connection

Analog Devices
AD521
Integrated Circuit Precision Instrumentation Amplifier

FEATURES
- Programmable gains from 0.1 to 1000
- Differential inputs
- High CMRR: 110 dB min.
- Low drift: 2 μV/°C max. (L)
- Complete input protection, power on and power off
- Functionally complete with the addition of two resistors
- Internally compensated
- Gain bandwidth product: 40 MHz
- Output current limited: 25 mA
- Very low noise: 0.5 μV p-p, 0.1 Hz to 10 Hz, RTI @ G=1000
- Chips are available

PRODUCT HIGHLIGHTS
1. The AD521 is a true instrumentation amplifier in integrated circuit form, offering the user performance that is comparable to many modular instrumentation amplifiers at a fraction of the cost.
2. The AD521 has low guaranteed input offset-voltage drift (2 μV/°C for L grade) and low noise for precision, high-gain applications.
3. The AD521 is functionally complete with the addition of two resistors. Gain can be preset from 0.1 to more than 1000.
4. The AD521 is fully protected for input levels up to 15 V beyond the supply voltages and 30 V differential at the inputs.
5. Internally compensated for all gains, the AD521 also offers the user the provision for limiting bandwidth.
6. Offset nulling can be achieved with an optional trim pot.
7. The AD521 offers superior dynamic performance with a gain-bandwidth product of 40 MHz, full peak response of 100 kHz (independent of gain) and a settling time of 5 μs to 0.1% of a 10-V step.

SPECIFICATIONS (typical @ $V_S = \pm 15$ V, $R_L = 2$ kΩ and $T_A = 25$ °C unless otherwise specified)

MODEL	AD521JD	AD521KD	AD521LD	AD521SD (AD521SD/883B)
GAIN				
Range (For Specified Operation, Note 1)	1 to 1000	•	•	•
Equation	G = R_S/R_G V/V	•	•	•
Error from Equation	($\pm$0.25 $-$0.004G)%	•	•	•
Nonlinearity (Note 2)				
$1 \leq G \leq 1000$	0.2% max	•	0.1% max	•
Gain Temperature Coefficient	$\pm$(3 $\pm$0.05G)ppm/°C	•	•	$\pm$(15 $\pm$0.4G)ppm/°C
OUTPUT CHARACTERISTICS				
Rated Output	$\pm$10V, $\pm$10mA min	•	•	•
Output at Maximum Operating Temperature	$\pm$10V @ 5mA min	•	•	•
Impedance	0.1Ω	•	•	•
DYNAMIC RESPONSE				
Small Signal Bandwidth ($\pm$3dB)				
G = 1	>2MHz	•	•	•
G = 10	300kHz	•	•	•
G = 100	200kHz	•	•	•
G = 1000	40kHz	•	•	•
Small Signal, $\pm$1.0% Flatness				
G = 1	75kHz	•	•	•
G = 10	26kHz	•	•	•
G = 100	24kHz	•	•	•
G = 1000	6kHz	•	•	•
Full Peak Response (Note 3)	100kHz	•	•	•
Slew Rate, $1 \leq G \leq 1000$	10V/μs	•	•	•
Settling Time (any 10V step to within 10mV of Final Value)				
G = 1	7μs	•	•	•
G = 10	5μs	•	•	•
G = 100	10μs	•	•	•
G = 1000	35μs	•	•	•

MODEL	AD521JD	AD521KD	AD521LD	AD521SD (AD521SD/883B)
Differential Overload Recovery (±30V Input to within 10mV of Final Value) (Note 4)				
G = 1000	50µs	*	*	**
Common Mode Step Recovery (30V Input to within 10mV of Final Value) (Note 5)				
G = 1000	10µs	*	*	**
VOLTAGE OFFSET (may be nulled)				
Input Offset Voltage (V_{OS_I})	3mV max (2mV typ)	1.5mV max (0.5mV typ)	1.0mV max (0.5mV typ)	**
vs. Temperature	15µV/°C max (7µV/°C typ)	5µV/°C max (1.5µV/°C typ)	2µV/°C max	**
vs. Supply	3µV/%	*	*	*
Output Offset Voltage (V_{OS_O})	400mV max (200mV typ)	200mV max (30mV typ)	100mV max	**
vs. Temperature	400µV/°C max (150µV/°C typ)	150µV/°C max (50µV/°C typ)	75µV/°C max	**
vs. Supply (Note 6)	$0.005 V_{OS_O}/\%$	*	*	*
INPUT CURRENTS				
Input Bias Current (either input)	80nA max	40nA max	**	**
vs. Temperature	1nA/°C max	500pA/°C max	**	**
vs. Supply	2%/V	*	*	*
Input Offset Current	20nA max	10nA max	**	**
vs. Temperature	250pA/°C max	125pA/°C max	**	**
INPUT				
Differential Input Impedance (Note 7)	$3 \times 10^9 \Omega \| 1.8pF$	*	*	*
Common Mode Input Impedance (Note 8)	$6 \times 10^{10} \Omega \| 3.0pF$	*	*	*
Input Voltage Range for Specified Performance (with respect to ground)	±10V	*	*	*
Maximum Voltage without Damage to Unit, Power ON or OFF Differential Mode (Note 9)	30V	*	*	*
Voltage at either input (Note 9)	V_S ±15V	*	*	*
Common Mode Rejection Ratio, DC to 60Hz with 1kΩ source unbalance				
G = 1	70dB min (74dB typ)	74dB min (80dB typ)	**	**
G = 10	90dB min (94dB typ)	94dB min (100dB typ)	**	**
G = 100	100dB min (104dB typ)	104dB min (114dB typ)	**	**
G = 1000	100dB min (110dB typ)	110dB min (120dB typ)	**	**
NOISE				
Voltage RTO (p-p) @ 0.1Hz to 10Hz (Note 10)	$\sqrt{(0.5G)^2 + (225)^2}$ µV	*	*	*
RMS RTO, 10Hz to 10kHz	$\sqrt{(1.2G)^2 + (50)^2}$ µV	*	*	*
Input Current, rms, 10Hz to 10kHz	15pA (rms)	*	*	*
REFERENCE TERMINAL				
Bias Current	3µA	*	*	*
Input Resistance	10MΩ	*	*	*
Voltage Range	±10V	*	*	*
Gain to Output	1	*	*	*
POWER SUPPLY				
Operating Voltage Range	±5V to ±18V	*	*	*
Quiescent Supply Current	5mA max	*	*	*
TEMPERATURE RANGE				
Specified Performance	0 to +70°C	*	*	-55°C to +125°C
Operating	-25°C to +85°C	*	*	-55°C to +125°C
Storage	-65°C to +150°C	*	*	*
PACKAGE OPTION				
Ceramic (D-14)	AD521JD	AD521KD	AD521LD	AD521SD

NOTES

*Specifications same as AD521JD.
**Specifications same as AD521KD.
Specifications subject to change without notice.

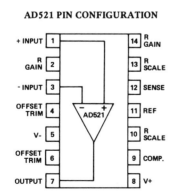

AD521 PIN CONFIGURATION

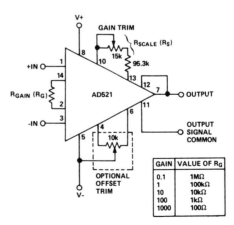

Operating Connections for AD521

Analog Devices AD521

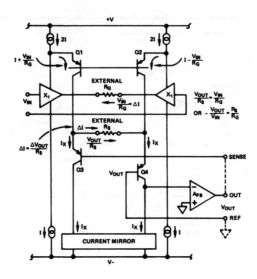

Simplified AD521 Schematic

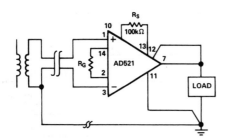

a). Transformer Coupled, Direct Return

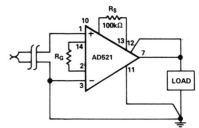

b). Thermocouple, Direct Return

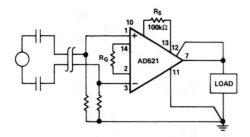

c). AC Coupled, Indirect Return

Ground Returns for "Floating" Transducers

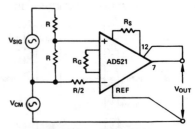

1. INCREASE R_G TO PICK UP GAIN LOST BY R DIVIDER NETWORK
2. INPUT SIGNAL MUST BE REDUCED IN PROPORTION TO POWER SUPPLY VOLTAGE LEVEL

Operating Conditions for $V_{IN} \approx V_S = 10V$

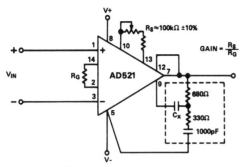

$$C_X = \frac{1}{100\pi f_t} \text{ when } f_t \text{ is the desired bandwidth.}$$

(f_t in kHz, C_X in μF)

Optional Compensation Circuit

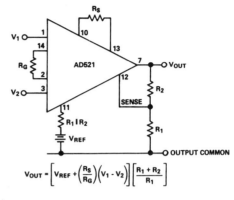

$$V_{OUT} = \left[V_{REF} + \left(\frac{R_S}{R_G} \right)(V_1 - V_2) \right]\left[\frac{R_1 + R_2}{R_1} \right]$$

Circuit for utilizing some of the unique features of the AD521. Note that gain changes introduced by changing R1 and R2 will have a minimum effect on output offset if the offset is carefully nulled at the highest gain setting.

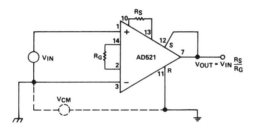

Ground loop elimination. The reference input, Pin 11, allows remote referencing of ground potential. Differences in ground potentials are attenuated by the high CMRR of the AD521.

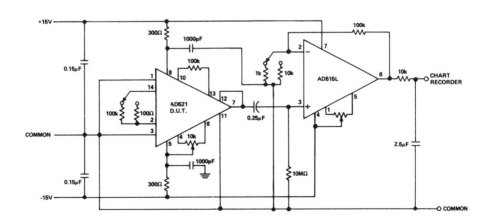

Test circuit for measuring peak to peak noise in the bandwidth 0.1Hz to 10Hz. Typical measurements are found by reading the maximum peak to peak voltage noise of the device under test (D.U.T.) for 3 observation periods of 10 seconds each.

Analog Devices
AD625
Programmable-Gain Instrumentation Amplifier

FEATURES
- User programmed gains of 1 to 10,000
- Low gain error: 0.02% max.
- Low gain TC: 5 ppm/°C max.
- Low nonlinearity: 0.001% max.
- Low offset voltage: 25 µV
- Low noise 4 nV/$\sqrt{Hz}$ (at 1 kHz)   RTI
- Gain bandwidth product: 25 MHz
- 16-pin ceramic or plastic DIP package
- MIL-standard parts available
- Low cost

PRODUCT HIGHLIGHTS

1. The AD625 affords up to 16-bit precision for user-selected fixed gains from 1 to 10,000. Any gain in this range can be programmed by 3 external resistors.

2. A 12-bit software-programmable gain amplifier can be configured using the AD625, a CMOS multiplexer and a resistor network. Unlike previous instrumentation amplifier designs, the ON resistance of a CMOS switch does not affect the gain accuracy.

3. The gain-accuracy and gain-temperature coefficient of the amplifier circuit are primarily dependent on the user-selected external resistors.

4. The AD625 provides totally independent input and output offset nulling terminals for high-precision applications. This minimizes the effects of offset voltage in gain-ranging applications.

5. The proprietary design of the AD625 provides input voltage noise of 4 nV/$\sqrt{Hz}$ at 1 kHz.

6. External resistor matching is not required to maintain high common-mode rejection.

SPECIFICATIONS (typical @ $V_S = \pm 15$ V, $R_L = 2$ kΩ and $T_A = +25$°C unless otherwise specified)

Model	AD625A/J/S Min	Typ	Max	AD625B/K Min	Typ	Max	AD625C Min	Typ	Max	Units
GAIN										
Gain Equation		$\frac{2 R_F}{R_G} + 1$			$\frac{2 R_F}{R_G} + 1$			$\frac{2 R_F}{R_G} + 1$		
Gain Range	1		10,000	1		10,000	1		10,000	
Gain Error[1]		±.035	±0.05		±0.02	±0.03		±0.01	±0.02	%
Nonlinearity, Gain = 1-256			±0.005			±0.002			±0.001	%
Gain>256			±0.01			±0.008			±0.005	%
Gain vs. Temp. Gain<1000[1]			5			5			5	ppm/°C
GAIN SENSE INPUT										
Gain Sense Current		300	500		150	250		50	100	nA
vs. Temperature		5	20		2	15		2	10	nA/°C
Gain Sense Offset Current		150	500		75	250		50	100	nA
vs. Temperature		2	15		1	10		1	5	nA/°C
VOLTAGE OFFSET (May be Nulled)										
Input Offset Voltage		50	200		25	50		10	25	µV
vs. Temperature		1	2/2		0.25	0.50/1		0.1	0.25	µV/°C
Output Offset Voltage		4	5		2	3		1	2	mV
vs. Temperature		20	50/50		10	25/40		10	15	µV/°C
Offset Referred to the Input vs. Supply										
G = 1	70	75		75	85		80	90		dB
G = 10	85	95		90	100		95	105		dB
G = 100	95	100		105	110		110	120		dB
G = 1000	100	110		110	120		115	140		dB
INPUT CURRENT										
Input Bias Current		±30	±50		±20	±25		±10	±15	nA
vs. Temperature		±50			±50			±50		pA/°C
Input Offset Current		±2	±35		±1	±15		±1	±5	nA
vs. Temperature		±20			±20			±20		pA/°C

Model	AD625A/J/S Min	AD625A/J/S Typ	AD625A/J/S Max	AD625B/K Min	AD625B/K Typ	AD625B/K Max	AD625C Min	AD625C Typ	AD625C Max	Units
INPUT										
Input Impedance										
Differential Resistance		1			1			1		GΩ
Differential Capacitance		4			4			4		pF
Common-Mode Resistance		1			1			1		GΩ
Common-Mode Capacitance		4			4			4		pF
Input Voltage Range										
Differ. Input Linear $(V_{DL})^2$			±10			±10			±10	V
Common-Mode Linear (V_{CM})		$12V - \left(\frac{G}{2} \times V_D\right)$			$12V - \left(\frac{G}{2} \times V_D\right)$			$12V - \left(\frac{G}{2} \times V_D\right)$		
Common-Mode Rejection Ratio dc to 60Hz with 1kΩ Source Imbalance										
G = 1	70	75		75	85		80	90		dB
G = 10	90	95		95	105		100	115		dB
G = 100	100	105		105	115		110	125		dB
G = 1000	110	115		115	125		120	140		dB
OUTPUT RATING		±10V @5mA			±10V @5mA			±10V @5mA		
DYNAMIC RESPONSE										
Small Signal −3dB										
G = 1 (R_F = 20kΩ)		650			650			650		kHz
G = 10		400			400			400		kHz
G = 100		150			150			150		kHz
G = 1000		25			25			25		kHz
Slew Rate		5.0			5.0			5.0		V/μs
Settling Time to 0.01%, 20V Step										
G = 1 to 200		15			15			15		μs
G = 500		35			35			35		μs
G = 1000		75			75			75		μs
NOISE										
Voltage Noise, 1kHz										
R.T.I.		4			4			4		nV/√Hz
R.T.O.		75			75			75		nV/√Hz
R.T.I., 0.1 to 10Hz										
G = 1		10			10			10		μV p-p
G = 10		1.0			1.0			1.0		μV p-p
G = 100		0.3			0.3			0.3		μV p-p
G = 1000		0.2			0.2			0.2		μV p-p
Current Noise 0.1Hz to 10Hz		60			60			60		pA p-p
SENSE INPUT										
R_{IN}		10			10			10		kΩ
I_{IN}		30			30			30		μA
Voltage Range	±10			±10			±10			V
Gain to Output		1 ± 0.01			1 ± 0.01			1 ± 0.01		%
REFERENCE INPUT										
R_{IN}		20			20			20		kΩ
I_{IN}		30			30			30		μA
Voltage Range	±10			±10			±10			V
Gain to Output		1 ± 0.01			1 ± 0.01			1 ± 0.01		%
TEMPERATURE RANGE										
Specified Performance										
J/K Grades	0		+70	0		+70				°C
A/B/C Grades	−25		+85	−25		+85	−25		+85	°C
S Grade	−55		+125							°C
Storage	−65		+150	−65		+150	−65		+150	°C
POWER SUPPLY										
Power Supply Range		±6 to ±18			±6 to ±18			±6 to ±18		V
Quiescent Current		3.5	5		3.5	5		3.5	5	mA
PACKAGE OPTIONS										
Ceramic (D-16)		AD625AD/SD			AD625BD			AD625CD		
Plastic DIP (N-16)		AD625JN			AD625KN					
Leadless Chip Carrier (E-20A)		AD625AE								

NOTES
[1]Gain Error and Gain TC are for the AD625 only. Resistor network errors will add to the specified errors.
[2]V_{DL} is the maximum differential input voltage at G = 1 for specified nonlinearity.
  V_{DL} at other gains = 10V/G.
  V_D = actual differential input voltage.
  Example: G = 10, V_D = 0.50
  V_{CM} = 12V−(10/2 × 0.50V) = 9.5V.

Specifications subject to change without notice.

All min and max specifications are guaranteed. Specifications shown in **boldface** are tested on all production units at final electrical test. Results from those tests are used to calculate outgoing quality levels.

ABSOLUTE MAXIMUM RATINGS[1]

Supply Voltage .... ±18V
Internal Power Dissipation 450mW
Input Voltage .... ±V_S
Differential Input Voltage[2] .... ±V_S
Output Short Circuit Duration .... Indefinite
Storage Temperature Range D ...... −65°C to +150°C
Storage Temperature Range N ...... −65°C to +125°C
Operating Temperature Range
  AD625J/K .... 0 to +70°C
  AD625A/B/C .... −25°C to +85°C
  AD625S .... −55°C to +125°C
Lead Temperature Range
  (Soldering, 60 seconds) .... +300°C

NOTE
[1]Stresses above those listed under "Absolute Maximum Ratings" may cause permanent damage to the device. This is a stress rating only and functional operation of the device at these or any other conditions above those indicated in the operational section of this specification is not implied. Exposure to absolute maximum rating conditions for extended periods may affect device reliability.

FUNCTIONAL BLOCK DIAGRAM

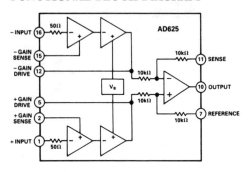

GAIN	R_F	R_G
1	20kΩ	∞
2	19.6kΩ	39.2kΩ
5	20kΩ	10kΩ
10	20kΩ	4.42kΩ
20	20kΩ	2.1kΩ
50	19.6kΩ	806Ω
100	20kΩ	402Ω
200	20.5kΩ	205Ω
500	19.6kΩ	78.7Ω
1000	19.6kΩ	39.2Ω
4	20kΩ	13.3kΩ
8	19.6kΩ	5.62kΩ
16	20kΩ	2.67kΩ
32	19.6kΩ	1.27kΩ
64	20kΩ	634Ω
128	20kΩ	316Ω
256	19.6kΩ	154Ω
512	19.6kΩ	76.8Ω
1024	19.6kΩ	38.3Ω

Common Gains Nominally within ±0.5% Error Using Standard 1% Resistors

PIN CONFIGURATIONS

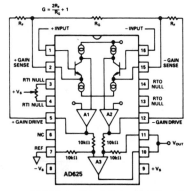

AD625 in Fixed Gain Configuration

Leadless Chip Carrier (E) Package

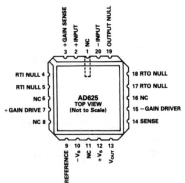

Ceramic and Plastic DIP (D and N) Packages

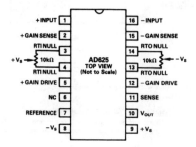

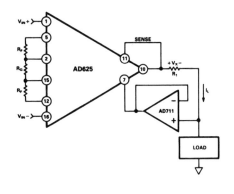

Voltage-to-Current Converter

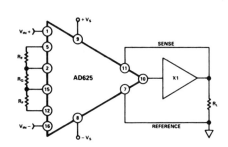

AD625 Instrumentation Amplifier with Output Current Booster

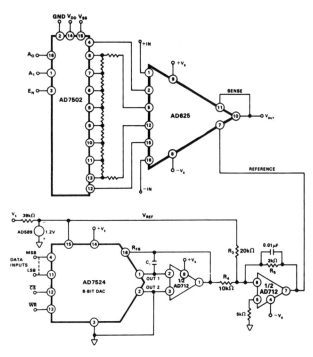

Software Controllable Offset

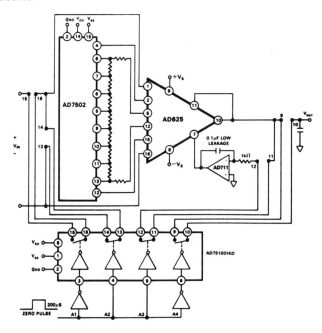

Auto-Zero Circuit

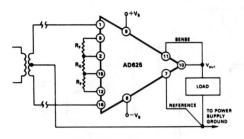

Ground Returns for Bias Currents with Transformer Coupled Inputs

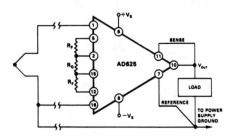

Ground Returns for Bias Currents with Thermocouple Input

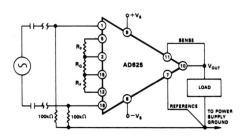

Ground Returns for Bias Currents with AC Coupled Inputs

The information included herein is believed to be accurate and reliable. However, LSI Computer Systems, Inc. assumes no responsibilities for inaccuracies, nor for any infringements of patent rights of others which may result from its use.

LSI
LS7100
BCD to 7-Segment Latch/Decoder/Driver for Liquid Crystal (Dynamic Scattering) Displays

FEATURES
- Up to -50-V segment output
- All inputs are TTL or CMOS compatible
- Internal pull-down resistors on all inputs
- Operating voltage range from -5 V to -60 V

ABSOLUTE MAXIMUM RATINGS
(All voltages referenced to V_{SS}, Pin 16)

	SYMBOL	VALUE	UNIT
dc supply voltage	V_{DD}	+0.3 to −60	V
Common in	V_{CI}	+0.3 to −60	V
All other inputs	V_{IN}	+0.3 to −30	V
Operating temperature	T_A	−40 to +70	°C
Storage temperature	T_{stg}	−65 to +125	°C

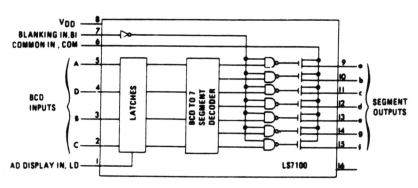

LS7100 FUNCTIONAL DIAGRAM

DISPLAY FORMAT

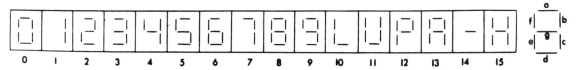

PIN ASSIGNMENT DIAGRAM

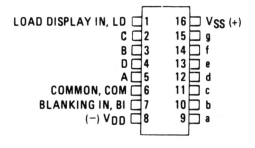

TOP VIEW
STANDARD 16 PIN DIP

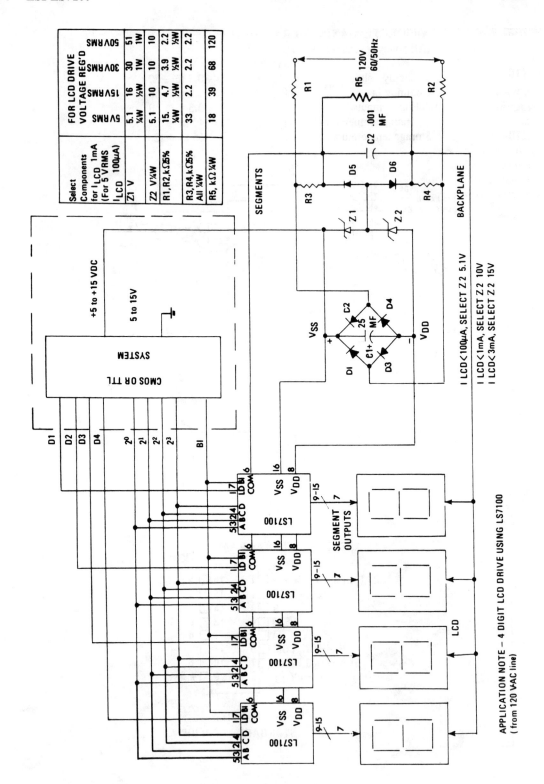

LSI
LS7110

Binary-Addressable Latched 8-Channel Demultiplexer/Driver for Liquid-Crystal (Dynamic-Scattering) Displays

FEATURES
- Up to -50-V output
- All inputs are TTL and CMOS compatible
- Internal pull-down resistors on all inputs
- Operating voltage range from -5 V to -60 V

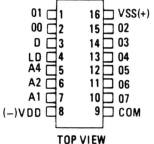

PIN ASSIGNMENT DIAGRAM

TOP VIEW
STANDARD 16 PIN DIP

ABSOLUTE MAXIMUM RATINGS
(All voltages referenced to V_{SS}, Pin 16)

PARAMETER	SYMBOL	VALUE	UNITS
dc supply voltage	V_{DD}	$+.3$ to -60	V
Common in	V_{CI}	$+.3$ to -60	V
All other inputs	V_{IN}	$+.3$ to -30	V
Operating temperature	T_A	-40 to $+85$	°C
Storage temperature	T_{stg}	$-65\ +125$	°C

FUNCTIONAL DIAGRAM

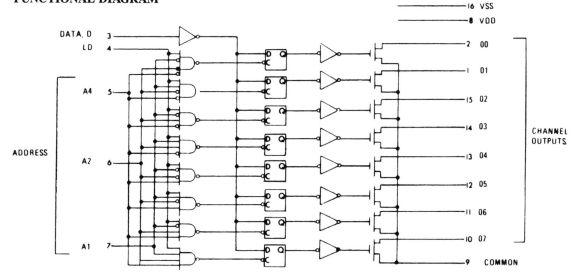

LSI LS7110

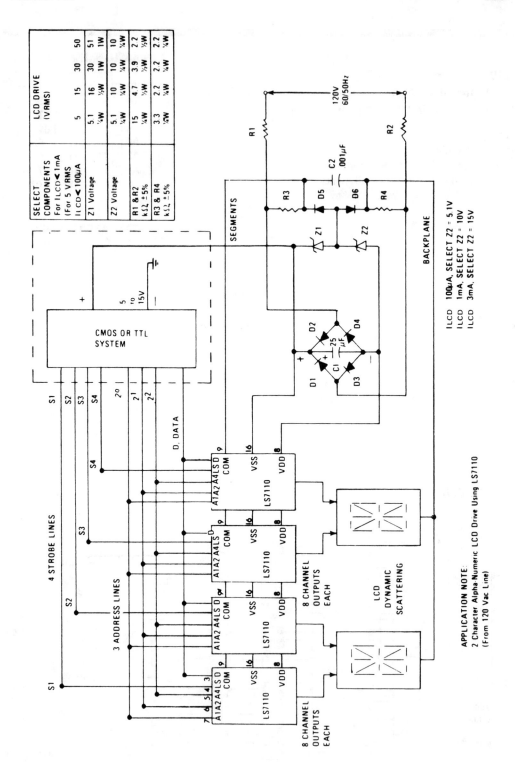

APPLICATION NOTE:
2 Character Alpha-Numeric LCD Drive Using LS7110 (From 120 Vac Line)

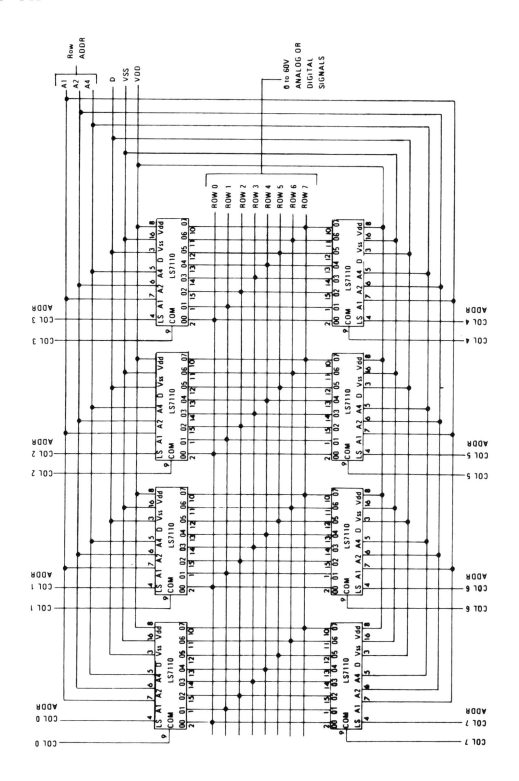

APPLICATION NOTE 8 × 8 SWITCHBOARD MATRIX USING LS7110

LSI
LS7310-LS7313
ac Power Controllers

FEATURES
- Phase-locked-loop (PLL) synchronization produces pure ac across the output load (no dc offset)
- 10 levels of output power, ranging from 37% to 97% of the rated load wattage
- Controls output power by controlling the ac duty cycle
- Operates on 50-Hz/60-Hz line frequency for PLL synchronization
- 10-V to 14-V supply voltage
- 10 I/O's for touch or mechanical switch inputs for power selection and LED driver outputs to indicate selected power
- Speed controller for universal and shaded pole motors

ABSOLUTE MAXIMUM RATINGS

PARAMETER	SYMBOL	VALUE	UNITS
dc supply voltage	V_{SS}	+20	V
Any input voltage	V_{IN}	$V_{SS}-20$ to $V_{SS}+0.5$	V
Operating temperature	T_A	0 to +80	°C
Storage temperature	T_{stg}	−65 to +150	°C

CONNECTION DIAGRAM— TOP VIEW
STANDARD 18 PIN PLASTIC DIP

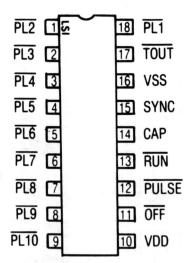

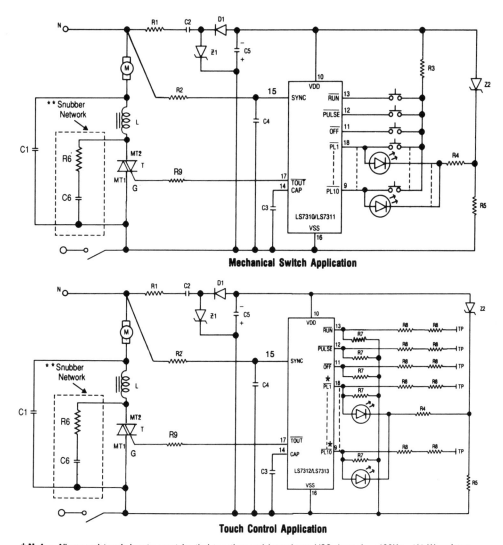

Mechanical Switch Application

Touch Control Application

*Note: All unused touch inputs must be tied together and brought to VSS through a 100KΩ, 1/4 W resistor.

115V

- C1 = 0.15μF/150VAC
- C2 = See C2 Value Table
- C3 = 0.047μF/25 V
- C4 = 470pF/25 V
- C5 = 220μF/25V
- **C6 = 0.47μF/150VAC
- R1 = 270 ohms/1W
- R2 = 1.5M ohms/ 1/4 W
- R3 = 10KΩ/ 1/4W
- R4 = 560Ω/ 1/4W
- R5 = 10KΩ/ 1/4W
- **R6 = 1.8KΩ/ 1W
- R7 = 1MΩ to 5MΩ/ 1/4 W (Select For Sensitivity)
- R8 = 2.7MΩ/ 1/4W
- R9 = 100Ω/ 1/4W
- Z1 = 13V/1W Zener (± 5%)
- *Z2 = 6.2 V/ 1/4W Zener (±5%)
- D1 = 1N4148
- T = Q4004L4 Triac (Typical)
- L = 100μH (Rfi Filter)

220V

- C1 = 0.15μF/300VAC
- C2 = See C2 Value Table
- C3 = 0.047μF/25 V
- C4 = 470pF/25 V
- C5 = 220μF/25V
- **C6 = 0.47μF/300VAC
- R1 = 270 ohms/2W
- R2 = 1.5M ohms/ 1/4 W
- R3 = 10KΩ/ 1/4W
- R4 = 560Ω/ 1/4 W
- R5 = 10KΩ/ 1/4W
- **R6 = 1.8KΩ/ 2W
- R7 = 1MΩ to 5MΩ/ 1/4 W (Select For Sensitivity)
- R8 = 4.7MΩ/ 1/4W
- R9 = 100Ω/ 1/4W
- Z1 = 13V/1W Zener (± 5%)
- *Z2 = 6.2 V/ 1/4W Zener (±5%)
- D1 = 1N4148
- T = Q4004L4 Triac (Typical)
- L = 200μH (Rfi Filter)

*Zener type should be that which produces its rated voltage at 500 microamperes or less such as part type MZ4627.
**R6-C6 snubber network may be required for some motor inductive loads. Note: Use LEDs requiring 5 mA or less.

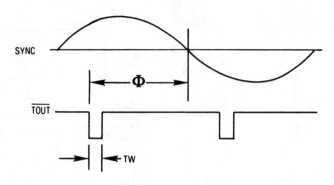

OUTPUT PHASE ANGLE Φ

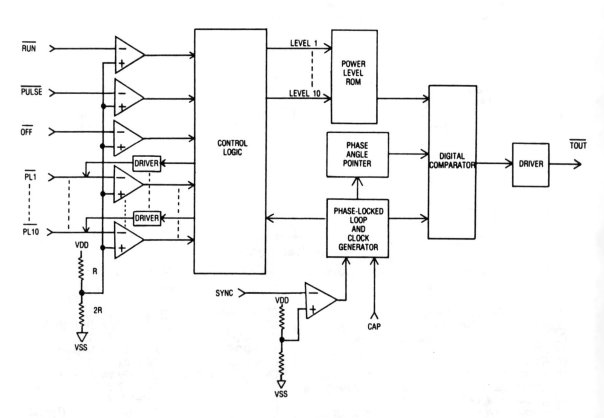

BLOCK DIAGRAM

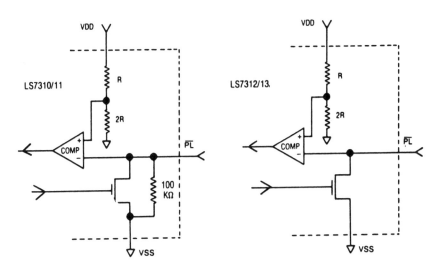

PL INPUT/OUTPUT CIRCUIT

LSI
LS7314/LS7315
ac Power Controllers

FEATURES
- Phase-locked-looped (PLL) synchronization produces pure ac across the output load (no dc offset)
- 10 levels of output power, ranging from 8% to 99% of rated load wattage
- Controls output power by controlling the ac duty cycle
- Operates on 50-Hz/60-Hz line frequency for PLL synchronization
- 10-V to 14-V supply voltage
- 10 I/O's for touch or mechanical switch inputs for power selection and LED driver outputs to indicate selected power
- Speed controller for universal and shaded-pole motors

ABSOLUTE MAXIMUM RATINGS

PARAMETER	SYMBOL	VALUE	UNITS
dc supply voltage	V_{SS}	+20	V
Any input voltage	V_{IN}	$V_{SS}-20$ to $V_{SS}+0.5$	V
Operating temperature	T_A	0 to +80	°C
Storage temperature	T_{stg}	−65 to +150	°C

LSI LS7314/LS7315

CONNECTION DIAGRAM — TOP VIEW
STANDARD 16 PIN PLASTIC DIP

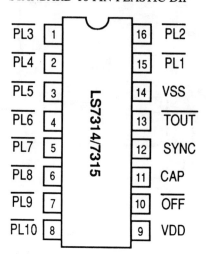

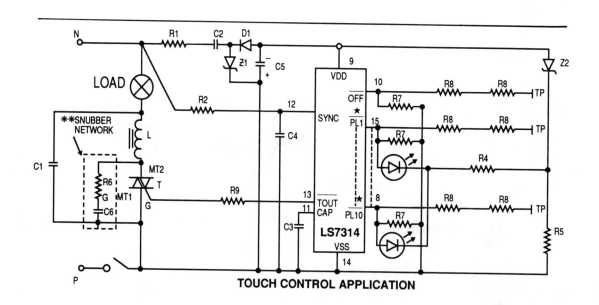

TOUCH CONTROL APPLICATION

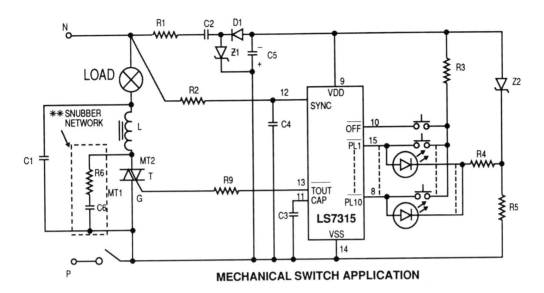

MECHANICAL SWITCH APPLICATION

* NOTE: ALL UNUSED TOUCH INPUTS MUST BE TIED TOGETHER AND BROUGHT TO VSS THROUGH A 100KΩ, 1/4W RESISTOR

115 VAC

C1	=	0.15µF/150VAC	** R6 =	1.8KΩ/ 1W
C2	=	See Value Table	R7 =	1MΩ to 5MΩ/ 1/4W
C3	=	0.047µF/25VDC		(Select For Sensitivity)
C4	=	470pF/25V	R8 =	2.7MΩ/ 1/4W
C5	=	220µF/25V	R9 =	100 Ω/ 1/4W
** C6	=	0.47µF/150VAC	Z1 =	13V/1W Zener (± 5%)
R1	=	270Ω/1W	* Z2 =	6.2 V/ 1/4W Zener (±5%)
R2	=	1.5M Ω/ 1/4W	D1 =	1N4148
R3	=	10KΩ/ 1/4W	T =	Q4004L4 Triac (Typical)
R4	=	560Ω/ 1/4W	L =	100 µH (Rfi Filter)
R5	=	10KΩ/ 1/4W		

220 VAC

C1	=	0.15µF/300VAC	** R6 =	1.8KΩ/2W
C2	=	See Value Table	R7 =	1MΩ to 5MΩ/ 1/4W
C3	=	0.047µF/25V		(Select For Sensitivity)
C4	=	470pF/25V	R8 =	4.7MΩ/ 1/4W
C5	=	220µF/25V	R9 =	100Ω/1/4W
** C6	=	0.47µF/300VAC	Z1 =	13V/ 1W Zener (±5%)
R1	=	1KΩ/2W	* Z2 =	6.2V/ 1/4W Zener (±5%)
R2	=	1.5M Ω/ 1/4W	D1 =	1N4148
R3	=	10KΩ/ 1/4W	T =	Q5004L4 Triac (Typical
R4	=	560Ω/ 1/4W	L	= 200 µH (Rfi Filter)
R5	=	10KΩ/ 1/4W		

*Zener type should be that which produces its rated voltage at 500 microamperes or less such as part type MZ4627.
R6-C6 snubber network may be required for some motor inductive loads. **Note: Use LEDs requiring 5mA or less.

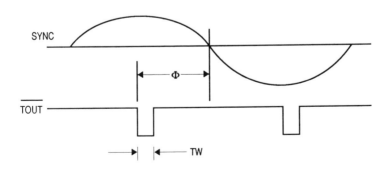

OUTPUT PHASE ANGLE Φ

400 LSI LS7314/LS7315

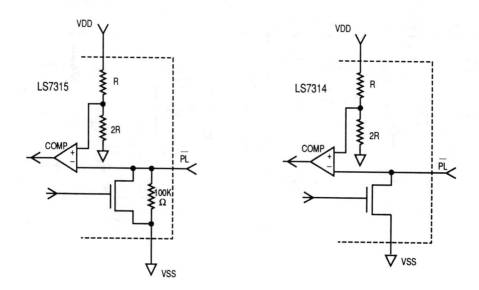

$\overline{PL}$ INPUT/OUTPUT CIRCUIT

BLOCK DIAGRAM

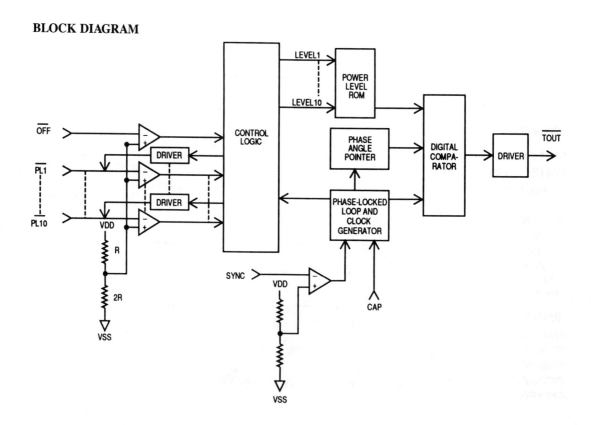

APPLICATION NOTE

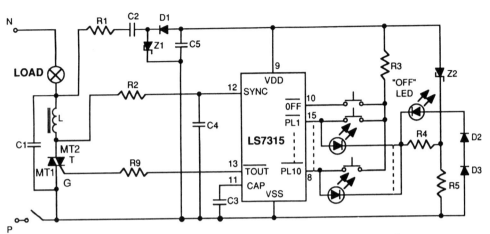

Mechanical Switch Multi-Level Wall Dimmer with LED Indicators
Parts List for 115 VAC Operation

C1	= 0.15µF/150VAC	R2	= 1.5MΩ/.25W	Z2	= 6.2V/.25W Zener (±5%)
**C2	= 1.0µF/150VAC	R3	= 10KΩ/.25W	D1	= 1N4148
C3	= 0.047µF/25V	R4	= 560Ω/.25W	D2	= 1N4148
C4	= 470pF/25V	R5	= 10KΩ/.25W	D3	= 1N4148
C5	= 100µF/25V	R9	= 100Ω/.25W	T	= Q4004L4 Triac (Typical)
R1	= 270Ω/1W	Z1	= 13V/1W Zener (±5%)	L	= 100µH (RFI Filter)

*Zener type should be that which produces its rated voltage at 500 microamperes or less such as part type MZ462.
**This value of C2 for 33µs wide output pulse. Use 1.8µF/150VAC for 1.0msec wide output pulse.
Note: Use LEDs requiring 5mA or less.

Raytheon
RV4143, 4144
Ground-Fault Interrupters

FEATURES
- Direct interface to SCR
- Supply voltage derived from ac line: 26-V shunt
- Adjustable sensitivity
- Grounded neutral fault detection
- Meets U.L. 943 standards

ABSOLUTE MAXIMUM RATINGS
Supply current	18 mA
Internal power dissipation	500 mW
Storage temperature range	−65C to +150°C
Operating temperature range	−35°C to +80°C
Lead soldering temperature (60 S)	+300°C

ORDERING INFORMATION

Part Number	Package	Operating Temperature Range
RV4143N	N	-35°C to +85°C
RV4144N	N	-35°C to +85°C

Notes:
N = 8-lead plastic DIP
Contact a Raytheon sales office or representative for ordering information on special package/temperature range combinations.

THERMAL CHARACTERISTICS

	8-Lead Plastic DIP
Max. Junction Temp.	125°C
Max. P_D T_A < 50°C	468mW
Therm. Res. θ_{JC}	—
Therm. Res. θ_{JA}	160°C/W
For T_A > 50°C Derate at	6.25mW per °C

ELECTRICAL CHARACTERISTICS (I_S = 5 mA and T_A = +25°C)

Parameters	Test Conditions	Min	Typ	Max	Units
Shunt Regulator					
Zener Shunt Voltage	Pin 6	25	26	29.2	V
Reference Voltage	Pin 3	12.5	13	14.6	V
Op Amp					
Input Offset Voltage	Pin 2 to Pin 3	-3	±1	+3	mV
Output Voltage Swing	Pin 7 to Pin 3	±11	±13.5		V
AC Output Voltage	A_V = 500, f_{IN} = 50kHz, V_{IN} = 1mV_{RMS}	50		180	mV_{RMS}
Resistors					
R3		3.8	4.7	5.7	kΩ
R1 RV4143		0.8	1.0	1.2	kΩ
R1 RV4144		0.6	0.75	0.9	kΩ
R2 RV4143		8.0	10.0	12.0	kΩ
R2 RV4144		2.0	2.5	3.0	kΩ
SCR Trigger					
V_{OH}	Across 4.7kΩ	1.5	2.8	6	V
V_{OL}			.001	.01	V

ELECTRICAL CHARACTERISTICS ($I_S = 5$ mA, over the specified temperature range)

Parameters	Test Conditions	Min	Typ	Max	Units
Shunt Regulator					
Zener Shunt Voltage	Pin 6	24	26	30	V
Reference Voltage	Pin 3	12	13	15	
Op Amp					
Input Offset Voltage	Pin 2 to Pin 3	−6	±2	+6	mV
Output Voltage Swing	Pin 7 to Pin 3	±10.5	±13		V
AC Output Voltage	$A_V = 500$, $f_{IN} = 50$kHz $V_{IN} = 1$mV$_{RMS}$	50		200	mV$_{RMS}$
Resistors					
R3		3.3	4.7	6.1	
R1 RV4143		0.7	1.0	1.3	
R1 RV4144		0.52	0.75	0.98	kΩ
R2 RV4143		7.0	10	13.0	
R2 RV4144		1.75	2.5	3.25	
SCR Trigger					
V_{OH}	Across 4.7kΩ	1.3	2.8	5	V
V_{OL}			.003	.05	

CONNECTION INFORMATION

8-Lead Dual In-Line Package (Top View)

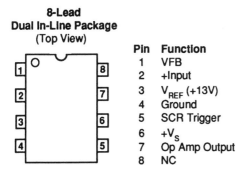

Pin	Function
1	VFB
2	+Input
3	V_{REF} (+13V)
4	Ground
5	SCR Trigger
6	+V_S
7	Op Amp Output
8	NC

FUNCTIONAL BLOCK DIAGRAM

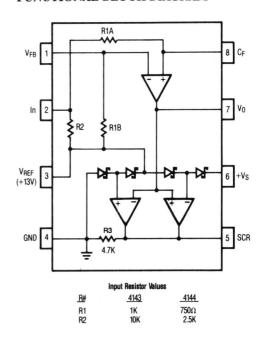

Input Resistor Values

R#	4143	4144
R1	1K	750Ω
R2	10K	2.5K

404 Raytheon RV4143/4144

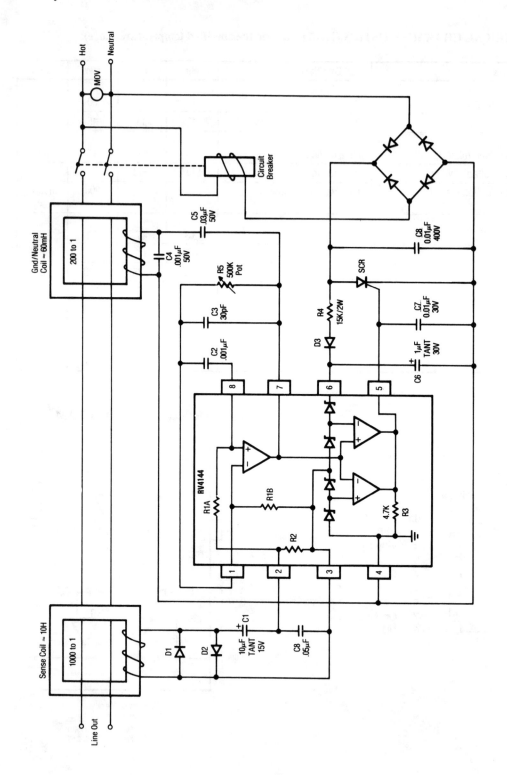

GFI Application Circuit (RV4144)

Raytheon RV4145
Low-Power Ground-Fault Interrupter

FEATURES
- No potentiometer required
- Direct interface to SCR
- Supply voltage derived from ac line: 26-V shunt
- Adjustable sensitivity
- Grounded neutral fault detection
- Meets U.L. 943 standards
- 450 µA quiescent current
- Ideal for 120-V or 220-V systems

ABSOLUTE MAXIMUM RATINGS

Supply current	18 mA
Internal power dissipation	500 mW
Storage temperature range	−65 °C to +150 °C
Operating temperature range	−35 °C to +80 °C
Lead soldering temperature	
(60 S, DIP)	+300 °C
(10 S, SO)	+260 °C

ORDERING INFORMATION

Part Number	Package	Operating Temperature Range
RV4145N	N	-35°C to +85°C
RV4145M	M	-35°C to +85°C

Notes:
N = 8-lead plastic DIP
M = 8-lead plastic SOIC
Contact a Raytheon sales office or representative for ordering information on special package/temperature range combinations.

THERMAL CHARACTERISTICS

	8-Lead Plastic SOIC	8-Lead Plastic DIP
Max. Junction Temp.	+125°C	+125°C
Max. P_D T_A <50°C	300 mW	468 mW
Therm. Res θ_{JC}	—	—
Therm. Res. θ_{JA}	240°C/W	160°C/W
For T_A >50°C Derate at	4.1 mW per °C	6.25 mW per °C

ELECTRICAL CHARACTERISTICS ($I_S = 1.5$ mA and $T_A = +25°C$)

Parameters	Test Conditions	Min	Typ	Max	Units
Shunt Regulator					
Zener Voltage (V_s)	Pin 6 to Pin 4	25	26	29.2	V
Reference Voltage	Pin 3 to Pin 4	12.5	13	14.6	V
Quiescent Current (I_s)	$+V_s = 24$V		450	750	µA
Operational Amplifier					
Offset Voltage	Pin 2 to Pin 3	-3.0	0.5	+3.0	mV
+Output Voltage Swing	Pin 7 to Pin 3	6.8	7.2	8.1	V
-Output Voltage Swing	Pin 7 to Pin 3	-9.5	-11.2	-13.5	V
+Output Source Current	Pin 7 to Pin 3		650		µA
-Output Sink Current	Pin 7 to Pin 3		1.0		mA
Gain Bandwidth Product	$f = 50$ kHz	1.0	1.8		MHz
Detector Reference Voltage	Pin 7 to Pin 3	6.8	7.2	8.1	±V
Resistors	$I_s = 0$ mA				
R1	Pin 2 to Pin 3		10		kΩ
R2	Pin 1 to Pin 3		10		kΩ
R5	Pin 5 to Pin 4	4.0	4.7	5.4	kΩ
SCR Trigger Voltage	Pin 5 to Pin 4				
Detector On		1.5	2.8		V
Detector Off		0	1	10	mV

ELECTRICAL CHARACTERISTICS ($I_S = 1.5$ mA, < over the specified temperature range)

Parameters	Test Conditions	Min	Typ	Max	Units
Shunt Regulator					
Zener Voltage (V_s)	Pin 6 to Pin 4	24	26	30	V
Reference Voltage	Pin 3 to Pin 4	12	13	15	V
Quiescent Current (I_s)	$+V_s = 23$V		500		µA
Operational Amplifier					
Offset Voltage	Pin 2 to Pin 3	-5.0	0.5	+5.0	mV
+Output Voltage Swing	Pin 7 to Pin 3	6.5	7.2	8.3	V
-Output Voltage Swing	Pin 7 to Pin 3	-9	-11.2	-14	V
Gain Bandwidth Product	$f = 50$ kHz		1.8		MHz
Detector Reference Voltage	Pin 7 to Pin 3	6.5	7.2	8.3	±V
Resistors	$I_s = 0$ mA				
R1	Pin 2 to Pin 3		10		kΩ
R2	Pin 1 to Pin 3		10		kΩ
R5	Pin 5 to Pin 4	3.8	4.7	5.6	kΩ
SCR Trigger Voltage	Pin 5 to Pin 4				
Detector On		1.3	2.8		V
Detector Off		0	3	50	mV

CONNECTION INFORMATION

8-Lead Plastic Dual In-Line SO-8
(Top View)

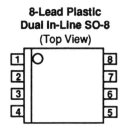

8-Lead Plastic Dual In-Line Package
(Top View)

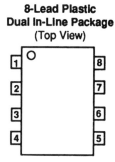

Pin	Function
1	VFB
2	+Input
3	V_{REF} (+13V)
4	Ground
5	SCR Trigger
6	$+V_S$
7	Op Amp Output
8	NC

FUNCTIONAL BLOCK DIAGRAM

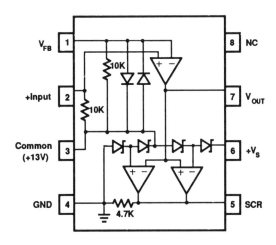

Raytheon RV4145

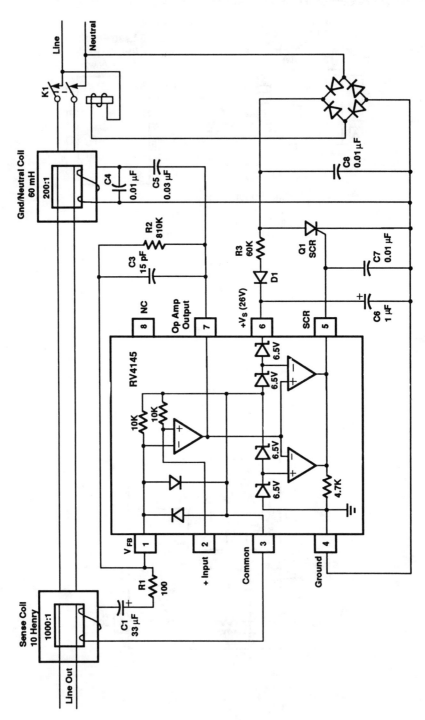

GFI Application Circuit

Raytheon XR-2207
Voltage-Controlled Oscillator

FEATURES
- Excellent temperature stability: 20 ppm/°C
- Linear frequency sweep
- Adjustable duty cycle: 0.1% to 99.9%
- Two- or four-level FSK capability
- Wide sweep range: 1000:1 min
- Logic compatible input and output levels
- Wide supply voltage range: ±4 V to ±13 V
- Low supply sensitivity: 0.15%/V
- Wide frequency range: 0.01 Hz to 1 MHz
- Simultaneous triangle- and square-wave outputs

APPLICATIONS
- FSK generation
- Voltage and current-to-frequency conversion
- Stable phase-locked loop
- Waveform generation triangle-, sawtooth-, pulse-, square-wave
- FM and sweep generation

ABSOLUTE MAXIMUM RATINGS
Supply voltage +26 V
Storage temperature range −65°C to +150°C

ORDERING INFORMATION

Part Number	Package	Operating Temperature Range
XR2207CN	N	0°C to +70°C
XR2207N	N	-25°C to +85°C
XR2207MD	D	-55°C to +125°C
XR2207MD/883B	D	-55°C to +125°C

Notes:
/883B suffix denotes Mil-Std-883, Level B processing
N = 14-lead plastic DIP
D = 14-lead ceramic DIP
Contact a Raytheon sales office or representative for ordering information on special package/temperature range combinations.

THERMAL CHARACTERISTICS

	14-Lead Plastic DIP	14-Lead Ceramic DIP
Max. Junction Temp.	125°C	175°C
Max. P_D $T_A < 50°C$	468mW	1042mW
Therm. Res. θ_{JC}	—	60°C/W
Therm. Res. θ_{JA}	160°C/W	120°C/W
For $T_A > 50°C$ Derate at	6.25mW per °C	8.33mW per °C

ELECTRICAL CHARACTERISTICS

$V_S = \pm 6$ V, $T_A = +25$ °C = 5000 pF, R1 = R2 = R3 = R4 = 20 kΩ, $R_L = 4.7$ kΩ, Binary inputs grounded, S1 and S2 closed unless otherwise specified

Parameters	Test Conditions	XR-2207 Min	XR-2207 Typ	XR-2207 Max	XR-2207C Min	XR-2207C Typ	XR-2207C Max	Units
General Characteristics								
Supply Voltage Single Supply Split Supplies	See Typical Performance Characteristics	+8.0 ±4	+12 ±6	+26 ±13	+8.0 ±4	+12 ±6	+26 ±13	V
Supply Current Single Supply	Measured at pin 1, S1 open (See Fig. 2)		5.0	7.0		5.0	8.0	mA
Split Supplies Positive	Measured at pin 1, S1 open (See Fig. 1)		5.0	7.0		5.0	8.0	
Negative	Measured at pin 12, S1, S2 open		4.0	6.0		4.0	7.0	
Binary Keying Inputs								
Switching Threshold	Measured at pins 8 and 9. Refer to pin 10	1.4	2.2	2.8	1.4	2.2	2.8	V
Input Resistance			5.0			5.0		kΩ
Oscillator Section — Frequency Characteristics								
Upper Frequency Limit	C = 500pF, R3 = 2kΩ	0.5	1.0		0.5	1.0		MHz
Lower Practical Frequency	C = 50μF, R3 = 2Ω		0.01			0.01		Hz
Frequency Accuracy			±1.0	±3.0		±1.0	±5.0	% of f_0
Frequency Matching			0.5			0.5		% of f_0
Frequency Stability Vs. Temperature (Note 1)	0°C < T_A < +75°C		20	50		30		ppm/°C
Vs. Supply Voltage			0.15			0.15		%/V
Sweep Range	R3 = 1.5kΩ for f_H R3 = 2MΩ for f_L	1000:1	3000:1		1000:1			f_H/f_L
Sweep Linearity 10:1 Sweep	C = 5000pF f_H = 10kHz, f_L = 1kHz		1.0	2.0		1.5		%
1000:1 Sweep	f_H = 100kHz, f_L = 100Hz		5.0			5.0		%
FM Distortion	±10% FM Deviation		0.1			0.1		%
Recommended Range of Timing Resistors	See Characteristic Curves	1.5		2000	1.5		2000	kΩ
Impedance at Timing Pins	Measured at pins 4, 5, 6 or 7		75			75		Ω
DC Level at Timing Terminals			10			10		mV

Parameters	Test Conditions	XR-2207			XR-2207C			Units
		Min	Typ	Max	Min	Typ	Max	
Output Characteristics								
Triangle Output Amplitude	Measured at pin 14	4	6		4	6		V_{p-p}
Impedance			10			10		Ω
DC Level	Referenced to pin 10		+100			+100		mV
Linearity	from 10% to 90% of swing		0.1			0.1		%
Squarewave Output Amplitude	Measured at pin 13, S2 Closed	11	12		11	12		V_{p-p}
Saturation Voltage	Referenced to pin 12		0.2	0.4		0.2	0.4	V
Rise Time	$C_L \leq 10pF$		200			200		nS
Fall Time	$C_L \leq 10pF$		20			20		nS

Note: 1. Guaranteed by design

CONNECTION INFORMATION

14-Lead Dual In-Line Package
(Top View)

Pin connections:
- 1: +Vs
- 2, 3: Timing Capacitor
- 4: R1 — Timing Resistors
- 5: R2
- 6: R3
- 7: R4
- 8: Binary Keying Inputs
- 9: Binary Keying Inputs
- 10: GND
- 11: Bias
- 12: -Vs
- 13: Squarewave Out
- 14: Trianglewave Out

Internal blocks: A1, A2 (VCO), Current Switches

Logic Table for Binary Keying Controls

Logic Level		Selected Timing Pins	Frequency	Definitions
8	9			
0	0	6	f_1	$f_1 = 1/R3C$, $\Delta f_1 = 1/R4C$
0	1	6 and 7	$f_1 + \Delta f_1$	$f_2 = 1/R2C$, $\Delta f_2 = 1/R1C$
1	0	5	f_2	Logic Levels: 0 = Ground
1	1	4 and 5	$f_2 + \Delta f_2$	Logic Levels: 1 = >3V

Note: For single-supply operation, logic levels are referenced to voltage at pin 10.

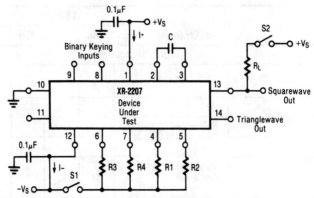

Note: This circuit is for Bench Tests only. DC testing is normally performed with automated test equipment using an equivalent circuit.

Test Circuit for Split Supply Operation

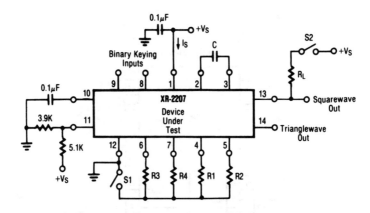

Test Circuit for Single Supply Operation

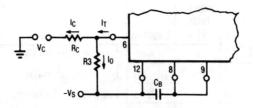

Frequency Sweep Operation

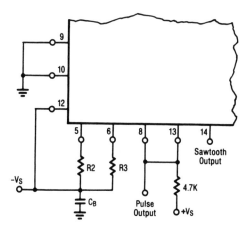

Pulse and Sawtooth Generation

SCHEMATIC DIAGRAM

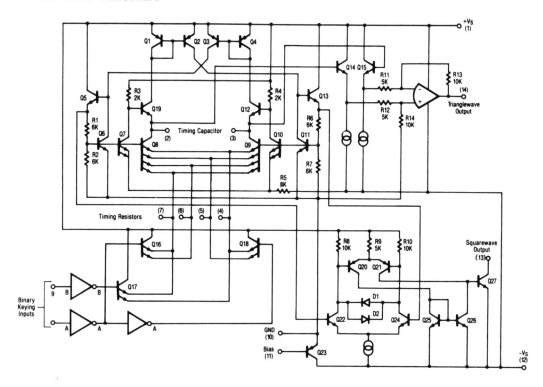

Suggested Additional Reading

Gibilisco, Stan, *International Encyclopedia of Integrated Circuits—2nd Edition*, TAB Books, 1992.

Gibilisco, Stan and Sclater, Neil, *Encyclopedia of Electronics—2nd Edition*, TAB Books, 1990.

Graf, Rudolf F., *Encyclopedia of Electronic Circuits*, Volume 1, TAB Books, 1985.

Graf, Rudolf F., *Encyclopedia of Electronic Circuits*, Volume 2, TAB Books, 1988.

Whitson, James A., *500 Electronic IC Circuits with Practical Applications*, TAB Books, 1987.

INDEX

A

ac power controllers, 394-401
address decoder/digital lock, 269-272
Allegro Microsystems, Inc., x
  AM receiver, 45-46
  automotive lamp monitors, 356-360
  automotive LCD clocks, 6-8
  automotive VF clocks, 3-5
  clocks, counters, and timers (CCT), 1-8
  communications circuits (COM), 39-53
  control circuits (CON), 109-120
  countdown power timers, 1-2
  fluid detector, 353-355
  FM decoder, 46-49
  Hall-effect sensor, linear-output, 361-364
  Hall-effect sensor, ratiometric linear, 364-367
  logic circuit, 215-223
  motor driver, BiMOS unipolar, 118
  motor driver, bipolar half-bridge, 119
  motor driver, dual full-bridge, 116
  motor driver, full-bridge, 115
  motor driver, linear, 117
  motor driver, voice-coil, 109-113
  noise blanker, 39-44
  optoelectronic switch, 49-53
  peripheral/power driver, 215-220
  power ICs for motor-drive applications, 114-120
  power supplies and test equipment (PTI), 353-370
  precision light sensor (PLS), with calibrated current amplifier, 367-370
  quad Darlington switches, 115
  relay driver applications, 120
  serial-input driver, 8-bit, 222-223
  serial-input driver, 10-bit, 221
  unipolar stepper-motor translator/driver, 114
AM noise blankers, 39-44
AM receiver, 45-46
amplifiers (*see* instrumentation amplifier; microphone amplifiers; operational amplifiers)
Analog Devices, Inc., x
  communications circuits (COM), 54-75
  control circuits (CON), 121-134
  data-conversion and processing circuits (DCP), 169-214
  instrumentation amplifier, 380-383
  instrumentation amplifier, programmable-gain, 384-388
  LVDT signal conditioner, 121-131
  microcomputer peripherals (MIC), 285-295
  modulator/demodulator, 169-174
  multiplier, 500-MHz four-quadrant, 371-375
  multiplier, IC, 174-177
  multiplier, internally trimmed IC, 376-379
  op amp, 54-56
  op amp, BiFET, 61-65
  op amp, IC, 57-60
  op amp, video, 66-69
  op amp, wideband, 70-75
  power supplies and test equipment (PTI), 371-375
  signal conditioner, bridge-transducer, 290-295
  signal conditioner, wide-bandwidth strain-gage, 285-289
  thermocouple conditioners and set-point controllers, 131-134
automotive
  clocks, LCD, 6-8
  clocks, VF, 3-5
  lamp monitors, 356-360

B

balance circuit, adaptive, 185-189
binary up counter, 20-22

C

clocks
  automotive LCD, 6-8
  automotive VF, 3-5
  quadrature converters, 23-27
  regenerator, 181-185

communications circuits, 1, 39-108
  AM noise blankers, 39-44
  AM receiver, 45-46
  DTMF generator, 78-79
  DTMF receiver, 95-101
  DTMF transceiver, 101-104
  DTMF transceiver and call
    progress detection, 105-108
  FM decoder, 46-49
  keypad pulse dialer, 81-83
  line-isolation device, 84-86
  microphone amplifier for
    telephone, 80-81
  op amp, 54-56
  op amp, BiFET, 61-65
  op amp, general purpose, 87-90
  op amp, IC, 57-60
  op amp, low-power, 91-94
  op amp, video, 66-69
  op amp, wideband, 70-75
  optoelectronic switches, 49-53
comparators
  high-speed latching, 273-276
  micropower programmable quad,
    280-284
  single-supply quad, 277-279
control circuits, 1, 109-168
  controller, buffer, 163-168
  controller, dc motor commutator,
    139-144
  controller, dc motor-speed, 145-148
  controller, light switch and ac
    motor-speed, 135-138
  controller, PC AT/XT combo,
    156-159
  controller, SCSI, 152-156
  controller/sequencer, 149-151
  controller, storage, 160-162
  LVDT signal conditioner, 121-131
  motor driver, BiMOS unipolar, 118
  motor driver, bipolar half-bridge,
    119
  motor driver, dual full-bridge, 116
  motor driver, full-bridge, 115
  motor driver, linear, 117
  motor driver, voice-coil, 109-113
  power ICs for motor-drive
    applications, 114-120
  quad Darlington switches, 115
  relay driver applications, 120
  thermocouple conditioners and
    set-point controllers, 131-134
  unipolar stepper-motor
    translator/driver, 114
  voltage-to-frequency converter,
    208-210
controllers
  ac power, 394-401
  buffer, 163-168
  dc motor commutator/controller,
    139-144
  dc motor-speed controller, 145-148
  light switch and ac motor-speed,
    135-138
  PC AT/XT combo, 156-159
  SCSI, 152-156
  sequencer, 149-151
  single-chip, 240-252
  storage, 160-162
converters
  8-bit D/A, 192-196
  8-bit D/A with microprocessor
    interface latches, 301-305
  10-bit D/A, 196-207
  12-bit D/A, 296-300
  polar-to-Cartesian, 233-236
  voltage-to-frequency, 208-210
countdown power timers, 1-2
counters, 1
  binary up, 20-22
  FET address generation counter,
    8-12
  up, 17-19
  up/down, 12-16

D

D/A converters
  8-bit, 192-196
  8-bit with microprocessor interface
    latches, 301-305
  10-bit, 196-207
  12-bit, 296-300
data synchronizer and
  write-precompensator device,
  348-352
data-conversion and processing
  circuits (DCP), 1, 169-214
  adaptive balance circuit, 185-189
  clock regenerator, 181-185
  converter, 8-bit D/A, 192-196
  converter, 10-bit D/A 192-196
  converter, voltage-to-frequency,
    208-210
  electronic filter, low-power
    programmable, 214
  electronic filter, programmable,
    210-213
  digital filter and detector (FAD),
    178-180
  IC multiplier, 174-177
  modulator/demodulator, 169-174
  PCM line interface circuit, 189-192
  dc motor commutator/controller,
    139-144
  dc motor-speed controller, 145-148
delay timer, 27-32
demultiplexer/driver for LCDs,
  391-393
digital filter and detector (FAD),
  178-180
digital lock circuit, 253-257
  and address decoder, 269-272
  keyboard-programmable, 257-261
  with tamper output, 263-269

E

electronic filter
  low-power programmable, 214
  programmable, 210-213

F

Fast Fourier Transforms (FETs), 8
fir filter, programmable, 223-230
floppy-disk drive, port-expander,
  344-348
floppy-disk read/write device
  2-channel, 338-341
  2- or 4-channel, 342-344
fluid detector, 353-355
FM decoder, 46-49

G

GEC Plessey Semiconductors, x
  adaptive balance circuit, 185-189
  clock regenerator, 181-185
  clocks, counters, and timers
    (CCT), 8-12
  communications circuits (COM),
    76-83
  data-conversion and processing
    circuits (DCP), 178-191
  digital filter and detector (FAD),
    178-180
  DTMF generator, 76-77, 78-79
  FET address generation counter,
    8-12
  fir filter, 223-230
  I/Q splitter/NCO, 236-240
  keypad pulse dialer, 81-83
  logic circuits, 223-252
  microphone amplifier for
    telephone, 80-81
  PCM line interface circuit, 189-192
  polar-to-Cartesian converter,
    233-236

pythagoras processor, 230-232
single-chip controller, 240-252
generator, DTMF, 76-77, 78-79
ground-fault interrupters, 401-404
low power, 405-408

H

Hall-effect sensor
linear-output, 361-364
ratiometric linear, 364-367

I

I/Q splitter/NCO, 236-240
IC multiplier, 174-177
ICs
application categories, xi
clocks, counters, and timers (CCT), 1-37
communications circuits (COM), 39-108
control circuits (CON), 109-168
data-conversion and processing circuits (DCP), 169-214
logic circuits (LOG), 215-284
microcomputer peripherals (MIC), 285-352
power supplies and test equipment (PTI), 353-413
instrumentation amplifier, 380-383
programmable-gain, 384-388

K

keypad pulse dialer, 81-83

L

lamp monitors, automotive, 356-360
latch/decoder/driver for LCDs, 388-390
light switch
and ac motor-speed controller, 135-138
timer, 32-37
line interface circuit, PCM, 189-192
line-isolation device, tune-activated, 84-86
logic circuits, 1, 215-284
address decoder/digital lock, 269-272
comparator, high-speed latching, 273-276
comparator, micropower programmable quad, 280-284
comparator, single-supply quad, 277-279
digital lock, 253-257
digital lock, keyboard programmable, 257-261
digital lock, with tamper output, 263-269
fir filter, 223-230
I/Q splitter/NCO, 236-240
polar-to-Cartesian converter, 233-236
program-mode lockout, 262
pythagoras processor, 230-232
serial-input driver, 8-bit, 222-223
serial-input driver, 10-bit, 221
single-chip controller, 240-252
LSI Computer Systems Inc., x
address decoder/digital lock, 269-272
binary up counter, 20-22
clocks, counters, and timers (CCT), 12-37
communications circuits (COM), 84-86
control circuits (CON), 135-151
controller, ac power, 394-401
controller, dc motor commutator, 139-144
controller, dc motor-speed, 145-148
controller, light switch and ac motor-speed, 135-138
controller/sequencer, 149-151
delay timer, 27-32
demultiplexer/driver for LCDs, 391-393
digital lock circuit, 253-257
digital lock circuit, keyboard-programmable, 257-261
digital lock circuit, with tamper output, 263-269
latch/decoder/driver for LCDs, 388-390
light switch timer, 32-37
line-isolation device, 84-86
logic circuits (LOG), 253-272
power supplies and test equipment (PTI), 388-401
program-mode lockout, 262
quadrature clock converters, 23-27
up counter, 17-19
up/down counter, 12-16

M

microcomputer peripherals, 1, 285-352
D/A converter, 8-bit, 301-305
D/A converter, 12-bit, 296-300
data synchronizer and write-precompensator device, 348
floppy-disk drive port-expander, 344-348
floppy-disk read/write device, 2-channel, 338-341
floppy-disk read/write device, 2- or 4-channel, 342-344
modem, 314-329, 330-337
modem, K-series, 337-338
modem, single-chip, 306-313
signal conditioner, bridge-transducer, 290-295
signal conditioner, wide-bandwidth strain-gage, 285-289
microphone amplifier, telephone, 80-81
modem, 314-322, 322-329, 330-337
K-series, 337-338
modem, single-chip, 306-313
modulator/demodulator, 169-174
motor driver
BiMOS unipolar, 118
bipolar half-bridge, 119
dual full-bridge, 116
full-bridge, 115
linear, 117
power ICs for, 114-120
quad Darlington switches, 115
relay applications, 120
unipolar stepper- translator/driver, 114
voice-coil, 109-113
multipliers
500-MHz four-quadrant, 371-375
IC, 174-177
internally trimmed IC, 376-379

N

numerically controlled oscillator (NCO), 236-240

O

operational amplifiers, 54-56
BiFET, 61-65
general-purpose, 87-90
IC, 57-60
low-power, 91-94
video, 66-69
wideband, 70-75
optoelectronic switches, 49-53
oscillators
numerically controlled (NCO), 236-240
voltage-controlled, 409-413

P

peripheral/power driver, 215-220
power supplies and test equipment, 1, 353-413
  ac power controllers, 394-401
  automotive lamp monitors, 356-360
  demultiplexer/driver for LCDs, 391-393
  fluid detector, 353-355
  ground-fault interrupters, 401-404, 405-408
  Hall-effect sensor, linear-output, 361-364
  Hall-effect sensor, ratiometric linear, 364-367
  instrumentation amplifier, 380-383
  instrumentation amplifier, programmable-gain, 384
  latch/decoder/driver for LCDs, 388-390
  multiplier, 500-MHz four quadrant, 371-375
  multiplier, internally trimmed IC, 376-379
  precision light sensor (PLS), with calibrated, 367
  voltage-controlled oscillator, 409-413
  precision light sensor (PLS), with calibrated current amplifier, 367-370
program-mode lockout, 262
pythagoras processor, 230-232

Q

quadrature clock converter, 23-27

R

RAM, 8
Raytheon Company, x
  communications circuits (COM), 87-94
  comparator, high-speed latching, 273-276
  comparator, micropower programmable quad, 280-284
  comparator, single-supply quad, 277-279
  D/A converter, 8-bit, 192-196, 301-305
  D/A converter, 10-bit, 196-207
  D/A converter, 12-bit, 296-300
  data-conversion and processing circuits (DCP), 192-210
  ground-fault interrupter, 401-404
  ground-fault interrupter, low-power, 405-408
  logic circuits, 273-284
  microcomputer peripherals (MIC), 296-305
  op amp, general-purpose, 87-90
  op amp, low-power, 91-94
  oscillator, voltage-controlled, 409-413
  power supplies and test equipment (PTI), 401-413
  receiver, DTMF, 95-101

S

serial-input driver
  8-bit, 222-223
  10-bit, 221
signal conditioner
  bridge-transducer, 290-295
  LVDT, 121-131
  wide-bandwidth strain-gage, 285-289
Silicon Systems, Inc., x
  buffer controller, 163-168
  communications circuits (COM), 95-108
  control circuits (CON), 152-168
  data-conversion and processing circuits (DCP), 210-214
  data synchronizer and write-precompensator device, 348-352
  DTMF receiver, 95-101
  DTMF transceiver, 101-104
  DTMF transceiver and call progress detection, 105-108
  electronic filter, low-power programmable, 214
  electronic filter, programmable, 210-213
  floppy-disk drive port-expander, 344-348
  floppy-disk read/write device, 2- or 4-channel, 342-344
  floppy-disk read/write device, 2-channel, 338-341
  microcomputer peripherals (MIC), 306-352
  modem, 314-329, 330-337
  modem, K-series, 337-338
  modem, single chip, 306-313
  PC AT/XT combo controller, 156-159
  SCSI controller, 152-156
  storage controller, 160-162
switches
  light and ac motor-speed controller, 135-138
  light timer, 32-37
  optoelectronic, 49-53
  quad Darlington, 115

T

test equipment (*see* power supplies and test equipment)
thermocouple conditioners and set-point controllers, 131-134
timers, 1
  countdown power, 1-2
  delay, 27-32
  light switch, 32-37
transceiver
  DTMF, 101-104
  DTMF and call progress detection, 105-108

U

up counter, 17-19
up/down counter, 12-16

V

voltage-controlled oscillator, 409-413